Electroactive Polymer-Based Smart Materials and Applications

Scrivener Publishing
100 Cummings Center, Suite 541J
Beverly, MA 01915-6106

Publishers at Scrivener
Martin Scrivener (martin@scrivenerpublishing.com)
Phillip Carmical (pcarmical@scrivenerpublishing.com)

Electroactive Polymer-Based Smart Materials and Applications

Edited by

Raneesh Balakrishnan
Department of Physics, Catholicate College, Pathanamthitta, Kerala, India

Sobi K. Chacko
Department of Physics, NSS Hindu College, Changanassery, Kerala, India

and

Visakh P.M.
Natural Bioactive Materials Laboratory, Department of Bioengineering, Ege University, Bornova/Izmir, Turkey

Scrivener Publishing

WILEY

This edition first published 2026 by John Wiley & Sons, Inc., 111 River Street, Hoboken, NJ 07030, USA and Scrivener Publishing LLC, 100 Cummings Center, Suite 541J, Beverly, MA 01915, USA

For more information about Scrivener publications please visit www.scrivenerpublishing.com.

Wiley Global Headquarters
111 River Street, Hoboken, NJ 07030, USA

For details of our global editorial offices, customer services, and more information about Wiley products visit us at www.wiley.com.

The manufacturer's authorized representative according to the EU General Product Safety Regulation is Wiley-VCH GmbH, Boschstr. 12, 69469 Weinheim, Germany, e-mail: Product_Safety@wiley.com.

Library of Congress Cataloging-in-Publication Data

ISBN 978-1-394-21435-8

Front cover images courtesy of Wikimedia Commons
Cover design by Russell Richardson

Set in size of 11pt and Minion Pro by Manila Typesetting Company, Makati, Philippines

Printed in the USA

10 9 8 7 6 5 4 3 2 1

Contents

Preface

The field of electroactive polymers has seen remarkable advancements over the past few decades and has emerged as a transformative area in materials science and nanotechnology. These polymers, known for their unique ability to respond to electrical stimuli with mechanical, thermal, or electrical changes, have unlocked new possibilities in energy harvesting, flexible electronics, biomedical applications, and smart materials research. With their increasing relevance, a comprehensive understanding of electroactive polymers is now more critical than ever for researchers, students, and professionals working in multidisciplinary fields.

This book bridges the gap between the fundamental principles and advanced applications of electroactive polymers, providing a detailed exploration of their thermal properties, structural characteristics, and multifunctional capabilities. It delves into their role in energy scavenging, electromagnetic interference shielding, energy storage, and biomedical applications, offering insights into both the theoretical aspects and experimental approaches. By compiling cutting-edge research and notable case studies, this book aims to equip readers with the necessary knowledge to innovate and contribute meaningfully to this evolving discipline.

For students, this book serves as a structured introduction to electroactive polymers, laying a strong foundation for understanding their material properties and potential applications in the field. Researchers and professionals, on the other hand, will find it a valuable resource for exploring the latest advancements, methodologies, and emerging trends in EAP-based technologies. The interdisciplinary nature of this subject makes it an essential reference for those in physics, materials science, electrical engineering, and biomedical engineering, fostering cross-disciplinary collaboration and novel innovation.

The increasing global demand for sustainable, flexible, and high-performance materials has made electroactive polymers an indispensable area of study. Whether for developing next-generation wearable sensors, designing self-powered devices, or advancing medical technologies, the knowledge encapsulated in this book will be instrumental in shaping future scientific and industrial breakthroughs.

We extend our deepest gratitude to the researchers, colleagues, and contributors whose insights and expertise have enriched this book. We are also thankful to Martin Scrivener and the team at Scrivener Publishing for their effort and guidance in this publication. We hope that this work will inspire new ideas, drive further research, and serve as a valuable guide for those eager to explore the vast potential of electroactive polymers.

The Editors
December 2025

1

An Introduction to Electroactive Polymers

Sobi K. Chacko[1], Raneesh Balakrishnan[2*] and Visakh P.M.[3]

[1]*Department of Physics, NSS Hindu College Changanassery, Kerala, India*
[2]*Department of Physics, Catholicate College, Pathanamthitta, Kerala, India*
[3]*Natural Bioactive Materials Laboratory, Department of Bioengineering, Ege University, Bornova/Izmir, Turkey*

Abstract

This chapter provides a comprehensive introduction to electroactive polymers (EAPs), a class of smart materials that exhibit electromechanical responses to external stimuli, such as electric fields, or mechanical stress. EAPs are broadly classified into two main categories: ionic and electronic EAPs, based on the mechanisms underlying their electromechanical behaviour. This chapter explores the classifications and fundamental properties of EAPs, including their electrical, mechanical, and thermal behaviors, highlighting their unique characteristics, such as flexibility, biocompatibility, and ease of processing. The chapter also briefly discusses various applications in which EAPs play a vital role, such as in sensors, actuators, artificial muscles, energy harvesting systems and flexible electronic devices. This overview aims to provide a foundational understanding of electroactive polymers, laying the groundwork for the further exploration of their diverse applications in advanced technologies.

Keywords: Electroactive polymers, smart polymers, piezoelectric polymers, dielectric elastomers, stimuli responsive, ionic polymer, electronic polymers

1.1 Introduction

Electroactive polymers (EAPs) represent a class of electroresponsive materials with a remarkable ability to experience mechanical deformation or changes in shape in response to an electric field [1]. This unique property

**Corresponding author*: raneesh.b@catholicatecollege.ac.in; ORCID: 0000-0003-0056-559X

Raneesh Balakrishnan, Sobi K. Chacko and Visakh P.M. (eds.) Electroactive Polymer-Based Smart Materials and Applications, (1–38) © 2026 Scrivener Publishing LLC

can cause significant shape or size changes in response to an electric field, known as electroactive actuation, and distinguishes EAPs from traditional polymers, making them highly attractive for a wide range of applications in fields, such as robotics, biomedical engineering, and smart materials [1–3]. EAPs have gained much interest in recent decades because of their unique properties and advantages over conventional ceramic actuators, such as flexibility, light weight, large deformation, responsiveness to various stimuli, biocompatibility, structural simplicity, low acoustic noise, low cost, scalability, and enhanced dielectric constant [4–7]. The properties and performance of EAPs depend on various factors, such as the chemical structure, morphology, fabrication method, environmental conditions, and applied stimuli. This ability makes them unique, leading to various potential applications in different fields including soft robotics, electrolytes, sensors and actuators, supercapacitor applications, energy harvesting and storage, smart textiles, tissue engineering, and wound healing [1, 3, 8–10]. EAPs can be categorized into two main types based on their actuation mechanisms: ionic EAPs (e.g., ionic polymer-metal composites and conductive polymers) and electronic EAPs (e.g., piezoelectric polymers, electrostrictive polymers, and dielectric elastomers). Each type offers distinct advantages and challenges depending on the specific application requirements. This chapter explores the underlying principles of electroactive polymers and their classification based on mechanisms, fundamental properties, recent advancements, and promising applications [1, 3, 11]. By understanding the unique capabilities of EAPs and their potential impact on emerging technologies, we can unlock new possibilities for innovation and design in responsive materials.

Table 1.1 outlines the key milestones in the development of electroactive polymers. The journey of EAPs began in 1880, when Wilhelm Röntgen conducted an experiment to test the effect of an electrostatic field on the mechanical properties of a strip of natural rubber [12]. But it did not directly involve the discovery of electroactive polymers as we understand them today. The piezoelectric response in natural rubber is fundamentally different from the engineered electroactive behaviors seen in synthetic polymers, which typically exhibit stronger and more tunable properties due to their crystalline structure and dipole alignment. Later, in 1899, Sacerdote demonstrated that a rubber band can produce a strain response when activated by an electric field [13, 20].

The discovery of electrets, a class of dielectric materials capable of retaining quasi-permanent electric charge or polarization, is an important milestone in the field of electroactive materials. The initial discovery was credited to Motojiro Eguchi, a Japanese physicist who, in 1925, found

Table 1.1 Key milestones in the development of electroactive polymers.

Event	Year	References
Wilhem Roentgen observed the electrically-induced deformation in a rubber sheet.	1880	[12]
Extended study on the relation between Electrical stimuli and Strain response by Sacerdote.	1899	[13]
The first piezoelectric polymer was discovered by Eguchi–Electret.	1925	[14]
Discovery of piezoelectric response in PVDF by Kawai.	1969	[15]
Development of conductive polymers by Alan MacDiarmid, Alan Heeger, and Hideki Shirakawa.	1977	[16–18]
Introduction of dielectric elastomer actuators (DEAs) by Pelrine *et al.*	1998	[19]
The first EAP Actuators and Devices (EAPAD) Conference by Bar-Cohen.	1999	[20]

that certain dielectric materials could hold an electric field-induced charge after being heated and cooled in the presence of a strong external electric field [14]. Eguchi's groundbreaking work, published in 1925, described the behavior of waxes and other materials that exhibited permanent electrical polarization, which he termed "electret." By subjecting these substances to high electric fields, Eguchi demonstrated that they could retain a stable electric charge even after the field was removed, and the material was cooled to room temperature. This significant milestone has expanded the scope of electroactive materials and laid the foundation for the development of various electret-based technologies [14]. While electrets exhibit permanent electrical polarization, the discovery of modern electroactive polymers, such as piezoelectric and conducting polymers emerged much later. Although Rontgen's natural rubber experiment laid the groundwork for understanding the piezoelectric effect, the discovery of electroactive polymers as synthetic materials with specific electromechanical properties occurred later. Polyvinylidene fluoride (PVDF) was discovered to exhibit piezoelectric behavior in 1969 by Tadashi Kawai [15]. PVDF is one of the earliest and most well-known electroactive polymers, owing to its piezoelectric properties [21, 22]. This discovery opens up new possibilities for the use of EAPs in various fields [1, 23, 24].

In the 1970s, remarkable advancements occurred in chemistry with the discovery of a novel category of polymers that demonstrated significant electronic conductivity when partially oxidized. This surprising innovation was credited to the joint work of three distinguished scientists, Alan J. Heeger, Alan G. MacDiarmid, and Hideki Shirakawa [16–18]. Their remarkable contributions led to the Nobel Prize in Chemistry being awarded to them in 2000. This pivotal research has opened up new possibilities for developing conducting polymers that hold immense potential for diverse applications across various fields.

1.2 Types of EAPs

In accordance with their operating mechanisms and working principles, electroactive polymers are fundamentally categorized as electronic/dielectric/dry polymers and ionic/wet polymers [3, 11]. Subsequently, the primary classifications were further stratified into discrete subcategories.

1.2.1 Piezoelectric Polymers

Piezoelectric polymers can generate an electrical charge when they experience mechanical stress or deformation. This fascinating process, called the direct piezoelectric effect, occurs because of the unique asymmetrical arrangement of the molecular dipoles within the polymer structure. Understanding this phenomenon opens exciting possibilities for various applications, such as sensors and energy harvesting technologies. When subjected to mechanical forces, such as compression or stretching, the dipoles become reorient, resulting in the accumulation of electric charge. Conversely, an inverse piezoelectric effect occurs when an electric field is applied to induce mechanical deformation in the polymer [22, 25]. There are several remarkable materials with diverse applications, including PVDF and its co-polymers, nonbiodegradable materials, and poly (L-lactic acid) (PLLA), which are biodegradable and often used in medical applications [21, 22, 26–28]. Other notable materials include cellulose, a natural polymer derived from plants that is eco-friendly and versatile; polyimide, renowned for its thermal stability and electrical insulation properties; and polyvinylidene chloride (PVDC), which offers exceptional barrier properties against gases and liquids [29–33]. Additionally, polyhydroxybutyrate-co-hydroxyvalerate (PHBV) is a biodegradable piezoelectric copolymer that combines flexibility and strength with odd-numbered nylons like nylon 11, which have unique structural properties and are used in engineering plastics and fibers, and silk fibroin, a protein that is highly valued for its biocompatibility and tensile strength [34–37]. These materials possess a wide range of properties that make them suitable for various applications, such as energy harvesting, where they convert different forms of energy into usable power, and sensors and actuators, which play critical roles in smart technology in tissue engineering, where they serve as scaffolds for cell growth and regeneration [38–47]. Furthermore, these materials can be fabricated in various forms, including fibers, foams, thin films, textiles, and coatings, allowing for innovative designs and functionalities in smart systems [48–51].

1.2.2 Conducting Polymers

Conducting polymers possess inherent electrical conductivity arising from the mobility of charge carriers such as electrons or ions within their polymeric backbone [1, 52]. This conductivity can be achieved by doping the polymer with dopant molecules, which introduces charge carriers and facilitates electron transport. Examples of conducting polymers include polypyrrole (PPy), polythiophene, and polyaniline (PANI). These materials have garnered significant attention because of their unique combination of electrical conductivity, processability, and tunable properties, making them promising candidates for a wide range of applications [18, 53–57]. These materials exhibit distinct characteristics that contribute to their unique electronic and optical properties. PPy, for instance, is known for its high electrical conductivity and stability, which make it particularly useful in applications such as organic electronics and sensors. Their ability to be doped with various ions further enhances their conductivity, allowing them to be tailored for specific applications [58–60]. On the other hand, polythiophene and its derivatives stands out for its exceptional light-emitting properties, making it suitable for use in organic light-emitting diodes and photovoltaic cells [61–63]. The conjugated structure of polythiophene enables effective charge transport and light absorption, which are critical for these applications. PANI is another notable conducting polymer owing to its versatility and relatively easy synthesis. This polymer can exist in multiple oxidation states, offering tunable conductivity depending on environmental conditions. This adaptability makes PANI suitable for various applications, including as a component of batteries and supercapacitors for energy storage [61–63]. Figure 1.1 shows the chemical structure of some conducting polymers.

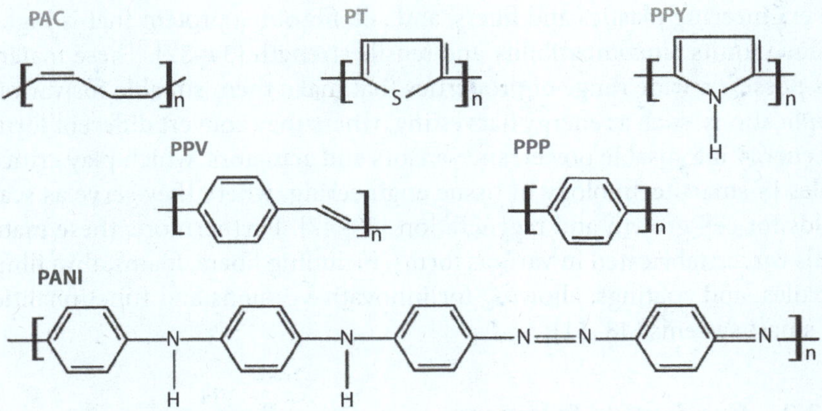

Figure 1.1 The chemical structures of some conducting polymers (CC-BY) [64].

1.2.3 Ionic Polymers

Ionic polymers contain ionic groups within their polymer chains, which enable them to conduct ions when exposed to an electric field or moisture. These materials exhibit electroactive behavior through the migration of ions within the polymer matrix, leading to changes in their mechanical, optical, and transport properties [65–68]. Ionic polymers are used in actuators, artificial muscles, and biomedical devices because of their responsive nature and biocompatibility [68–72]. Various ionic polymers include Polymer-Ionic Liquid Gels (IL Gels), which are composed of a polymer matrix impregnated with an ionic liquid. This combination results in materials with high ionic conductivities, excellent electrochemical stabilities, and good mechanical properties. IL gels have applications in diverse fields, including energy storage devices, sensors, actuators, and drug delivery systems [73–76]. IPMCs are thin polymer films coated with noble-metal electrodes. When an electric field is applied, the ions migrate within the polymer matrix, causing the material to bend or deform. This property makes IPMCs suitable for application in soft robotics, artificial muscles, and microelectromechanical systems (MEMS) [77–83]. Porous ionic polymers (PiPs) possess porous structures with ionic functional groups. This unique architecture enables them to exhibit high ionic conductivity, a large surface area, and good mechanical stability. PiPs have applications in ion-exchange membranes, sensors, and catalysts [84–87]. Poly(ionic liquids) (PILs) are polymers with ionic liquid moieties that are directly incorporated into their molecular structures. This design endows the PILs with high ionic conductivity, thermal stability, and good mechanical properties. They are used in various applications, including electrolytes, sensors, and catalysts [88–90].

1.2.4 Dielectric Elastomers

The inherent softness of dielectric elastomers (DEs) is a fundamental characteristic that enables a substantial deformation in response to an applied electric field. This inherent compliance is a key factor in their potential applications, particularly in fields like soft robotics and artificial muscles where flexibility and adaptability are paramount [91]. Composed of an insulating elastomer film sandwiched between compliant electrodes, DEs offer a unique blend of mechanical and electrical properties. When a voltage is applied across electrodes, the generated electrostatic forces induce considerable strain in the elastomer, enabling a diverse array of applications. The application of voltage across the electrodes causes the elastomer to experience electrostatic forces, resulting in contraction in thickness and lateral expansion. This electromechanical response allows dielectric elastomers to function as both sensors and actuators, rendering them highly versatile for various applications [92–94]. DEs offer several advantages, including a large deformation, high-energy density, fast response times, and low operating voltages. These properties make them promising candidates for soft robotics, artificial muscles, haptic devices, and energy harvesting applications. However, challenges, such as electrical breakdown and fatigue limitations, must be addressed for widespread practical applications. Some examples include Acrylates, such as VHB 4905 and 4910, silicones, polyurethanes [95–103].

Each type of electroactive polymer offers distinct advantages and challenges. Table 1.2 shows a comparison between the different types of EAPs, and Figure 1.2 gives the broad classification of Ionic and Electronic EAPs. By understanding their underlying mechanisms researchers can tailor their designs and syntheses to meet the demands of diverse technological applications.

Ionic-conducting polymers can be broadly categorized into two primary types: intrinsically conducting polymers and extrinsically. Intrinsic polymers possess a conjugated backbone structure characterized by alternating single and double bonds. This arrangement allows for the delocalization of π-electrons along the polymer chain, contributing to their inherent conductivity [3, 11, 52]. When subjected to an electric field, these delocalized electrons facilitate the charge transport within the polymer matrix.

Table 1.2 A comparison between the different types of EAPs.

Classification	Properties	Advantages	Differences
Piezoelectric Polymers (Polyurethane, polyvinylidene fluoride)	• Generate electric voltage in response to mechanical stress • Exhibit direct and inverse piezoelectric effects • High mechanical flexibility and strength	• Transform mechanical energy into electrical energy and vice versa • Used in sensors, actuators, and energy-harvesting devices	• Mechanism relies on molecular dipole alignment • Can exhibit strong piezoelectric response at ambient conditions
Conducting Polymers (PPy, PANI)	• Conductivity due to doping with dopant molecules • Tunable electronic and optical properties • High charge carrier mobility	• Versatile materials for electronic devices, sensors, and energy storage • Compatibility with flexible and stretchable substrates	• Mechanism involves charge carrier movement within the polymer matrix • Can exhibit metallic or semiconducting behaviour based on the doping level
Ionic Polymers (Nafion, Flemion)	• Ionic groups facilitate ion conduction within the polymer matrix • Responsive to electric fields or moisture • Biocompatible and environmentally friendly	• Suitable for applications in actuators, artificial muscles, and biomedical devices • Responsive to external stimuli, such as pH or temperature	• Mechanism involves ion migration rather than electronic conduction • Can exhibit swelling or changes in mechanical properties in response to ion movement

(*Continued*)

Table 1.2 A comparison between the different types of EAPs. (*Continued*)

Classification	Properties	Advantages	Differences
Dielectric Elastomers (Polyacrylate, silicone rubber)	• Deform in response to an applied electric field • High energy density and mechanical compliance • Rapid response times	• Potential for use in actuators, soft robotics, and haptic interfaces • Simple fabrication and low cost compared to other actuators	• Mechanism relies on electrostatic forces between charged electrodes and elastomer • Require compliant electrodes to maintain electrical conductivity

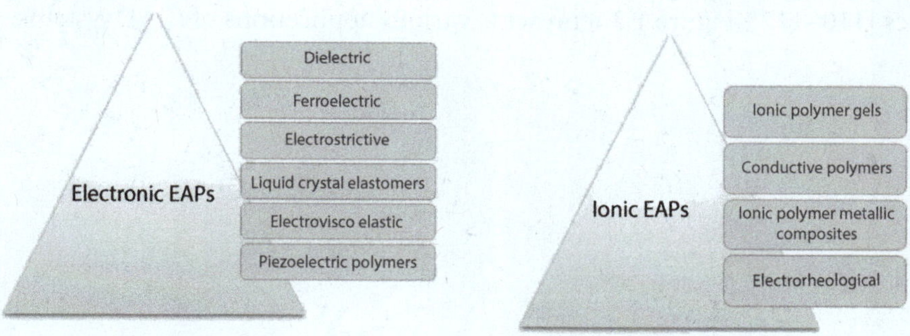

Figure 1.2 Classification of ionic and electronic EAPs.

In contrast, extrinsic polymers acquire conductive properties through the incorporation of external components, such as carbon black, carbon nanotubes, or conductive oxides. These additives modify the polymer matrix, creating pathways for the charge flow [104–109]. Ionic electroactive polymers (IEAPs) exhibit significant shape changes, such as bending, twisting, or expansion/contraction in response to an applied electric field. Key examples include conductive polymers such as PANI, despite its moderate environmental stability and decent electrical conductivity, is sensitive to pH and dopant leaching, and poly(3,4-ethylenedioxythiophene) (PEDOT), which is often doped with poly(styrenesulfonate) (PSS) to enhance its conductivity and enable applications in various devices. PPy exhibits fast response times and high actuation strains owing to the intercalation and deintercalation of ions within its polymer matrix. Although polythiophenes, such as poly(3-hexylthiophene) (P3HT) and poly(3-octylthiophene) (P3OT), are primarily utilized in organic electronics, their limited ionic conductivity typically restricts their use in practical IEAP applications. Ionic polymer–metal composites (IPMCs), consisting of a thin ionomeric polymer membrane sandwiched between two noble metal electrodes, exhibit large deformations at low voltages, making them promising for applications in robotics and soft actuators. Other materials, such as poly(vinyl alcohol) (PVA), can be used as support materials in electroactive polymer composites. Polyacetylene, the first conductive polymer discovered, has limitations owing to its instability and processability. Poly(phenylene vinylene) (PPV) and its derivatives are known for their electronic properties and extensive applications in organic electronics [110–117]. Figure 1.3 represents various applications of iEAP systems.

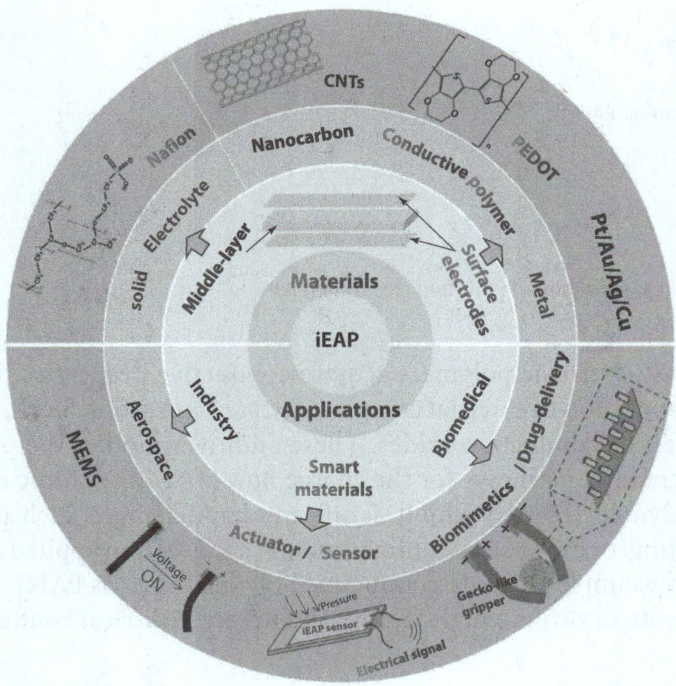

Figure 1.3 Diverse applications of iEAP systems demonstrating their versatility across multiple industries (CC-BY) [118].

1.3 Properties of EAPs

EAPs exhibit a unique combination of physical, chemical, and mechanical properties, rendering them suitable for a diverse array of applications, including actuators, sensors, and artificial muscles. The characteristics of EAPs can be systematically classified based on their underlying response mechanisms and material attributes. These properties are broadly categorized into the mechanical, electrical, thermal, and chemical domains, each exerting a significant influence on their overall performance across various applications.

1.3.1 Mechanical Properties

The mechanical properties of the EAPs are central to their functionality. They exhibit a unique combination of flexibility, strength, and responsiveness, which allow them to actuate or sense mechanical changes with high efficiency. One of the most remarkable aspects of EAPs is their ability to undergo significant deformations while sustaining large forces. This characteristic is in stark contrast to traditional ceramic piezoelectric materials, which, while able to withstand large forces, typically deform only a fraction of a percentage. The mechanical properties that are of particular interest in the study and application of EAPs include modulus of elasticity, fatigue limit, tensile strength, flexibility, and hardness [119, 120].

These properties are instrumental in dictating the effectiveness of EAPs in applications that require significant deformation and dynamic responses, such as actuators and artificial muscles [46, 121, 122]. EAPs are well known for their exceptional flexibility and capacity to undergo substantial deformations, often replicating the movements of biological muscles. Electronic EAPs, such as dielectric elastomers, can achieve remarkable strains of up to 300% under high electric fields, highlighting their impressive performance in terms of deformation. In contrast, ionic EAPs, such as ionic polymer-metal composites, exhibit bending and deformation at lower voltages, with a slower response time [119, 120]. The superior flexibility and elasticity of materials such as silicone-based dielectric elastomers and Nafion-based IPMCs make them well suited for applications requiring mechanical adaptability, such as in soft robotics and prosthetics [123–125]. The Young's modulus of the EAPs varies across different categories, reflecting their stiffness. For example, dielectric elastomers generally have a low modulus (0.1 to 10 MPa), which makes them highly stretchable under electrical fields [92]. Conversely, ferroelectric polymers, such as PVDF and its family, have a higher modulus, offering increased stiffness and structural integrity [22, 26]. This characteristic is particularly important for applications requiring specific mechanical stiffness, such as energy-harvesting devices or precision actuators [46, 126].

The ability of EAPs, especially electronic types, to withstand fatigue is crucial for maintaining their long-term stability during operation. For example, VHB acrylic elastomers exhibit exceptional durability by enduring numerous cycles of deformation with minimal degradation. Similarly, conductive polymers, such as PPy, have demonstrated excellent fatigue resistance, particularly in demanding biomedical applications that require continuous actuation [127–129].

1.3.2 Electrical Properties

The electrical properties of electroactive polymers (EAPs) determine their response to electric fields, affecting their ability to store and transfer electrical energy. These properties are crucial for various applications such as sensors, energy storage devices, and actuators, where electrical input and output are essential functions [3, 93, 121, 123, 130, 131]. EAPs with high dielectric constants can store more electrical energy, making them more sensitive to external electric fields. For example, dielectric elastomers, such as PDMS, have moderate dielectric constants, whereas ferroelectric polymers, such as PVDF and its copolymers, have higher dielectric constants owing to their polarizable molecular structures [91, 132]. This high dielectric constant makes them well suited for energy-harvesting applications and capacitive sensors [93, 133]. In the case of ionic EAPs, the electrical conductivity depends on ion mobility, which is influenced by the hydration levels and ionic properties of the polymer [1, 11, 134]. For instance, conductive polymers, such as PANI, have intrinsic electrical conductivity, allowing them to function as actuators and electrochemical sensors. This conductivity can also be modulated based on the redox states, enabling their use in dynamic environments [135, 136]. Different categories of EAPs require different actuation voltages. For example, dielectric elastomers typically require high voltages (around 100 V/μm) for actuation, whereas ionic EAPs such as IPMCs operate at much lower voltages (1–3 V), making them suitable for bio-applications where low-voltage operation is crucial for safety [137–142].

1.3.3 Thermal Properties

The thermal properties of EAPs play a crucial role in their operational stability across different temperature ranges, making them essential for applications under various environmental conditions. These properties include the thermal conductivity, thermal expansion, and temperature stability. For instance, materials such as ferroelectric PVDF demonstrate high thermal stability, allowing them to maintain their functional properties over a wide temperature range [143]. This makes them valuable in demanding environments such as the aerospace and automotive industries, and the temperature dependence of various characteristics, such as piezoelectricity, dielectric behavior, and flexibility, makes them potential candidates in various fields [144–146]. On the other hand, dielectric elastomers, despite their flexibility, may undergo thermal degradation at high temperatures, which restricts their application under thermally unstable conditions [147–149].

The coefficient of thermal expansion (CTE) of EAPs is often low, which is a crucial feature for applications requiring dimensional stability over temperature changes. However, in certain ionic EAPs, thermal expansion may lead to swelling owing to the movement of water or other ions within the polymer matrix, affecting the actuation performance of the material [150–152].

1.3.4 Chemical Properties

EAPs exhibit several notable chemical properties such as chemical stability, biocompatibility, and ion-exchange capacity [6, 22, 153]. These properties are vital for determining the compatibility of a material in various environments, including biological systems and corrosive settings. Biocompatibility is particularly important for ionic EAPs, such as Nafion-based IPMCs and hydrogel-based actuators, which are often used in biomedical applications, such as drug delivery and artificial muscles [154, 155]. These materials are designed to interact safely with biological tissues without causing toxic effects or significant immune responses [156, 157]. Chemical stability is crucial for EAPs used in environments exposed to moisture, chemicals, or solvents. Conductive polymers, such as PPy and PANI, demonstrate excellent chemical stability, making them suitable for applications in harsh chemical environments, including sensors and anti-corrosion coatings. Ionic EAPs rely heavily on their ion-exchange capacities for actuation. Materials, such as Nafion and polyacrylamide gels, exhibit high ion-exchange capacities, allowing efficient ion movement under an applied electric field. This property directly influences the ability of the material to actuate and respond to external stimuli, making it crucial in soft robotics and biomedical actuators [154, 155, 158, 159].

1.4 Applications of EAPs

Research on electroactive polymers began in the latter half of the 20th century and remained relatively limited in scope initially. However, towards the end of the century, there has been a significant expansion in research on electroactive polymers, leading to the development of new electroactive polymers. Figure 1.4 illustrates the key characteristics and functionalities of electroactive polymers. These polymers are now used in a variety of applications, ranging from biomedical devices to robotics [1, 160–163]. In 1999, the EAPAD Conference was held, marking the first major international forum for communicating the developments in the field of electroactive polymers. The first SPIE EAPAD Conference was held at Newport Beach, CA from March 1-5, 1999 [20]. Since then, the EAPAD conference has flourished and expanded in scope and influence within the scientific community. Its annual occurrence has attracted an increasing number of presentations and attendees, reflecting the growing interest and investment in electroactive polymer research and development. Research in the field of electroactive polymers is rapidly advancing, with new discoveries and applications emerging every year. Researchers are currently exploring the potential of these materials in fields such as soft robotics, energy harvesting, and artificial muscles. Electroactive polymers have numerous

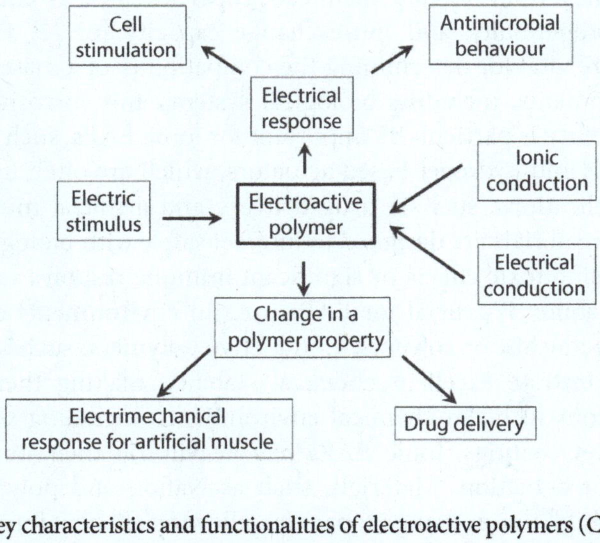

Figure 1.4 Key characteristics and functionalities of electroactive polymers (CC-BY) [23].

applications in the biomedical field. They are similar to biological muscles in their ability to undergo large deformations in response to electrical stimulation, which has inspired researchers to explore their potential in various fields. One of the main applications of EAPs is in tissue engineering, which can be used to develop artificial tissues and organs. By mimicking the properties of biological muscles, EAPs can provide a more realistic and efficient approach for building tissues and organs [9, 23, 39, 43, 122, 153, 158, 164–166].

A composite electrospun mat composed of gold nanoparticles and polyvinylidene fluoride (PVDF) was fabricated with enhanced piezoelectricity for nerve tissue engineering [167]. PVDF-TrFE scaffolds have been designed and developed by researchers to serve dual purposes as biological scaffolds and sensors [42]. These scaffolds can provide functional measurements of contractile function, which is an essential characteristic of tissue engineering applications [42]. Hasti Ghaedsharafi *et al.* reported core-shell fiber mats made of PVDF-ZnO/PCL that possess enhanced cell growth and cell proliferation [168]. These fiber mats are potential candidates for bone regeneration. A hydrogel composite was fabricated for bone tissue engineering and energy harvesting using a freeze-drying technique with a blend of cross-linked polyvinylidene fluoride-sodium alginate as the base material, calcium copper titanate nanowires as the piezoelectric agent, and hydroxyapatite nanoparticles to enhance its mineral content [169]. Composite scaffolds formed by integrating piezoelectric materials with biodegradable polymers such as PLLA exhibit enhanced piezoelectric responsiveness. Such scaffolds create a supportive microenvironment that promotes bone regeneration and contributes to functional recovery. A conductive polymer known as PEDOT: PSS has demonstrated the potential for promoting wound healing in animals because of its heightened angiogenic effect [170]. A conductive hydrogel, characterized by flexibility, biodegradability, and electroactivity, was successfully created using a method inspired by mussels. The hydrogel was synthesized by triggering the *in-situ* polymerization of PEDOT within a polydopamine-functionalized gelatin hydrogel, supported by the Hofmeister effect. This breakthrough is significant for managing different biological processes, particularly antioxidant function and cell movement [171]. Wu *et al.* developed a hydrogel composed of sulfonated hyaluronic acid and doped it with PANI. This unique hydrogel mimics the electrical conductivity of the human skin and has been shown to have antibacterial properties. Conductive polymers also enhance electrical stimulation, which contributes to the promotion of chronic wound healing [172]. Additionally, EAPs have potential applications as antimicrobials and biosensors. EAPs based composites can be used to develop medical devices to prevent infection and improve patient outcomes. Biosensors using EAPs can EAPs to detect environmental changes and provide feedback to users [55, 166, 173, 174]. Although the discovered electroactive polymers were not initially very effective in exhibiting strain, new strategies and methodologies have been developed to fabricate EAPs with high efficiency and potential for use in biomedical applications. Furthermore, the emergence of lead-free piezoelectric polymers has made these systems more biocompatible and sustainable. Figure 1.5 represents the various biomedical applications of electroactive polymer composites.

EAPs for electromagnetic interference (EMI) shielding are a growing area of research, particularly because of their adaptability, lightweight nature, and tunable properties compared to traditional metallic materials. The unique properties of some EAPs, including such as their conductivity, dielectric permittivity, and piezoelectric behavior, can be optimized to absorb or reflect electromagnetic waves, making them promising

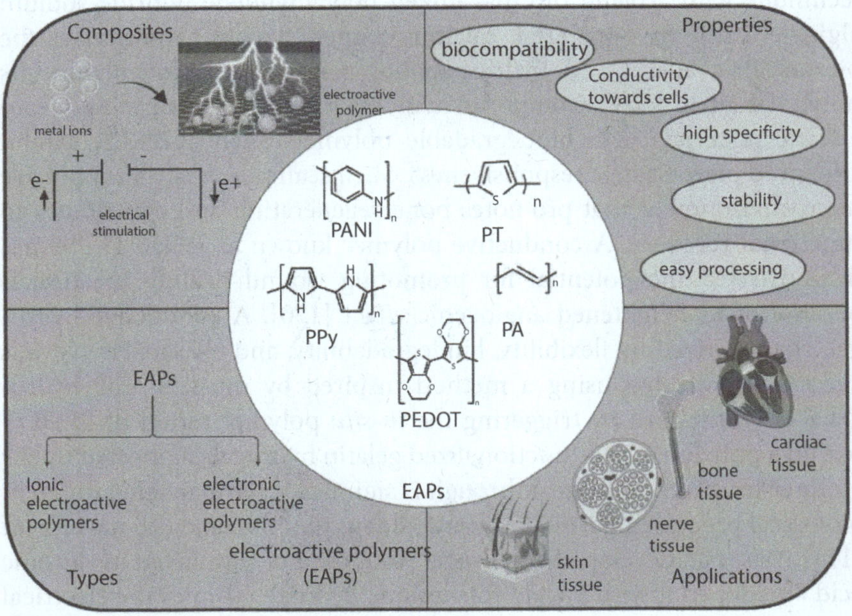

Figure 1.5 Diagrammatic representation of the various biomedical applications of electroactive polymer composites (CC-BY) [175].

candidates for advanced EMI shielding applications. EAPs can be engineered to have high dielectric constants, allowing them to absorb electromagnetic waves and dissipate energy as heat through dielectric relaxation [176–180]. The dielectric properties of polymers can be adjusted by adding fillers such as carbon nanotubes (CNTs), graphene, or metal oxide nanoparticles, enhancing their ability to attenuate EMI. Conductive polymers or polymer composites, such as PPy or PANI, function by creating a conductive path that reflects and absorbs EMI waves. Conductive fillers form a percolation network within the polymer matrix, providing a percolated conductive network that facilitates reflection and partial absorption of EMI through Joule heating and impedance matching [178, 179, 181, 182]. Some EAPs exhibit piezoelectric or ferroelectric properties, allowing them to interact with electric fields and convert electromagnetic energy into mechanical energy. This interaction can be tailored to enhance EMI shielding, especially at specific frequencies, by tuning the response of the material to external stimuli, such as mechanical strain or electric fields. Incorporating graphene quantum dot-decorated graphene into a polyvinylidene fluoride matrix achieved an EMI shielding effectiveness of 31 dB at 8 GHz, primarily through reflection and electron tunneling absorption. Enhancing this composite with conductive AgNPs further increased the total shielding effectiveness (SE_T) to 43 dB, improving the electrical conductivity. Consequently, the PVDF/G-D-GQDsAg nanocomposite can effectively shield 99.9% of the electromagnetic radiation in the X-band range, making it a promising material for EMI shielding in consumer electronics [180]. A ternary polymer blend of PC, PVDF, and PMMA was developed for EMI shielding, which achieved >99% EM radiation blocking. MWNTs in the PVDF phase improved the conductivity, whereas barium ferrite nanoparticles in the PC phase enhanced the microwave absorption. The blend achieved a shielding effectiveness (SE) of −37 dB at 18 GHz, with PMMA contributing to the mechanical strength, making it ideal for durable microwave absorbers [183]. A layered PVDF-based foam/film nanocomposite was developed for lightweight high-efficiency EMI shielding with minimal reflection. The structure combines a PVDF/SiC nanowire/MXene foam for effective absorption and a conductive PVDF/MWCNT/GnP film for reflection, achieving an EMI shielding effectiveness of 32.6 dB across the Ku-band and a reflection efficiency of only 0.0022%. This design offers a promising solution for advanced low-reflection EMI-shielding applications [184]. Rayan *et al.* published a review article highlighting the growing interest in ferrite-based composites, particularly when combined with polymers such as PANI, for electromagnetic interference (EMI) shielding. This study focuses on ferrite-based nanocomposites, emphasizing the influence of particle size, distribution, and various synthesis techniques to enhance EMI shielding through both reflection and absorption [181]. A review paper by Yangyong Wang *et al.* provides a comprehensive overview of the research dedicated to EMI shielding utilizing intrinsically conducting polymers. It specifically emphasizes the properties and applications of PANI and PPY, which are two of the most studied conducting polymers in the field. This paper explores how these materials, along with their composites, perform across a wide range of frequencies, highlighting their effectiveness and potential in improving EMI shielding solutions [185]. A study measured the EMI shielding effectiveness of various conductive polymer blends, such as PVC or nylon compounded with an inherently conducting polymer (ICP), observing that samples transitioned from electrically thin to thick across a frequency range of 1 MHz to 3 GHz. The shielding efficiency reached 70 dB at 1 GHz for pure organic composites with intrinsically conductive polymers. To address the limitations of existing models for near- and far-field shielding, the authors developed new expressions to accurately describe the material behavior in the crossover frequency region [177].

Hybrid polymer composites of liquid-exfoliated MoS_2 nanosheets and PANI nanoparticles dispersed in a polystyrene (PS) matrix demonstrated high dielectric and EMI shielding performances. The addition of MoS_2 to the PS/PANI blends formed an interconnected network, significantly enhancing the electrical conductivity and dielectric behavior, making the material suitable for EMI shielding. At 1 wt% MoS_2 in PS/PANI (5 wt%), the composite achieved an estimated shielding effectiveness of ~92 dB at 100 Hz, owing to an increased dielectric constant and loss. This improvement is attributed to the accumulation of charged dipoles at the insulator-conductor interface [179]. Polypropylene-based composites with Ag-coated glass microfibers and tin microparticles exhibit a significant EMI shielding enhancement of over 60 dB across a 0.1 MHz to 18 GHz range, attributed to a guided electrical percolation effect. This enhancement is attributed to effective electrical percolation, driven by a morphological transformation from a uniform to a highly heterogeneous structure, which facilitated increased conductivity and lowered the percolation threshold [186].

EAPs are advanced materials that convert mechanical energy from their surroundings into electrical energy [10, 130]. This characteristic makes EAPs ideal for various energy-scavenging applications, particularly low-power devices. EAPs are revolutionizing energy-scavenging technologies by efficiently converting mechanical energy from ambient sources to electrical energy. These polymers can be integrated into wearable devices, such as smart clothing and shoes, to harvest energy from body movements and provide a sustainable power source for portable electronics [188–195]. Additionally, EAPs are employed in structural health monitoring systems, where they harvest energy from environmental vibrations and mechanical stresses to power sensors that detect the structural integrity of buildings and bridges. In the medical field, EAPs enable the development of self-sustaining medical implants by harvesting energy from physiological movements, such as heartbeats, thus reducing the dependency on traditional batteries [196–199].

Another area in which electroactive polymers are making significant strides is energy scavenging, capacitors, and battery applications [10, 200–209]. Figure 1.6 shows the response mechanism in piezoelectric polymer. A comprehensive review by Shacklette examined the use of electroactive conductive polymers as electrode materials in non-aqueous secondary batteries. Their research evaluated the performance of various oxidized polymers, such as polyacetylene, PPy, and PANI, when employed as positive electrode materials. These polymers are noted for their exceptional ability to efficiently conduct electricity and undergo stable redox reactions, making them highly suitable for secondary battery applications that require long-term durability and cycling stability. By exploring the electrochemical properties of these materials, the authors provided a comprehensive understanding of how conductive polymers can contribute to enhancing the performance of non-aqueous secondary batteries. Their findings underscore the potential of these materials to improve the energy density, cycling life, and charge–discharge efficiency, thereby advancing the design and functionality of modern rechargeable battery systems [210].

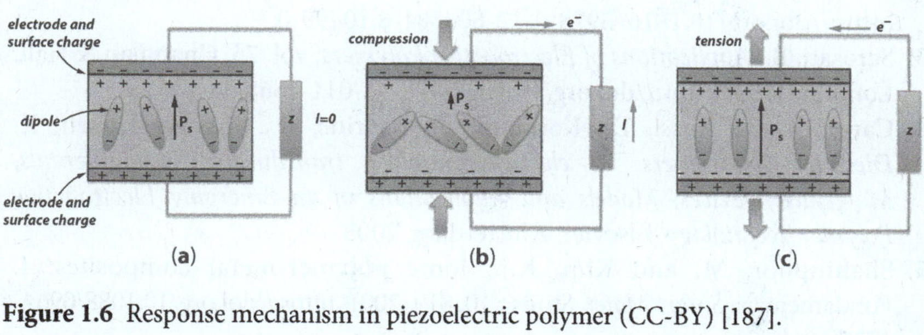

Figure 1.6 Response mechanism in piezoelectric polymer (CC-BY) [187].

1.5 Conclusion

Electroactive polymers (EAPs) are highly flexible materials that have widespread applications, ranging from electronics and healthcare to soft robotics. These materials offer a promising future for new technologies such as artificial muscles, flexible electronics, biosensors, and drug delivery systems. This chapter has laid the conceptual foundation by outlining the fundamental principles and characteristics of electroactive polymers, providing a basis for understanding their unique electromechanical behavior. Building on this groundwork, the following chapters will explore advanced research directions and real-world applications, highlighting how EAP technologies are evolving to meet future demands. These discussions will clarify the roles of EAPs across various sectors and demonstrate how they are driving innovation and contributing to progress in both scientific and industrial domains.

AI disclosure statement: The author acknowledges the use of AI-based tools for minor language editing or grammatical correction. However, all intellectual contributions, analyses, and conclusions presented in this chapter are entirely the author's own.

References

1. Rahman, M.H., Werth, H., Goldman, A., Hida, Y., Diesner, C., Lane, L., Menezes, P.L., Recent Progress on Electroactive Polymers: Synthesis, Properties and Applications. *Ceramics*, 4, 516–541, 2021, https://doi.org/10.3390/CERAMICS4030038.
2. Ribeiro, C., Costa, C.M., Martins, P., Correia, V., Lanceros-Mendez, S., Piezoelectric Polymers and Polymer Composites for Sensors and Actuators, in: *Reference Module in Materials Science and Materials Engineering*, 2018, https://doi.org/10.1016/B978-0-12-803581-8.10499-0.
3. Scrosati, B., *Applications of Electroactive Polymers*, vol. 75, Chapman & Hall, London, 1993, https://doi.org/10.1007/978-94-011-1568-1.
4. Carpi, F., De Rossi, D., Kornbluh, R., Pelrine, R., Sommer-Larsen, P., *Dielectric elastomers as electromechanical transducers: Fundamentals, Materials, Devices, Models and Applications of an Emerging Electroactive Polymer Technology*, Elsevier, Amsterdam, 2008.
5. Shahinpoor, M. and Kim, K.J., Ionic polymer-metal composites: I. Fundamentals. *Smart Mater. Struct.*, 10, 819, 2001, https://doi.org/10.1088/0964-1726/10/4/327.

6. Bar-Cohen, Y., Electroactive Polymers as Artificial Muscles: A Review. *J. Spacecr. Rockets*, 39, 822–827, 2002, https://doi.org/10.2514/2.3902.
7. Peng, C., Chen, Y., Yang, B., Jiang, Z., Liu, Y., Liu, Z., Zhou, L., Tang, L., Recent Advances of Soft Actuators in Smart Wearable Electronic-Textile. *Adv. Mater. Technol.*, 9, 2400079, 2024, https://doi.org/10.1002/ADMT.202400079.
8. Wax, S.G. and Sands, R.R., Electroactive polymer actuators and devices. *Proc. SPIE*, 3669, 2–10, 1999, https://doi.org/10.1117/12.349666.
9. Carpi, F. and De Rossi, D., Electroactive polymer-based devices for e-textiles in biomedicine. *IEEE Trans. Inform. Technol. Biomed.*, 9, 574, 2005, https://doi.org/10.1109/titb.2005.858429.
10. Mangold, K., Shaw, J.A., Vollmer, M., García Alía, R., Brugger, M., Jean-Mistral, C., Basrour, S., Chaillout, J.-J., Comparison of electroactive polymers for energy scavenging applications. *Smart Mater. Struct.*, 19, 085012, 2010, https://doi.org/10.1088/0964-1726/19/8/085012.
11. Şensoy, K.G., Properties of Electroactive Polymers, in: *Electroactive Polymeric Materials*, pp. 51–62, 2022, https://doi.org/10.1201/9781003173502-3.
12. Röntgen, W.C., Ueber die durch Electricität bewirkten Form- und Volumenänderungen von dielectrischen Körpern. *Ann. Phys.*, 247, 771–786, 1880, https://doi.org/10.1002/ANDP.18802471304.
13. Sacerdote, M.P., On the electrical deformation of isotropic dielectric solids. *J. Phys.*, 3, 282–285, 1899.
14. Eguchi, M., XX. On the permanent electret. *Lond. Edinb. Dublin Philos. Mag. J. Sci.*, 49, 178–192, 1925, https://doi.org/10.1080/14786442508634594.
15. Kawai, H., The Piezoelectricity of Poly (vinylidene Fluoride). *Jpn. J. Appl. Phys.*, 8, 975, 1969, https://doi.org/10.1143/JJAP.8.975/META.
16. Heeger, A.J., Semiconducting and metallic polymers: The fourth generation of polymeric materials. *J. Phys. Chem. B*, 105, 8475–8491, 2001, https://doi.org/10.1021/JP011611W.
17. MacDiarmid, A.G., Nobel lecture: "Synthetic metals": A novel role for organic polymers. *Rev. Mod. Phys.*, 73, 701–712, 2001, https://doi.org/10.1103/REVMODPHYS.73.701.
18. Shirakawa, H., Nobel lecture: The discovery of polyacetylene film - The dawning of an era of conducting polymers. *Rev. Mod. Phys.*, 73, 713–718, 2001, https://doi.org/10.1103/REVMODPHYS.73.713.
19. Pelrine, R.E., Kornbluh, R.D., Joseph, J.P., Electrostriction of polymer dielectrics with compliant electrodes as a means of actuation. *Sens. Actuators A Phys.*, 64, 77–85, 1998, https://doi.org/10.1016/S0924-4247(97)01657-9.
20. Bar-Cohen, Y., Smart Structures and Materials 1999: Electroactive Polymer Actuators and Devices, 3669,1999.
21. Ameduri, B., From Vinylidene Fluoride (VDF) to the Applications of VDF-Containing Polymers and Copolymers: Recent Developments and Future Trends. *Chem. Rev.*, 109, 6632–6686, 2009, https://doi.org/10.1021/cr800187m.

22. Martins, P., Lopes, A.C., Lanceros-Mendez, S., Electroactive phases of poly(vinylidene fluoride): Determination, processing and applications. *Prog. Polym. Sci.*, 39, 683–706, 2014, https://doi.org/10.1016/j.progpolymsci.2013.07.006.
23. Palza, H., Zapata, P.A., Angulo-Pineda, C., Electroactive smart polymers for biomedical applications. *Materials*, 12, 277, 2019, https://doi.org/10.3390/MA12020277.
24. Ribeiro, C., Costa, C.M., Correia, D.M., Nunes-Pereira, J., Oliveira, J., Martins, P., Gonçalves, R., Cardoso, V.F., Lanceros-Méndez, S., Electroactive poly(vinylidene fluoride)-based structures for advanced applications. *Nat. Protoc.*, 13, 681–704, 2018, https://doi.org/10.1038/nprot.2017.157.
25. Bera, B. and Das Sarkar, M., Piezoelectricity in PVDF and PVDF Based Piezoelectric Nanogenerator: A Concept. *IOSR J. Appl. Phys. (IOSR-JAP)*, 9, 95–99, 2017. https://doi.org/10.9790/4861-0903019599.
26. Xia, W. and Zhang, Z., PVDF-based dielectric polymers and their applications in electronic materials. *IET Nanodielectr.*, 1, 17–31, 2018, https://doi.org/10.1049/IET-NDE.2018.0001.
27. Dallaev, R., Pisarenko, T., Sobola, D., Orudzhev, F., Ramazanov, S., Trčka, T., Brief Review of PVDF Properties and Applications Potential. *Polymers*, 14, 4793, 2022, https://doi.org/10.3390/POLYM14224793.
28. Nalwa, H.S., *Ferroelectric polymers: chemistry: physics, and applications.* CRC Press, Boca Raton, 1995.
29. Li, Y.Y., Wang, B., Ma, M.G., Wang, B., Review of recent development on preparation, properties, and applications of cellulose-based functional materials. *Int. J. Polym. Sci.*, 2018, 1, 8973643, 2018.
30. Nevstrueva, D., Murashko, K., Vunder, V., Aabloo, A., Pihlajamäki, A., Mänttäri, M., Pyrhönen, J., Koiranen, T., Torop, J., Natural cellulose ionogels for soft artificial muscles. *Colloids Surf., B*, 161, 244–251, 2018, https://doi.org/10.1016/J.COLSURFB.2017.10.053.
31. Kim, J., Yun, S., Ounaies, Z., Discovery of cellulose as a smart material. *Macromolecules*, 39, 12, 4202–4206. 2006, https://doi.org/10.1021/ma060261e.
32. Chao, D., Cui, L., Lu, X., Mao, H., Zhang, W., Wei, Y., Electroactive polyimide with oligoaniline in the main chain *via* oxidative coupling polymerization. *Eur. Polym. J.*, 43, 2641–2647, 2007, https://doi.org/10.1016/J.EURPOLYMJ.2007.03.048.
33. Huang, T.C., Yeh, L.C., Lai, G.H., Huang, B.S., Yang, T.I., Hsu, S.C., Lo, A.Y., Yeh, J.M., Advanced superhydrophobic electroactive fluorinated polyimide and its application in anticorrosion coating. *Int. J. Green Energy*, 14, 113–120, 2017, https://doi.org/10.1080/15435075.2016.1233495.
34. Amaro, L., Correia, D.M., Marques-Almeida, T., Martins, P.M., Pérez, L., Vilas, J.L., Botelho, G., Lanceros-Mendez, S., Ribeiro, C., Tailored Biodegradable and Electroactive Poly(Hydroxybutyrate-Co-Hydroxyvalerate) Based

Morphologies for Tissue Engineering Applications. *Int. J. Mol. Sci.*, 19, 2149, 2018, https://doi.org/10.3390/IJMS19082149.

35. Najjar, R., Luo, Y., Jao, D., Brennan, D., Xue, Y., Beachley, V., Hu, X., Xue, W., Biocompatible Silk/Polymer Energy Harvesters Using Stretched Poly (vinylidene fluoride-co-hexafluoropropylene) (PVDF-HFP) Nanofibers. *Polymers (Basel)*, 9, 479, 2017, https://doi.org/10.3390/POLYM9100479.
36. Sarkar, L., Kandala, A.B., Bonam, S., Mohanty, S., Singh, S.G., Krishna Vanjari, S.R., Flexible polymer-based triboelectric nanogenerator using Poly(vinylidene fluoride) and bombyx mori silk. *Mater. Today Sustain.*, 20, 100230, 2022, https://doi.org/10.1016/J.MTSUST.2022.100230.
37. Lee, J.W., Takase, Y., Newman, B.A., Scheinbeim, J.I., Effect of annealing on the ferroelectric behavior of nylon - 11 and nylon - 7. *J. Polym. Sci. B Polym. Phys.*, 29, 3, 279–286, 1991, https://doi.org/10.1002/polb.1991.090290303.
38. Yang, R., Zhou, Y., Ren, Y., Xu, D., Guan, L., Guo, X., Zhang, R., Zhao, B., Promising PVDF-CNT-Graphene-NiCo chains composite films with excellent electromagnetic interference shielding performance. *J. Alloys Compd.*, 908, 164538, 2022, https://doi.org/10.1016/j.jallcom.2022.164538.
39. Gimenes, R., Zaghete, M.A., Bertolini, M., Varela, J.A., Coelho, L.O., Silva, Jr., N.F., Composites PVDF-TrFE/BT used as bioactive membranes for enhancing bone regeneration. *Smart Structures and Materials 2004: Electroactive Polymer Actuators and Devices (EAPAD)*, vol. 5385, p. 539, 2004, https://doi.org/10.1117/12.548647.
40. Jiang, R., Bian, T., Zheng, X., Cui, Y., Zhang, Y., Li, M., Li, Z., Corrosion-resistant porous hydrophobic PVDF-CBC foam for the treatment of oil-water separation. *Mater. Chem. Phys.*, 273, 125080, 2021, https://doi.org/10.1016/j.matchemphys.2021.125080.
41. Wang, X., Xiao, C., Liu, H., Huang, Q., Fu, H., Fabrication and properties of PVDF and PVDF-HFP microfiltration membranes. *J. Appl. Polym. Sci.*, 135, 46711, 2018, https://doi.org/10.1002/APP.46711.
42. Adadi, N., Yadid, M., Gal, I., Asulin, M., Feiner, R., Edri, R., Dvir, T., Electrospun Fibrous PVDF-TrFe Scaffolds for Cardiac Tissue Engineering, Differentiation, and Maturation. *Adv. Mater. Technol.*, 5, 1900820, 2020, https://doi.org/10.1002/ADMT.201900820.
43. Lee, J.C., Suh, I.W., Park, C.H., Kim, C.S., Polyvinylidene fluoride/silk fibroin-based bio-piezoelectric nanofibrous scaffolds for biomedical application. *J. Tissue Eng. Regener. Med.*, 15, 869–877, 2021, https://doi.org/10.1002/TERM.3232.
44. Guo, H.F., Li, Z.S., Dong, S.W., Chen, W.J., Deng, L., Wang, Y.F., Ying, D.J., Piezoelectric PU/PVDF electrospun scaffolds for wound healing applications. *Colloids Surf., B*, 96, 29–36, 2012, https://doi.org/10.1016/J.COLSURFB.2012.03.014.

45. Zhang, Q., Mochalin, V.N., Neitzel, I., Knoke, I.Y., Han, J., Klug, C.A., Zhou, J.G., Lelkes, P.I., Gogotsi, Y., Fluorescent PLLA-nanodiamond composites for bone tissue engineering. *Biomaterials*, 32, 87–94, 2011, https://doi.org/10.1016/J.BIOMATERIALS.2010.08.090.
46. Bar-Cohen, Y. and Anderson, I.A., Electroactive polymer (EAP) actuators—background review. *Mech. Soft Mater.*, 1, 5, 2019, https://doi.org/10.1007/S42558-019-0005-1.
47. Maksimkin, A.V., Dayyoub, T., Telyshev, D.V., Gerasimenko, A.Y., Electroactive Polymer-Based Composites for Artificial Muscle-like Actuators: A Review. *Nanomaterials*, 12, 2272, 2022, https://doi.org/10.3390/NANO12132272.
48. Lee, J.E., Shin, Y.E., Lee, G.H., Kim, J., Ko, H., Chae, H.G., Polyvinylidene fluoride (PVDF)/cellulose nanocrystal (CNC) nanocomposite fiber and triboelectric textile sensors. *Compos. B Eng.*, 223, 109098, 2021, https://doi.org/10.1016/j.compositesb.2021.109098.
49. Yang, T., Wan, C., Zhang, X., Liu, T., Niu, L., Fang, J., Liu, Y., High-efficiency preparation of multifunctional conjugated electrospun graphene doped PVDF/CF yarns for energy harvesting and human movement monitoring in TENG textile. *Nano Res.*, 17, 4478–4488, 2024, https://doi.org/10.1007/S12274-023-6373-8/METRICS.
50. Kundu, C.K., Song, L., Hu, Y., Nanoparticles based coatings for multifunctional Polyamide 66 textiles with improved flame retardancy and hydrophilicity. *J. Taiwan Inst. Chem. Eng.*, 112, 15–19, 2020, https://doi.org/10.1016/J.JTICE.2020.07.013.
51. Du Toit, L.C., Kumar, P., Choonara, Y.E., Pillay, V., Electroactive Polymers and Coatings, in: *Industrial Applications for Intelligent Polymers and Coatings*, pp. 51–89, 2016, https://doi.org/10.1007/978-3-319-26893-4_3.
52. Jaiswal, A. and Gupta, K., History and Progress of Electroactive Polymers, in: *Electroactive Polymeric Materials*, pp. 75–84, CRC Press, 2022, https://doi.org/10.1201/9781003173502-5
53. Inzelt, G., Recent advances in the field of conducting polymers. *J. Solid State Electrochem.*, 21, 1965–1975, 2017, https://doi.org/10.1007/S10008-017-3611-6.
54. Nakamura, N., Yokoshima, T., Nara, H., Zhen Jin, Z., Bin Qiu, Y., Peng Guo, X., Vahid Ebadi, S., Fashandi, H., Semnani, D., Rezaei, B., Fakhrali, A., Electroactive actuator based on polyurethane nanofibers coated with polypyrrole through electrochemical polymerization: a competent method for developing artificial muscles. *Smart Mater. Struct.*, 29, 045008, 2020, https://doi.org/10.1088/1361-665X/AB73E5.
55. Dhand, C., Das, M., Datta, M., Malhotra, B.D., Recent advances in polyaniline based biosensors. *Biosens. Bioelectron.*, 26, 2811–2821, 2011, https://doi.org/10.1016/J.BIOS.2010.10.017.

56. Pathiranage, T.M.S.K., Dissanayake, D.S., Niermann, C.N., Ren, Y., Biewer, M.C., Stefan, M.C., Role of polythiophenes as electroactive materials. *J. Polym. Sci. A Polym. Chem.*, 55, 3327–3346, 2017, https://doi.org/10.1002/POLA.28726.
57. Kaloni, T.P., Giesbrecht, P.K., Schreckenbach, G., Freund, M.S., Polythiophene: from fundamental perspectives to applications. *Chem. Mater.*, 29, 24, 10248–10283, 2017. https://doi.org/10.1021/acs.chemmater.7b03035.
58. Pan, Y., Zhang, J., Guo, X., Li, Y., Li, L., Pan, L., Recent Advances in Conductive Polymers-Based Electrochemical Sensors for Biomedical and Environmental Applications. *Polymers*, 16, 1597, 2024, https://doi.org/10.3390/POLYM16111597.
59. Mahmood, J., Arsalani, N., Naghash-Hamed, S., Hanif, Z., Geckeler, K.E., Preparation and characterization of hybrid polypyrrole nanoparticles as a conducting polymer with controllable size. *Sci. Rep.*, 14, 11653, 2024, https://doi.org/10.1038/S41598-024-61587-1.
60. Thadathil, A., Pradeep, H., Joshy, D., Ismail, Y.A., Periyat, P., Polyindole and polypyrrole as a sustainable platform for environmental remediation and sensor applications. *Mater. Adv.*, 3, 2990–3022, 2022, https://doi.org/10.1039/D2MA00022A.
61. Del-Oso, J.A., Maldonado, J.L., Ramos-Ortíz, G., Rodríguez, M., Güizado-Rodríguez, M., Escalante, J., Frontana-Uribe, B.A., Pérez-Gutiérrez, E., Santillan, R., New polythiophene derivatives and enhanced photovoltaic effect by a boron compound blended with them in OPVs cells. *Synth. Met.*, 196, 83–91, 2014, https://doi.org/10.1016/J.SYNTHMET.2014.07.019.
62. Bai, Q., Cheng, Y., Wang, W., Chen, J., Sun, H., Polythiophene and its derivatives for all-polymer solar cells. *J. Mater. Chem. A Mater.*, 12, 16251–16267, 2024, https://doi.org/10.1039/D4TA02245A.
63. Perepichka, I.F., Perepichka, D.F., Meng, H., Wudl, F., Light-Emitting Polythiophenes. *Adv. Mater.*, 17, 2281–2305, 2005, https://doi.org/10.1002/ADMA.200500461.
64. Kolasińska, E. and Kolasiński, P., A Review on Electroactive Polymers for Waste Heat Recovery. *Materials*, 9, 485, 2016, https://doi.org/10.3390/MA9060485.
65. Potaufeux, J.-E., Odent, J., Notta-Cuvier, D., Lauro, F., Raquez, J.-M., A comprehensive review of the structures and properties of ionic polymeric materials. *Polym. Chem.*, 11, 5914–5936, 2020, https://doi.org/10.1039/D0PY00770F.
66. Capek, I., Dispersions of polymer ionomers: I. *Adv. Colloid Interface Sci.*, 112, 1–29, 2004, https://doi.org/10.1016/J.CIS.2004.06.001.
67. Choi, U.H., Lee, M., Wang, S., Liu, W., Winey, K.I., Gibson, H.W., Colby, R.H., Ionic conduction and dielectric response of poly(imidazolium acrylate) ionomers. *Macromolecules*, 45, 3974–3985, 2012, https://doi.org/10.1021/ma202784e

68. Zhang, H., Lin, Z., Hu, Y., Ma, S., Liang, Y., Ren, L., Ren, L., Low-Voltage Driven Ionic Polymer-Metal Composite Actuators: Structures, Materials, and Applications. *Adv. Sci.*, 10, 2206135, 2023, https://doi.org/10.1002/ADVS.202206135.
69. Bekaroğlu, M.G., Alemdar, A., İşçi, S., Comparison of ionic polymers in the targeted drug delivery applications as the coating materials on the Fe3O4 nanoparticles. *Mater. Sci. Eng., C*, 103, 109838, 2019, https://doi.org/10.1016/J.MSEC.2019.109838.
70. Schacht, E.H., Ionic Polymers as Drug Carriers, in: *Controlled Drug Delivery*, pp. 149–173, 2019, https://doi.org/10.1201/9780429262258-6.
71. Nagao, Y., Proton-Conducting Polymers: Key to Next-Generation Fuel Cells, Electrolyzers, Batteries, Actuators, and Sensors. *ChemElectroChem*, 11, e202300846, 2024, https://doi.org/10.1002/CELC.202300846.
72. Kim, K.J. and Shahinpoor, M., A novel method of manufacturing three-dimensional ionic polymer–metal composites (IPMCs) biomimetic sensors, actuators and artificial muscles. *Polymer (Guildf)*, 43, 797–802, 2002, https://doi.org/10.1016/S0032-3861(01)00648-6
73. Lebedeva, O., Kultin, D., Kustov, L., Advanced research and prospects on polymer ionic liquids: trends, potential and application. *Green Chem.*, 25, 9001–9019, 2023, https://doi.org/10.1039/D3GC02131A.
74. Fuller, J., Breda, A.C., Carlin, R.T., Ionic Liquid-Polymer Gel Electrolytes. *J. Electrochem. Soc.*, 144, L67–L70, 1997, https://doi.org/10.1149/1.1837555/XML.
75. Snedden, P., Cooper, A.I., Scott, K., Winterton, N., Cross-Linked Polymer–Ionic Liquid Composite Materials. *Macromolecules*, 36, 4549–4556, 2003, https://doi.org/10.1021/MA021710N.
76. Shaplov, A.S., Marcilla, R., Mecerreyes, D., Recent Advances in Innovative Polymer Electrolytes based on Poly(ionic liquid)s. *Electrochim. Acta*, 175, 18–34, 2015, https://doi.org/10.1016/J.ELECTACTA.2015.03.038.
77. Lee, J.W. and Yoo, Y.T., Preparation and performance of IPMC actuators with electrospun Nafion®-MWNT composite electrodes. *Sens. Actuators B Chem.*, 159, 103–111, 2011, https://doi.org/10.1016/J.SNB.2011.06.057.
78. Lufrano, F., Staiti, P., Minutoli, M., Evaluation of nafion based double layer capacitors by electrochemical impedance spectroscopy. *J. Power Sources*, 124, 314–320, 2003, https://doi.org/10.1016/S0378-7753(03)00589-5.
79. Park, S.W., Kim, S.J., Park, S.H., Lee, J., Kim, H., Kim, M.K., Recent Progress in Development and Applications of Ionic Polymer–Metal Composite. *Micromachines*, 13, 1290, 2022, https://doi.org/10.3390/MI13081290.
80. Marr, P.C. and Marr, A.C., Ionic liquid gel materials: applications in green and sustainable chemistry. *Green Chem.*, 18, 105–128, 2015, https://doi.org/10.1039/C5GC02277K.
81. Phillips, A.K. and Moore, R.B., Ionic actuators based on novel sulfonated ethylene vinyl alcohol copolymer membranes. *Polymer (Guildf)*, 46, 7788–7802, 2005, https://doi.org/10.1016/J.POLYMER.2005.02.123.

82. Lai, J., Zeng, B., Liu, J., Zhang, J., Pei, W., Zhou, Y., Liu, Y., Li, J., Tong, Y., Current research status of ionic polymer–metal composites in applications of low-voltage actuators. *Mater. Adv.*, 5, 4601–4617, 2024, https://doi.org/10.1039/D4MA00040D.
83. Yang, L., Wang, H., Zhang, X., Recent progress in preparation process of ionic polymer-metal composites. *Results Phys.*, 29, 104800, 2021, https://doi.org/10.1016/J.RINP.2021.104800.
84. Alshubramy, M.A., Alamry, K.A., Hussein, M.A., Alorfi, H.S., Reda Aouad, M., Rezki, N., Alosaimi, A.M., Effective synergy of ionic capacity and C3-symmetry porosity in porous ionic polymers based imidazolium/pyridinium-benzene and/or triazine molecular hybrids: A comprehensive review. *Results Chem.*, 9, 101673, 2024, https://doi.org/10.1016/J.RECHEM.2024.101673.
85. Wang, F., Hu, Z., Fan, Y., Bai, W., Wu, S., Sun, H., Zhu, Z., Liang, W., Li, A., Salt-Rejection Solar Absorbers Based on Porous Ionic Polymers Nanowires for Desalination. *Macromol. Rapid Commun.*, 42, 2000536, 2021, https://doi.org/10.1002/MARC.202000536.
86. Bai, X., Su, Z., Wei, J., Ma, L., Duan, S., Wang, N., Zhang, X., Li, J., Zinc(II) porphyrin-Based Porous Ionic Polymers (PIPs) as Multifunctional Heterogeneous Catalysts for the Conversion of CO2 to Cyclic Carbonates. *Ind. Eng. Chem. Res.*, 61, 5093–5102, 2022, https://doi.org/10.1021/acs.iecr.2c00161
87. Xu, D., Guo, J., Yan, F., Porous ionic polymers: Design, synthesis, and applications. *Prog. Polym. Sci.*, 79, 121–143, 2018, https://doi.org/10.1016/J.PROGPOLYMSCI.2017.11.005.
88. Kesküla, A., Peikolainen, A.L., Kilmartin, P.A., Kiefer, R., Solvent effect in imidazole-based poly(Ionic liquid) membranes: Energy storage and sensing. *Polymers (Basel)*, 13, 3466, 2021, https://doi.org/10.3390/POLYM13203466/S1.
89. Miralles-Comins, S., Zanatta, M., Sans, V., Advanced Formulations Based on Poly(ionic liquid) Materials for Additive Manufacturing. *Polymers*, 14, 5121, 2022, https://doi.org/10.3390/POLYM14235121.
90. Zhu, M. and Yang, Y., Poly(ionic liquid)s: an emerging platform for green chemistry. *Green Chem.*, 26, 5022–5102, 2024, https://doi.org/10.1039/D4GC00202D.
91. Billah, S.M.R., Dielectric Polymers, pp. 241–288, 2019, https://doi.org/10.1007/978-3-319-95987-0_8.
92. Pelrine, R. and Kornbluh, R., Dielectric Elastomers as Electroactive Polymers (EAPs): Fundamentals, in: *Electromechanically Active Polymers*, pp. 671–686, 2016, https://doi.org/10.1007/978-3-319-31530-0_30.
93. Wu, X., Chen, X., Zhang, Q.M., Tan, D.Q., Advanced dielectric polymers for energy storage. *Energy Storage Mater.*, 44, 29–47, 2022, https://doi.org/10.1016/J.ENSM.2021.10.010.

94. Liu, Z., Liu, Y.D., Shi, Q., Liang, Y., Electroactive dielectric polymer gels as new-generation soft actuators: a review. *J. Mater. Sci.*, 56, 27, 14943–14963, 2021, https://doi.org/10.1007/S10853-021-06260-Y.
95. Pan, X., Pu, W., Liu, Y., Xiao, Y., Pu, J., Shi, Y., Wu, H., Wang, H., Self-Perceptional Soft Robotics by a Dielectric Elastomer. *ACS Appl. Mater. Interfaces*, 16, 26797–26807, 2024, https://doi.org/10.1021/ACSAMI.4C04700.
96. Kofod, G., Sommer-Larsen, P., Kornbluh, R., Pelrine, R., Actuation response of polyacrylate dielectric elastomers. *J. Intell. Mater. Syst. Struct.*, 14, 787–793, 2003, https://doi.org/10.1177/104538903039260.
97. Anderson, I.A., Gisby, T.A., McKay, T.G., O'Brien, B.M., Calius, E.P., Multi-functional dielectric elastomer artificial muscles for soft and smart machines. *J. Appl. Phys.*, 112, 2012, https://doi.org/10.1063/1.4740023.
98. Jun, K., Kim, D., Ryu, S., Oh, I.K., Surface Modification of Anisotropic Dielectric Elastomer Actuators with Uni-A nd Bi-axially Wrinkled Carbon Electrodes for Wettability Control. *Sci. Rep.*, 7, 6091, 2017, https://doi.org/10.1038/S41598-017-06274-0.
99. Zhao, Y., Yin, L.J., Zhong, S.L., Zha, J.W., Dang, Z.M., Review of dielectric elastomers for actuators, generators and sensors. *IET Nanodielectr.*, 3, 99–106, 2020, https://doi.org/10.1049/IET-NDE.2019.0045.
100. Wang, K., Ouyang, G., Chen, X., Jakobsen, H., Engineering Electroactive Dielectric Elastomers for Miniature Electromechanical Transducers. *Polym. Rev.*, 57, 369–396, 2017, https://doi.org/10.1080/15583724.2016.1268156.
101. Chiba, S., Dielectric Elastomers, in: *Soft Actuators: Materials, Modeling, Applications, and Future Perspectives: Second Edition*, pp. 259–273, 2019, https://doi.org/10.1007/978-981-13-6850-9_14.
102. Michel, S., Zhang, X., Wissler, M., C.L.-P., A comparison between silicone and acrylic elastomers as dielectric materials in electroactive polymer actuators, 2010, Wiley Online Library, https://scijournals.onlinelibrary.wiley.com/doi/abs/10.1002/pi.2751
103. Ouyang, Z., Li, J., Wang, J., Li, Q., Ni, T., Zhang, X., Wang, H., Li, Q., Su, Z., Wei, G., Fabrication, characterization and sensor application of electrospun polyurethane nanofibers filled with carbon nanotubes and silver nanoparticles. *J. Mater. Chem. B*, 1, 2415–2424, 2013, https://doi.org/10.1039/C3TB20316F.
104. Le, T.H., Kim, Y. and Yoon, H., Electrical and electrochemical properties of conducting polymers. *Polymers*, 9, 4, 150, 2017. https://doi.org/10.3390/polym9040150
105. Sharma, S., Verma, A., Rangappa, S.M., Siengchin, S., Ogata, S., Recent progressive developments in conductive-fillers based polymer nanocomposites (CFPNC's) and conducting polymeric nanocomposites (CPNC's) for multifaceted sensing applications. *J. Mater. Res. Technol.*, 26, 5921–5974, 2023, https://doi.org/10.1016/J.JMRT.2023.08.300.

106. Tan, H.X. and Xu, X.C., Conductive properties and mechanisms of different polymers doped by carbon nanotube/polypyrrole 1D hybrid nanotubes. *RSC Adv.*, 5, 61383–61389, 2015, https://doi.org/10.1039/C5RA09816E.
107. Gómez, I.J., Sulleiro, M.V., Mantione, D., Alegret, N., Carbon Nanomaterials Embedded in Conductive Polymers: A State of the Art. *Polymers*, 13, 745, 2021, https://doi.org/10.3390/POLYM13050745.
108. Huang, Y., Kormakov, S., He, X., Gao, X., Zheng, X., Liu, Y., Sun, J., Wu, D., Conductive Polymer Composites from Renewable Resources: An Overview of Preparation, Properties, and Applications. *Polymers (Basel)*, 11, 187, 2019, https://doi.org/10.3390/POLYM11020187.
109. Barth, S., Bässler, H., Rost, H., Hörhold, H., Extrinsic and intrinsic dc photoconductivity in a conjugated polymer. *Phys. Rev. B: Condens. Matter Mater. Phys.*, 56, 3844–3851, 1997, https://doi.org/10.1103/PHYSREVB.56.3844.
110. Shirakawa, H., The discovery of polyacetylene film – the dawning of an era of conducting polymers. *Curr. Appl. Phys.*, 1, 281–286, 2001, https://doi.org/10.1016/S1567-1739(01)00052-9.
111. Bobrowska, D.M., Gdula, K., Breczko, J., Basa, A., Markiewicz, K.H., Winkler, K., Poly(p-phenylene vinylene) incorporated into carbon nanostructures. *J. Nanopart. Res.*, 24, 1–18, 2022, https://doi.org/10.1007/s11051-022-05589-y
112. Gruber, J., Chia Li, R.W., Hümmelgen, I.A., Synthesis, properties, and applications of poly(p-phenylene vinylene)S, in: *Handbook of Advanced Electronic and Photonic Materials and Devices*, pp. 163–184, 2001, https://doi.org/10.1016/B978-012513745-4/50067-6.
113. Goyal, M., Singh, K., Bhatnagar, N., Conductive polymers: A multipurpose material for protecting coating. *Prog. Org. Coat.*, 187, 108083, 2024, https://doi.org/10.1016/J.PORGCOAT.2023.108083.
114. Arenas, M.C., Mendoza, N., Cortina, H., Nicho, M.E., Hu, H., Influence of poly3-octylthiophene (P3OT) film thickness and preparation method on photovoltaic performance of hybrid ITO/CdS/P3OT/Au solar cells. *Sol. Energy Mater. Sol. Cells*, 94, 29–33, 2010, https://doi.org/10.1016/J.SOLMAT.2009.04.013.
115. Nicho, M.E., Hernández, F., Hu, H., Medrano, G., Güizado, M., Guerrero, J.A., Physicochemical and morphological properties of spin-coated poly(3-alkylthiophene) thin films. *Sol. Energy Mater. Sol. Cells*, 93, 37–40, 2009, https://doi.org/10.1016/J.SOLMAT.2008.02.016.
116. Bajpai, A.K., Bajpai, J., Soni, S.N., Designing polyaniline (PANI) and polyvinyl alcohol (PVA) based electrically conductive nanocomposites: Preparation, characterization and blood compatible study. *J. Macromol. Sci. Part A Pure Appl. Chem.*, 46, 774–782, 2009, https://doi.org/10.1080/10601320903004533.
117. Popov, A., Brasiunas, B., Mikoliunaite, L., Bagdziunas, G., Ramanavicius, A., Ramanaviciene, A., Comparative study of polyaniline (PANI), poly(3,4-ethylenedioxythiophene) (PEDOT) and PANI-PEDOT films electrochemically deposited on transparent indium thin oxide based

electrodes. *Polymer (Guildf)*, 172, 133–141, 2019, https://doi.org/10.1016/J.POLYMER.2019.03.059.

118. Dong, Y., Yeung, K.W., Tang, C.Y., Law, W.C., Tsui, G.C.P., Xie, X., Development of ionic liquid-based electroactive polymer composites using nanotechnology. *Nanotechnol. Rev.*, 10, 99–116, 2021, https://doi.org/10.1515/ntrev-2021-0009
119. Bar-Cohen Y., Sherrit, S., Lih, S.S., Characterization of the electromechanical properties of EAP materials, in: *Proc. SPIE 4329, Smart Structures and Materials 2001: Electroactive Polymer Actuators and Devices*, SPIE, pp. 319–327, 2001.
120. Zhao, X. and Suo, Z., Theory of dielectric elastomers capable of giant deformation of actuation. *Phys. Rev. Lett.*, 104, 178302, 2010, https://doi.org/10.1103/PhysRevLett.104.178302
121. Bar-Cohen, Y. and Zhang, Q., Electroactive polymer actuators and sensors. *MRS Bull.*, 33, 173–177, 2008, https://doi.org/10.1557/mrs2008.42
122. Bar-Cohen, Y., *Electroactive Polymer (EAP) Actuators as Artificial Muscles: Reality, Potential, and Challenges*, SPIE, Washington, 2004, https://doi.org/10.1117/3.547465.
123. Shahinpoor, M., Ionic polymer–conductor composites as biomimetic sensors, robotic actuators and artificial muscles—a review. *Electrochim. Acta*, 48, 2343–2353, 2003, https://doi.org/10.1016/S0013-4686(03)00224-X
124. Bernat, J., Gajewski, P., Kołota, J., Marcinkowska, A., Silicone-Based Membranes as Potential Materials for Dielectric Electroactive Polymer Actuators. *Energies (Basel)*, 15, 6324, 2022, https://doi.org/10.3390/EN15176324.
125. Yamakita, M., Kamamichi, N., Luo, Z.W., Asaka, K., Robotic application of IPMC actuators with redoping capability, in: *Electroactive Polymers for Robotic Applications: Artificial Muscles and Sensors*, pp. 199–225, 2007, https://doi.org/10.1007/978-1-84628-372-7_8.
126. Panwar, V., Cha, K., Park, J.O., Park, S., High actuation response of PVDF/PVP/PSSA based ionic polymer metal composites actuator. *Sens. Actuators B Chem.*, 161, 1, 460–470, 2012, https://doi.org/10.1016/j.snb.2011.10.062.
127. Chen, Y., Kang, G., Yuan, J., Li, T., Experimental study on pure-shear-like cyclic deformation of VHB 4910 dielectric elastomer. *J. Polym. Res.*, 26, 186, 2019, https://doi.org/10.1007/S10965-019-1858-6.
128. Tan, M.W.M., Wang, H., Gao, D., Huang, P., Lee, P.S., Towards high performance and durable soft tactile actuators. *Chem. Soc. Rev.*, 53, 3485–3535, 2024, https://doi.org/10.1039/D3CS01017A.
129. Ha, S.M., Yuan, W., Pei, Q., Pelrine, R., Stanford, S., Interpenetrating networks of elastomers exhibiting 300% electrically-induced areastrain. *Smart Mater. Struct.*, 16, S280, 2007, https://doi.org/10.1088/0964-1726/16/2/S12.
130. Li, Q. and Wang, Q., Ferroelectric Polymers and Their Energy-Related Applications. *Macromol. Chem. Phys.*, 217, 11, 1228–1244, 2016, https://doi.org/10.1002/macp.201500503.

131. Sharafkhani, S. and Kokabi, M., Ultrathin-shell PVDF/CNT nanocomposite aligned hollow fibers as a sensor/actuator single element. *Compos. Sci. Technol.*, 200, 108425, 2020, https://doi.org/10.1016/j.compscitech.2020.108425.
132. Zhang, Q.M., Li, H., Poh, M., Xia, F., Cheng, Z.Y., Xu, H., Huang, C., An all-organic composite actuator material with a high dielectric constant. *Nature*, 419, 6904, 284–287, 2002, https://doi.org/10.1038/nature01021.
133. Prateek, Thakur, V.K., Gupta, R.K., Recent Progress on Ferroelectric Polymer-Based Nanocomposites for High Energy Density Capacitors: Synthesis, Dielectric Properties, and Future Aspects. *Chem. Rev.*, 116, 4260–4317, 2016, https://doi.org/10.1021/ACS.CHEMREV.5B00495.
134. Linford, R.G., Electrical and electrochemical properties of ion conducting polymers, in: *Applications of Electroactive Polymers*, pp. 1–28, 1993, https://doi.org/10.1007/978-94-011-1568-1_1.
135. Hatchett, D.W. and Josowicz, M., Composites of intrinsically conducting polymers as sensing nanomaterials. *Chem. Rev.*, 108, 746–769, 2008, https://doi.org/10.1021/cr068112h
136. Wang, H., Lin, J., Shen, Z.X., Polyaniline (PANi) based electrode materials for energy storage and conversion. *J. Sci.: Adv. Mater. Devices*, 1, 225–255, 2016, https://doi.org/10.1016/J.JSAMD.2016.08.001.
137. Duduta, M., Wood, R.J., Clarke, D.R., Multilayer Dielectric Elastomers for Fast, Programmable Actuation without Prestretch. *Adv. Mater.*, 28, 8058–8063, 2016, https://doi.org/10.1002/ADMA.201601842.
138. Plante, J.S. and Dubowsky, S., Large-scale failure modes of dielectric elastomer actuators. *Int. J. Solids Struct.*, 43, 7727–7751, 2006, https://doi.org/10.1016/J.IJSOLSTR.2006.03.026.
139. Panwar, V., Lee, C., Ko, S.Y., Park, J.O., Park, S., Dynamic mechanical, electrical, and actuation properties of ionic polymer metal composites using PVDF/PVP/PSSA blend membranes. *Mater. Chem. Phys.*, 135, 928–937, 2012, https://doi.org/10.1016/J.MATCHEMPHYS.2012.05.081.
140. Romasanta, L.J., Lopez-Manchado, M.A., Verdejo, R., Increasing the performance of dielectric elastomer actuators: A review from the materials perspective. *Prog. Polym. Sci.*, 51, 188–211, 2015, https://doi.org/10.1016/J.PROGPOLYMSCI.2015.08.002.
141. Chen, Z., Um, T., Bart-Smith, H., Chen, Z., Um, T., Bart-Smith, H., Ionic Polymer-Metal Composite Artificial Muscles in Bio-Inspired Engineering Research: Underwater Propulsion, in: *Smart Actuation and Sensing Systems - Recent Advances and Future Challenges*, 2012, https://doi.org/10.5772/51292.
142. Palmre, V., Pugal, D., Kim, K.J., Leang, K.K., Asaka, K., Aabloo, A., Nanothorn electrodes for ionic polymer-metal composite artificial muscles. *Sci. Rep.*, 4, 1–10, 2014, https://doi.org/10.1038/srep06176.
143. Chacko, S.K., Rahul, M.T., Raneesh, B., Kalarikkal, N., Enhanced magnetoelectric coupling and dielectric constant in flexible ternary composite electrospun fibers of PVDF-HFP loaded with nanoclay and NiFe 2 O 4

nanoparticles. *New J. Chem.*, 44, 11356–11364, 2020, https://doi.org/10.1039/D0NJ02494E.

144. Sun, F.C., Dongare, A.M., Asandei, A.D., Pamir Alpay, S., Nakhmanson, S., Temperature dependent structural, elastic, and polar properties of ferroelectric polyvinylidene fluoride (PVDF) and trifluoroethylene (TrFE) copolymers. *J. Mater. Chem. C Mater.*, 3, 8389–8396, 2015, https://doi.org/10.1039/C5TC01224D.
145. Sherman, J.D., Elloian, J., Jadwiszczak, J., Shepard, K.L., On the Temperature Dependence of the Piezoelectric Response of Prepoled Poly(vinylidene fluoride) Films. *ACS Appl. Polym. Mater.*, 2, 5110–5120, 2020, https://doi.org/10.1021/acsapm.0c00902
146. Laiarinandrasana, L., Besson, J., Lafarge, M., Hochstetter, G., Temperature dependent mechanical behaviour of PVDF: Experiments and numerical modelling. *Int. J. Plast.*, 25, 1301–1324, 2009, https://doi.org/10.1016/J.IJPLAS.2008.09.008.
147. Sheng, J., Chen, H., Li, B., Wang, Y., Qiang, J., Effect of temperature on electromechanical instability of dielectric elastomers, in: *Electroactive Polymer Actuators and Devices (EAPAD).* vol. 8340, pp. 512–520, SPIE, 2012, https://doi.org/10.1117/12.917089.
148. Sheng, J., Chen, H., Li, B., Effect of temperature on the stability of dielectric elastomers. *J. Phys. D Appl. Phys.*, 44, 365406, 2011, https://doi.org/10.1088/0022-3727/44/36/365406.
149. Mehnert, M., Hossain, M., Steinmann, P., A complete thermo–electro–viscoelastic characterization of dielectric elastomers, Part I: Experimental investigations. *J. Mech. Phys. Solids*, 157, 104603, 2021, https://doi.org/10.1016/J.JMPS.2021.104603.
150. Abd-Elsalam, A., Badr, H.O., Abdel-Rehim, A.A., El-Mahallawi, I.S., Structure and thermoelectric behavior of polyaniline-based/ CNT-composite. *Curr. Appl Phys.*, 36, 88–92, 2022, https://doi.org/10.1016/J.CAP.2021.11.012.
151. Karbownik, I., Rac-Rumijowska, O., Rybicki, T., Suchorska-Wozniak, P., Teterycz, H., The Effect of Temperature on Electric Conductivity of Polyacrylonitrile-Polyaniline Fibers. *IEEE Access*, 9, 74017–74027, 2021, https://doi.org/10.1109/ACCESS.2021.3078835.
152. Rudajevova, A., Prokes, J., Varga, M., Effect of water release on thermal properties of polyaniline. *Chem. Pap.*, 71, 393–400, 2017, https://doi.org/10.1007/S11696-016-0104-5.
153. Bar-Cohen, Y., Electroactive polymers as artificial muscles - Reality and challenges. *19th AIAA Applied Aerodynamics Conference*, 2001, https://doi.org/10.2514/6.2001-1492.
154. Olvera, D. and Monaghan, M.G., Electroactive material-based biosensors for detection and drug delivery. *Adv. Drug Deliv. Rev.*, 170, 396–424, 2021, https://doi.org/10.1016/J.ADDR.2020.09.011.

155. Neuhaus, R., Zahiri, N., Petrs, J., Tahouni, Y., Siegert, J., Kolaric, I., Dahy, H., Bauernhansl, T., Integrating Ionic Electroactive Polymer Actuators and Sensors Into Adaptive Building Skins – Potentials and Limitations. *Front. Built Environ.*, 6, 95, 2020, https://doi.org/10.3389/fbuil.2020.00095
156. Pohanka, M., The piezoelectric biosensors: Principles and applications, a review. *Int. J. Electrochem. Sci.*, 12, 496–506, 2017, https://doi.org/10.20964/2017.01.44.
157. Rasouli, H., Naji, L., Hosseini, M.G., Electrochemical and electromechanical behavior of Nafion-based soft actuators with PPy/CB/MWCNT nanocomposite electrodes. *RSC Adv.*, 7, 3190–3203, 2017, https://doi.org/10.1039/C6RA25771B.
158. Carpi, F. and Smela, E., *Biomedical applications of electroactive polymer actuators*, John Wiley & Sons, Ltd., Chichester, UK, 2009.
159. Dewang, Y., Sharma, V., Baliyan, V.K., Soundappan, T., Singla, Y.K., Research Progress in Electroactive Polymers for Soft Robotics and Artificial Muscle Applications. *Polymers*, 17, 6, 746, 2025.
160. Wang, T., Farajollahi, M., Choi, Y.S., Lin, I.T., Marshall, J.E., Thompson, N.M., Kar-Narayan, S., Madden, J.D.W., Smoukov, S.K., Electroactive polymers for sensing. *Interface Focus*, 6, 20160026, 2016, https://doi.org/10.1098/RSFS.2016.0026.
161. Bar-Cohen, Y., Electroactive polymers as actuators, in: *Advanced Piezoelectric Materials: Science and Technology*, pp. 287–317, 2010, https://doi.org/10.1533/9781845699758.1.287.
162. Bashir, M. and Rajendran, P., A review on electroactive polymers development for aerospace applications. *J. Intell. Mater. Syst. Struct.*, 19, 29, 3681–3695, 2018, https://doi.org/10.1177/1045389X18798951.
163. Bar-Cohen, Y., Biomimetics using electroactive polymers (EAP) as artificial muscles-: A review. *J. Adv. Mater.*, 2006, Pascal-Francis.Inist.Fr, https://pascal-francis.inist.fr/vibad/index.php?action=getRecordDetail&idt=18150105 (accessed October 9, 2024).
164. Hu, S., Shi, Z., Zhao, W., Wang, L., Yang, G., Multifunctional piezoelectric elastomer composites for smart biomedical or wearable electronics. *Compos. B Eng.*, 160, 595–604, 2019, https://doi.org/10.1016/J.COMPOSITESB.2018.12.077.
165. Mullaveettil, F.N., Dauksevicius, R., Wakjira, Y., Strength and elastic properties of 3D printed PVDF-based parts for lightweight biomedical applications. *J. Mech. Behav. Biomed. Mater.*, 120, 104603, 2021, https://doi.org/10.1016/J.JMBBM.2021.104603.
166. Guarino, V., Zuppolini, S., Borriello, A., Ambrosio, L., Electro-Active Polymers (EAPs): A Promising Route to Design Bio-Organic/Bioinspired Platforms with on Demand Functionalities. *Polymers (Basel)*, 8, 185, 2016, https://doi.org/10.3390/POLYM8050185.

167. Motamedi, A.S., Mirzadeh, H., Hajiesmaeilbaigi, F., Bagheri-Khoulenjani, S., Shokrgozar, M.A., Piezoelectric electrospun nanocomposite comprising Au NPs/PVDF for nerve tissue engineering. *J. Biomed. Mater. Res. A*, 105, 1984–1993, 2017, https://doi.org/10.1002/JBM.A.36050.
168. Ghaedsharafi, H., Sherafat, Z., Sani, M., Azarpira, N., Piezoelectric core-shell fibrous scaffolds of PVDF-ZnO/PCL for bone regeneration. *Mater. Today Chem.*, 37, 102017, 2024, https://doi.org/10.1016/J.MTCHEM.2024.102017.
169. Khazani, Y., Rafiee, E., Samadi, A., Mahmoodi, M., Alginate-PVDF piezoelectric hydrogel containing calcium copper titanate- hydroxyapatite as a self-powered scaffold for bone tissue engineering and energy harvesting. *Colloids Surf., A*, 687, 133537, 2024, https://doi.org/10.1016/J.COLSURFA.2024.133537.
170. Chung, Y.L., Chou, P.Y., Sheu, M.J., The Wound-Healing Activity of PEDOT-PSS in Animals. *Int. J. Mol. Sci.*, 24, 12539, 2023, https://doi.org/10.3390/IJMS241612539.
171. Li, Y., Peng, W., Dong, Y., Fan, B., Qian, W., Ji, X., Lu, X., Gan, D., Liu, P., Mussel-Inspired PEDOT-Incorporated Gelatin-Based Conductive Hydrogel with Flexibility and Electroactivity to Accelerate Wound Healing *In Vitro*. *ACS Appl. Polym. Mater.*, 5, 4233–4243, 2023, https://doi.org/10.1021/acsapm.3c00445
172. Wu, C., Shen, L., Lu, Y., Hu, C., Liang, Z., Long, L., Ning, N., Chen, J., Guo, Y., Yang, Z., Hu, X., Zhang, J., Wang, Y., Intrinsic Antibacterial and Conductive Hydrogels Based on the Distinct Bactericidal Effect of Polyaniline for Infected Chronic Wound Healing. *ACS Appl. Mater. Interfaces*, 13, 52308–52320, 2021, https://doi.org/10.1021/acsami.1c14088
173. Cancelliere, R., Di Tinno, A., Cataldo, A., Bellucci, S., Kumbhat, S., Micheli, L., Nafion-based label-free immunosensor as a reliable warning system: The case of AFB1 detection in cattle feed. *Microchem. J.*, 191, 108868, 2023, https://doi.org/10.1016/J.MICROC.2023.108868.
174. Chacko, S.K., Balakrishnan, R., Kalarikkal, N., Thomas, N.G., Ternary Fiber Mats of PVDF-HFP/Cellulose/LiFe5O8 Nanoparticles with Enhanced Electric, Magnetoelectric, and Antibacterial Properties: A Promising Approach for Magnetic and Electric Field-Responsive Antibacterial Coatings. *ACS Appl. Polym. Mater.*, 6, 1429–1438, 2024, https://doi.org/10.1021/acsapm.3c02588
175. Acharya, R., Dutta, S.D., Patil, T.V., Ganguly, K., Randhawa, A., Lim, K.T., A Review on Electroactive Polymer–Metal Composites: Development and Applications for Tissue Regeneration. *J. Funct. Biomater.*, 14, 523, 2023, https://doi.org/10.3390/JFB14100523.
176. Mäkelä, T., Pienimaa, S., Taka, T., Jussila, S., H.I., Thin polyaniline films in EMI shielding. *Synth. Met.*, 85, 1335–1336, 1997, https://doi.org/10.1016/S0379-6779(97)80259-7.

177. Shacklette, L.W. and Colaneri, N.F., EMI shielding measurements of conductive polymer blends. *Conference Record - IEEE Instrumentation and Measurement Technology Conference*, pp. 72–78, 1991, https://doi.org/10.1109/IMTC.1991.161543.
178. Saini, P., Choudhary, V., Singh, B., R.M., Polyaniline–MWCNT nanocomposites for microwave absorption and EMI shielding. *Mater. Chem. Phys.*, 113, 919–926, 2009, https://doi.org/10.1016/j.matchemphys.2008.08.065
179. Saboor, A., Khalid, S.M., Jan, R., Khan, A.N., Zia, T., Farooq, M.U., Afridi, S., Sadiq, M., Arif, M., PS/PANI/MoS2 Hybrid Polymer Composites with High Dielectric Behavior and Electrical Conductivity for EMI Shielding Effectiveness. *Materials*, 12, 2690, 2019, https://doi.org/10.3390/MA12172690.
180. Lakshmi, N.V. and Tambe, P., EMI shielding effectiveness of graphene decorated with graphene quantum dots and silver nanoparticles reinforced PVDF nanocomposites. *Compos. Interfaces*, 24, 861–882, 2017, https://doi.org/10.1080/09276440.2017.1302202.
181. Rayar, A., Naveen, C.S., Onkarappa, H.S., Betageri, V.S., Prasanna, G.D., EMI shielding applications of PANI-Ferrite nanocomposite materials: A review. *Synth. Met.*, 295, 117338, 2023, https://doi.org/10.1016/J.SYNTHMET.2023.117338.
182. Dutta, B., Bose, N., Kar, E., Das, S., Mukherjee, S., Smart, lightweight, flexible NiO/poly(vinylidene flouride) nanocomposites film with significantly enhanced dielectric, piezoelectric and EMI shielding properties. *J. Polym. Res.*, 24, 1–15, 2017, https://doi.org/10.1007/s10965-017-1396-z
183. Biswas, S., Kar, G.P., Bose, S., Tailor-Made Distribution of Nanoparticles in Blend Structure toward Outstanding Electromagnetic Interference Shielding. *ACS Appl. Mater. Interfaces*, 7, 25448–25463, 2015, https://doi.org/10.1021/acsami.5b08333
184. Ma, L., Hamidinejad, M., Zhao, B., Liang, C., Park, C.B., Layered Foam/Film Polymer Nanocomposites with Highly Efficient EMI Shielding Properties and Ultralow Reflection. *Nano Micro Lett.*, 14, 1–18, 2022, https://doi.org/10.1007/s40820-021-00759-4
185. Wang, Y. and Jing, X., Intrinsically conducting polymers for electromagnetic interference shielding. *Polym. Adv. Technol.*, 16, 344–351, 2005, https://doi.org/10.1002/PAT.589.
186. Lecocq, H., Sudre, G., Alcouffe, P., Lhost, O., Cassagnau, P., Serghei, A., Enhanced electromagnetic interference shielding effectiveness of polypropylene/hybrid metallic fillers composite materials by coalescence-driven guided electrical percolation. *Polymer (Guildf)*, 246, 124740, 2022, https://doi.org/10.1016/J.POLYMER.2022.124740.
187. Wan, C. and Bowen, C.R., Multiscale-structuring of polyvinylidene fluoride for energy harvesting: the impact of molecular-, micro- and macro-structure. *J. Mater. Chem. A Mater.*, 5, 3091–3128, 2017, https://doi.org/10.1039/C6TA09590A.

188. Sun, C., Shi, J., Bayerl, D.J., Wang, X., PVDF microbelts for harvesting energy from respiration. *Energy Environ. Sci.*, 4, 4508–4512, 2011, https://doi.org/10.1039/C1EE02241E.
189. Bhattacharya, D., Bayan, S., Mitra, R.K., Ray, S.K., Flexible biomechanical energy harvesters with colossal piezoelectric output (~2.07 V/kPa) based on transition metal dichalcogenides-poly(vinylidene fluoride) nanocomposites. *ACS Appl. Electron. Mater.*, 2, 3327–3335, 2020, https://doi.org/10.1021/acsaelm.0c00632
190. Szewczyk, P.K., Gradys, A., Kim, S.K., Persano, L., Marzec, M., Kryshtal, A., Busolo, T., Toncelli, A., Pisignano, D., Bernasik, A., Kar-Narayan, S., Sajkiewicz, P., Stachewicz, U., Enhanced Piezoelectricity of Electrospun Polyvinylidene Fluoride Fibers for Energy Harvesting. *ACS Appl. Mater. Interfaces*, 12, 11, 13575–13583, 2020, https://doi.org/10.1021/acsami.0c02578.
191. Zhang, Q., Liu, Q., Xue, W., Xiang, Y., Hu, X., Synchronous Sound Recognition and Energy Harvesting by Flexible Piezoelectric PLLA/VB2 Composites. *Polymers (Basel)*, 16, 1071, 2024, https://doi.org/10.3390/POLYM16081071/S1.
192. Singh, V. and Singh, B., PDMS/PVDF- MoS2 based flexible triboelectric nanogenerator for mechanical energy harvesting. *Polymer (Guildf)*, 274, 125910, 2023, https://doi.org/10.1016/J.POLYMER.2023.125910.
193. Khandelwal, G., Min, G., Karagiorgis, X., Dahiya, R., Aligned PLLA electrospun fibres based biodegradable triboelectric nanogenerator. *Nano Energy*, 110, 108325, 2023, https://doi.org/10.1016/J.NANOEN.2023.108325.
194. Rocha, J.G., Goncalves, L.M., Rocha, P.F., Silva, M.P., Lanceros-Mendez, S., Energy Harvesting From Piezoelectric Materials Fully Integrated in Footwear. *IEEE Trans. Ind. Electron.*, 57, 813–819, 2010, https://doi.org/10.1109/TIE.2009.2028360.
195. Lee, D.W., Jeong, D.G., Kim, J.H., Kim, H.S., Murillo, G., Lee, G.H., Song, H.C., Jung, J.H., Polarization-controlled PVDF-based hybrid nanogenerator for an effective vibrational energy harvesting from human foot. *Nano Energy*, 76, 105066, 2020, https://doi.org/10.1016/j.nanoen.2020.105066.
196. Roy, K., Ghosh, S.K., Sultana, A., Garain, S., Xie, M., Bowen, C.R., Henkel, K., Schmeiβer, D., Mandal, D., A Self-Powered Wearable Pressure Sensor and Pyroelectric Breathing Sensor Based on GO Interfaced PVDF Nanofibers. *ACS Appl. Nano Mater.*, 2, 2013–2025, 2019, https://doi.org/10.1021/acsanm.9b00033
197. Mannsfeld, S.C.B., Tee, B.C.K., Stoltenberg, R.M., Chen, C.V.H.H., Barman, S., Muir, B.V.O., Sokolov, A.N., Reese, C., Bao, Z., Highly sensitive flexible pressure sensors with microstructured rubber dielectric layers. *Nat. Mater.*, 9, 10, 859–864, 2010, https://doi.org/10.1038/nmat2834.

198. Lee, J.-H., Yoon, H.-J., Kim, T.Y., Gupta, M.K., Lee, J.H., Seung, W., Ryu, H., Kim, S.-W., Micropatterned P(VDF-TrFE) Film-Based Piezoelectric Nanogenerators for Highly Sensitive Self-Powered Pressure Sensors. *Adv. Funct. Mater.*, 25, 3203–3209, 2015, https://doi.org/10.1002/adfm.201500856.
199. Lee, J.S., Shin, K.Y., Cheong, O.J., Kim, J.H., Jang, J., Highly Sensitive and Multifunctional Tactile Sensor Using Free-standing ZnO/PVDF Thin Film with Graphene Electrodes for Pressure and Temperature Monitoring. *Sci. Rep.*, 5, 1–8, 2015, https://doi.org/10.1038/srep07887.
200. Alluri, N.R., Chandrasekhar, A., Jeong, J.H., Kim, S.J., Enhanced electroactive β-phase of the sonication-process-derived PVDF-activated carbon composite film for efficient energy conversion and a battery-free acceleration sensor. *J. Mater. Chem. C Mater.*, 5, 4833–4844, 2017, https://doi.org/10.1039/c7tc00568g.
201. Yang, L., Wang, S., Mao, J., Deng, J., Gao, Q., Tang, Y., Schmidt, O.G., Hierarchical MoS2/Polyaniline Nanowires with Excellent Electrochemical Performance for Lithium-Ion Batteries. *Adv. Mater.*, 25, 8, 1180–1184. 2013, Infona.Pl, https://doi.org/10.1002/adma.201203999
202. Asghar, M.R., Anwar, M.T., Rasheed, T., Naveed, A., Yan, X., Zhang, J., Lithium Salt Doped Poly(Vinylidene Fluoride)/Cellulose Acetate Composite Gel Electrolyte Membrane for Lithium Ion Battery. *IOP Conf. Ser. Mater. Sci. Eng.*, 654, 012017, 2019, https://doi.org/10.1088/1757-899X/654/1/012017.
203. Song, J.Y., Wang, Y.Y., Wan, C.C., Review of gel-type polymer electrolytes for lithium-ion batteries. *J. Power Sources*, 77, 2, 183–197, 1999, https://doi.org/10.1016/S0378-7753(98)00193-1.
204. Bicy, K., Suriyakumar, S., Anu Paul, P., Anu, A.S., Kalarikkal, N., Stephen, A.M., Geethamma, V.G., Rouxel, D., Thomas, S., Highly lithium ion conductive, Al2O3 decorated electrospun P(VDF-TrFE) membranes for lithium ion battery separators. *New J. Chem.*, 42, 24, 19505–19520, 2018, https://doi.org/10.1039/c8nj01907j.
205. Singh, D., Choudhary, A., Garg, A., Flexible and Robust Piezoelectric Polymer Nanocomposites Based Energy Harvesters. *ACS Appl. Mater. Interfaces*, 10, 2793–2800, 2018, https://doi.org/10.1021/acsami.7b16973.
206. Song, J., Zhao, G., Li, B., Wang, J., Design optimization of PVDF-based piezoelectric energy harvesters. *Heliyon*, 3, e00377, 2017, https://doi.org/10.1016/j.heliyon.2017.e00377.
207. Chen, J.X., Li, J.W., Cheng, C.C., Chiu, C.W., Piezoelectric Property Enhancement of PZT/Poly(vinylidenefluoride-co-trifluoroethylene) Hybrid Films for Flexible Piezoelectric Energy Harvesters. *ACS Omega*, 7, 793–803, 2022, https://doi.org/10.1021/ACSOMEGA.1C05451.
208. Chinya, I., Sasmal, A., Sen, S., Conducting polyaniline decorated *in-situ* poled Ferrite nanorod-PVDF based nanocomposite as piezoelectric energy harvester. *J. Alloys Compd.*, 815, 152312, 2020, https://doi.org/10.1016/j.jallcom.2019.152312.

209. Kurra, N., Jiang, Q., Syed, A., Xia, C., Alshareef, H.N., Micro-Pseudocapacitors with Electroactive Polymer Electrodes: Toward AC-Line Filtering Applications. *ACS Appl. Mater. Interfaces*, 8, 12748–12755, 2016, https://doi.org/10.1021/acsami.5b12784
210. Shacklette, L.W., Maxfield, M., Gould, S., Wolf, J.F., Jow, T.R., Baughman, R.H., Secondary batteries with electroactive polymer electrodes. *Synth. Met.*, 18, 611–618, 1987, https://doi.org/10.1016/0379-6779(87)90949-0.

2

Structural, Chemical, and Electrical Properties of Dielectric Polymers

Rohit Nisthala[1], Asrar Rafiq Bhat[2], Ekta Jagtiani[2], Laxmeesha Somappa[3] and Prasanna Kumar Mural[2*]

[1]*Center for Research in Nano Technology and Science, Indian Institute of Technology Bombay (IITB), Powai, Mumbai, Maharashtra, India*
[2]*Department of Metallurgical Engineering and Materials Science, Indian Institute of Technology Bombay (IITB), Powai, Mumbai, Maharashtra, India*
[3]*Department of Electrical Engineering, Indian Institute of Technology Bombay (IITB), Powai, Mumbai, Maharashtra, India*

Abstract
Electroactive polymers with dielectric properties possess great potential to solve various intelligent engineering solutions in electronics applications, like actuators, generators, and sensors. Exceptional electrical insulation with flexibility, low specific weight, and ease of processing makes them highly desirable for various applications. The chapter discusses dielectric polymer's structural, chemical, and electrical properties. The structural properties have a significant effect on electrical and mechanical properties. By understanding their preparation methods, structural properties, chemical properties, electrical properties, and other characteristics, engineers can customize dielectric polymers to meet the requirements of various applications. Ongoing research and development efforts seek to enhance dielectric polymers further, enabling their application in emerging technologies and expanding their capabilities. Additionally, dielectric polymers serve as a crucial foundation for the efficient and dependable operation of electrical and electronic systems, contributing to the development of numerous industries and enhancing our quality of life.

Keywords: Dielectric polymers, electroactive polymers, polarization mechanisms, dielectric relaxation, piezoelectric materials, polymer-ceramic composites, energy storage, multilayer dielectric composites

**Corresponding author*: prasannamural@iitb.ac.in

Raneesh Balakrishnan, Sobi K. Chacko and Visakh P.M. (eds.) Electroactive Polymer-Based Smart Materials and Applications, (39–58) © 2026 Scrivener Publishing LLC

2.1 Introduction

Electroactive polymers are gaining interest in various applications like actuators, generators, and sensors. Electroactive polymers with dielectric properties are a class of material that converts one form of energy to another [1]. A dielectric material permits the electric field or electric flux to travel through it without being a charge carrier [2]. Further dielectric materials consist of insulators and other nonconductive substances. However, the vital aspect of dielectric materials in contemporary electronics and electric power systems is storing and controlling electrical charges and energies [3]. The dielectric capacitor is an energy-storing material with high power density, reliability, and operating voltage [4].

Dielectric materials store electrical energy *via* polarization of constituent molecules or atoms when subjected to an external electric field [5]. Dielectric materials require charge separation *via* polarizing the electron distributions around the constituent atoms or molecules in response to an electric field [6]. They are further applying an electric field to the dielectric material, resulting in the rearrangement of charge with no long-range transports of charge—this rearrangement of order with relative dielectric displacement results in material polarization.

Polarization under an electric field can be due to the composite's individual contributions of electronic, ionic, dipole, and interfacial polarization or a combination of these types. Figure 2.1 depicts the schematic of various possible polarization mechanisms. Electronic polarization is a displacement of electrons within atoms under an electric field. The displacement of cation/anion ions within a crystal structure or molecules under an applied electric field leads to ionic polarization. An asymmetric molecule's orientation permanent dipole in an electric field leads to orientation polarization. Interfacial polarization arises due to localized charge accumulation in the presence of conductivity phases in the insulating matrix subjected to an electric field. Individuals or a combination of these mechanisms can contribute to its polarization. Further, the polymers with a polar backbone offer higher dielectric constant owing to their ferroelectric nature.

The orientation of dipoles requires a certain amount of time to orient in the applied electric field. The time needed for dipoles to align in an electric field is known as relaxation time. The relaxation process is affected by dipole, ionic, and dielectric relaxation. The dipole relaxation comprises the alignment of permanent dipoles. Similarly, ionic relaxation encompasses interfacial relaxation and ionic conductivity. In contrast, dielectric relaxation results from the movement of dipoles caused by dipole relaxation

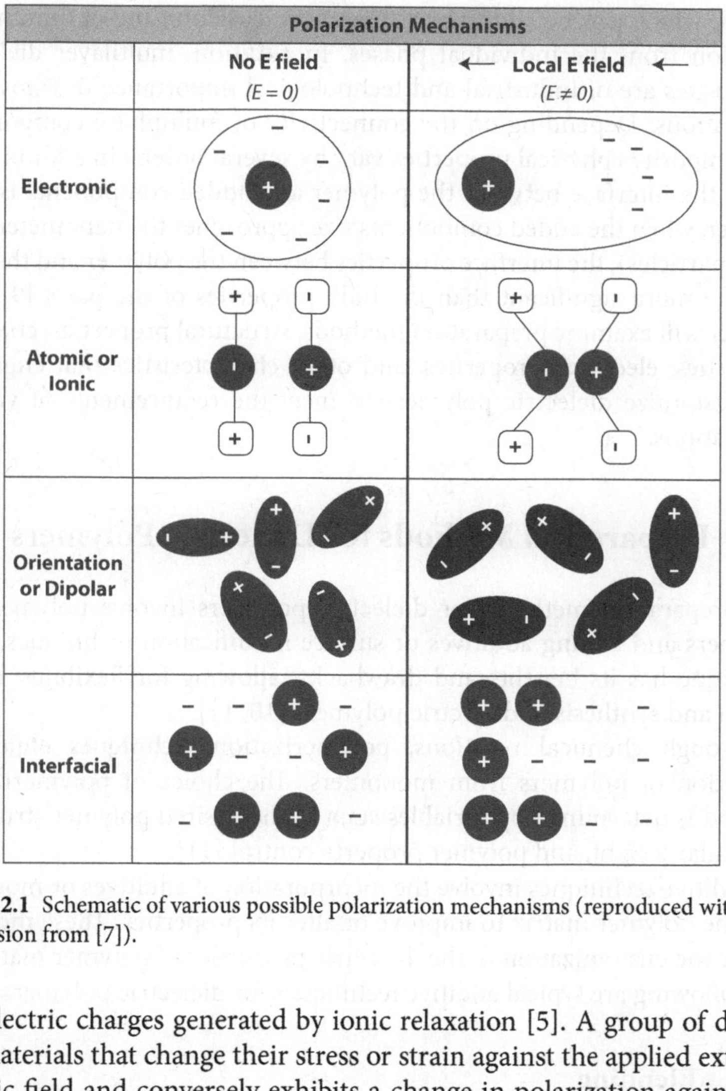

Figure 2.1 Schematic of various possible polarization mechanisms (reproduced with permission from [7]).

and electric charges generated by ionic relaxation [5]. A group of dielectric materials that change their stress or strain against the applied external electric field and conversely exhibits a change in polarization against the applied mechanical excitation can be categorized as piezoelectric material [8].

The combination of ceramic and polymer materials complements each other in overcoming individual limitations and provides superior dielectric properties. Combining ceramic and polymer materials with the optimum and appropriate components is pivotal as it will control the material's electric flux patterns and mechanical properties. Further physical and chemical properties of composites depend on the contribution from individual phases, which can be additive or synergistic based on the optimum contribution from the individual phases. In addition, multilayer dielectric composites are of industrial and technological importance in transducer applications. Depending on the connectivity of multiphase components, the composite's physical properties vary by several orders. In addition, tailoring the interface between the polymer and added components is vital. Further, when the added components size approaches the nanometer scale (nanoparticles), the interface properties between the polymer and the filler become more significant than the bulk properties of the parts [9]. This chapter will examine preparation methods, structural properties, chemical properties, electrical properties, and other characteristics that engineers can customize dielectric polymers to meet the requirements of various applications.

2.2 Preparation Methods for Dielectric Polymers

The preparation methods for dielectric polymers involve polymerizing polymers and adding additives or surface modification techniques. Each technique has its benefits and drawbacks, allowing for flexibility in the design and synthesis of dielectric polymers [10, 11].

Through chemical reactions, polymerization techniques entail the formation of polymers from monomers. The choice of polymerization method is determined by variables such as the desired polymer structure, molecular weight, and polymer property control [11].

Additive techniques involve the incorporation of additives or modifiers into the polymer matrix to improve or alter its properties. These methods enable the customization of the dielectric properties of polymer materials. The following are typical additive techniques for dielectric polymers:

2.2.1 Blending

Blending is the process of combining two or more polymers or adding additives to alter the dielectric properties. By combining various polymers, one can obtain a combination of the desirable properties of each constituent. Fillers, including nanoparticles and ceramics, can improve properties such as dielectric constant, thermal stability, and mechanical strength.

2.2.2 Copolymerization

Copolymerization is a method for producing novel materials with tailored dielectric properties by covalently bonding two or more distinct monomer units. The copolymers of these materials produce improved insulating materials appropriate for use in electronic packaging. The dielectric constant decreases as the number of siloxane units within the copolymers rises. The presence of silicon, which is larger than carbon, and the pliability of the silicon-oxygen bond inhibit the movement of the polyimide network as a whole, making it more difficult for dipoles to respond to polarity changes [12–14]. Additionally, a higher concentration of unbound electrons reduces molar polarization significantly. It entails polymerizing two or more monomers to create copolymers with particular properties. Copolymerization permits the tailoring of dielectric properties, mechanical strength, and thermal stability by adjusting the monomer ratio.

2.2.3 Cross-Linking of Polymers

Cross-linking entails the formation of chemical bonds between polymer chains to form a three-dimensional network. Cross-linked polymers possess enhanced mechanical strength, thermal stability, and deformation resistance [14, 15].

2.3 Structural Properties

Tailoring of molecular structure, molecular weight, chain architecture, polydispersity index, crosslinking, and morphology of dielectric polymers can affect dielectric properties, such as dielectric constant, breakdown strength, loss tangent, and mechanical strength. Further, the presence of polar or nonpolar groups, the nature and strength of chemical bonds, and the overall symmetry of polymer chains are crucial in determining the dielectric properties. The crystallinity of dielectric polymer will have a lower dielectric constant than the amorphous polymer. Additionally, tailoring the processing conditions and compositing leads to achieving the desired structural properties.

2.3.1 Crosslinking

Crosslinking of polymers increases the chain rigidity, which is governed by crosslinking density [16]. An increase in crosslink density increases the relaxation time due to enhanced interchain entanglements. In addition, as the cross-link density increases, the dielectric constant decreases, and the maximum dielectric loss shifts toward lower frequencies, indicating the polymer transformation from a rubbery to a vitreous state.

2.3.2 Polar Groups

Increasing fluorination in polyimide films generally decreases the dielectric constant and increases free volume [17]. Further, adding bulky molecules, such as adamantane, to a polyimide chain increases the available volume and decreases the dielectric constant. This strategy is also used in commercially available dielectric polymers to accomplish low dielectric loss and maintain desirable electrical properties at high frequencies [18]. Introducing polar urea or thiourea groups in the main chain increases dielectric response and minimizes dipole coupling [19, 20]. Further introduction of meta-aromatic polyurea and meta-aromatic polyurea into the chain results in dipole-dipole interactions and prevents ferroelectric loss [21–23].

Substitution of the amine group in the main chain of polyimides results in the doubling of capacitance and minimizing dissipation factors [24, 25]. Further incorporating functional groups such as carbonyl and ether in the bridge of the dianhydride and the polyether segment in the diamine, the polar group density in the polymer backbone was significantly increased, thereby increasing the dielectric constant [26]. The polyimide with the longest cross-conjugated system and the longest polyether segment (polyimide B4) displayed the highest dielectric constant of 7.8 while maintaining low dissipation factors and attaining a high energy density [24]. Also, incorporating polar nitrile groups into the main chain of polyimides exhibited marginally greater dielectric constants than polyimides without nitrile groups. The presence of an asymmetric structure in the repeating unit improved the processability of the polymers and increased their dielectric constants while keeping their dielectric loss relatively low [24]. To study the impact of dipole moment density and molecular asymmetry on the dielectric properties, polyimides with three nitrile groups per repeating unit were synthesized by polymerizing novel diamines with common dianhydrides [27, 28]. The increase in dipole moments and the para–para linkage in the diamine led to a higher permittivity and discharged energy density while the dielectric loss remained low (Figure 2.2). These results indicated that polyimides containing nitrile, particularly those with a high

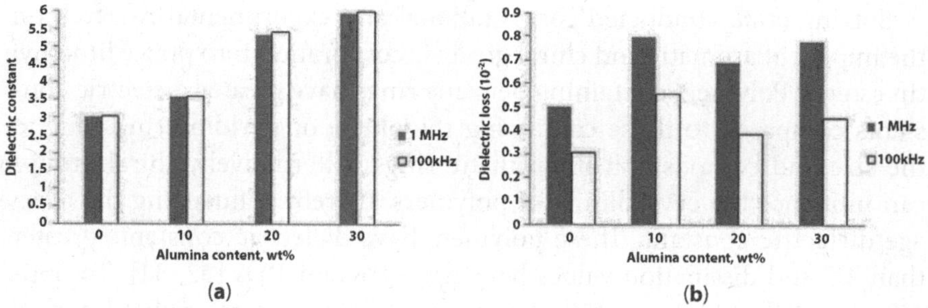

Figure 2.2 (a) Dielectric constant and (b) dielectric loss, as a function of weight percentage and frequency (reproduced with permission from [27]).

nitrile content, are promising dielectric materials for high-temperature film capacitors [29, 30].

Incorporating metal ions into a polymer backbone through covalent bonding can significantly improve dielectric properties. In contrast to polymer nanocomposites that entail adding inorganic nanofillers to a polymer matrix, directly incorporating metal atoms into the polymer structure allows for improved dispersion without adverse aggregation effects. Compared to carbon-heteroatom bonds, metal-heteroatom bonds in the polymer backbone increase dipolar and atomic polarization due to a greater electronegativity difference [31–33]. Sotzing *et al.* researched a series of related poly (dimethyl tin esters) with linker lengths ranging from 0 to 10 methylene units. By varying the linker length in the diacid monomers, aliphatic poly (dimethyl tin esters) with varying tin weight percentages were produced by varying the linker length in the diacid monomers. In general, increasing the linker length caused the dielectric constant to diminish and the bandgap to widen [34, 35]. Nevertheless, there were "optimal" linker lengths (4-7 methylene units) in which both the bandgap and the dielectric constant were high. Among these polymers, Poly (Dimethyl tin suberate) (pDMT Sub) exhibited a maximum dielectric constant of 7 and a bandgap of nearly 7 eV. The morphology of the film, which was determined by the length of the methylene spacer, affected the dielectric constant and bandgap and could be precisely controlled by merging with another homopolymer such as Poly (dimethyl tin 3,3-dimethylglutarate) (p(DMTDMG)). In addition, concurrent first-principal calculations were performed alongside experimental work in order to gain a comprehensive comprehension of the properties of this family of organotin polymers [34, 35].

Sotzing *et al.* conducted computational and experimental research on the impact of aromatic and chiral group incorporation into poly (dimethyl tin esters). Polymers containing benzene rings have greater dielectric constants compared to those containing thiophene or pyridine rings due to the size and composition of aromatic rings. Alternatively, chiral groups can influence the crystallinity of polymers, thereby influencing the average dielectric constant. These polymers have dielectric constants greater than 4.5 and dissipation values between 0.1% and 10% [32, 34]. To optimize the dielectric properties of poly (dimethyl tin esters), poly (dimethyl tin suberate) (pDMT sub), and poly (dimethyl tin 3,3-dimethylglutarate) (p(DMTDMG)) were selected for further investigation *via* blending and copolymerization. With a poly (dimethyl tin suberate) (pDMT Sub) concentration of 20%, both blend (BP20) and copolymer (CP20) films exhibited high dielectric constants of 6.6 and approximately 1% loss. The energy density of CP20 was calculated to be approximately 6 J cm^3 compared to 4 J cm^3 for BP20, indicating that copolymerization is more suitable than blending for energy storage. These findings contribute to the viability of Poly (dimethyl tin esters) as polymer dielectrics of the next generation [32, 34].

Adding a small number of polar OH groups to a nonpolar polymer could significantly increase the dielectric constant while preserving a low dielectric loss. The H-bonded clusters with the OH groups prevent dielectric loss. This finding suggests that functionalizing nonpolar polymers such as PP and PE with polar groups is a promising strategy for enhancing their dielectric properties and energy storage capacities, warranting further investigation *via* experiments and computer simulations [36, 37].

Ion-containing polymers could not attain high energy density and breakdown strength due to significant conduction loss at high electric fields. The dielectric properties of novel ion-containing (co)polymers will increase energy density, decrease energy loss, withstand high operating temperatures, and enhance thermal management at elevated temperatures [37–39]. Ionic cross-linking is effective at improving the dielectric performance and energy storage capacity. The cross-linked ionic polymer exhibits an enhanced dielectric constant, superior frequency, and temperature stability [39, 40].

Oligothiophene-functionalized side-chain polymers have been shown to exhibit high polarizability and are therefore, considered suitable for use in dielectric materials. These polymers may be identified by sulfur atoms and extended-electrons. Polymers containing conjugated pendant oligothiophene in the side chain were looked into by Tang *et al.* as potential candidates for high-performance dielectrics [41, 42]. Polymer brush architecture employing oligothiophene-containing polymers to improve the quality and processability of films. Even though the resulting films have enhanced mechanical durability, their dielectric properties are average. By customizing the polymeric backbone and side chain, Tang *et al.* reported in a follow-up study a synergistic macromolecular architecture strategy. They utilized polynorbornene as the polymeric backbone and oligothiophene as the polar groups attached to the polymeric backbone. This strategy yielded transparent and flexible dielectric polymer films with substantially increased electric field breakdown strength and minimal dielectric loss. Due to the intrinsic amorphous properties of polynorbornene-based films; however, the relative permittivity decreased. These results suggest that the arrangement of polar groups in the side chain plays a dominant role in dielectric permittivity. In contrast, the backbone influences the thermal and mechanical properties of breakdown strength and dielectric loss. The optimal equilibrium between the polymeric backbone and the polar conjugated group in the side chains is essential for optimizing dielectric performance at low and high electric fields [43–46].

2.4 Chemical Properties

The chemical properties of dielectric polymers that contribute to their ability to electrically insulate and provide additional desirable qualities. Dielectric polymers are chemically inert and resistant to solvents, acids, and bases, among other common compounds. This durability assures their functionality and stability in harsh environments. Chemical resistance is an essential characteristic of dielectric polymers, which refers to their ability to sustain exposure to a variety of chemicals without suffering significant degradation or damage. Frequently, dielectric polymers are used in environments where they may come into contact with solvents, acids, bases, and other compounds. Therefore, chemical resistance is necessary for maintaining their functionality, performance, and durability [47, 48]. Dielectric polymers with a high chemical resistance are typically inert to a wide variety of common chemicals. They are resistant to chemical reaction and dissolution, assuring their stability and integrity when exposed to aggressive or corrosive substances.

Chemical resistance prevents the polymer from deteriorating when subjected to corrosive substances. It protects the dielectric polymer against degradation, discoloration, softening, and loss of physical properties. Dielectric polymers with superior chemical resistance typically have an extended lifespan. They can retain their electrical and physical properties for extended periods of time, even in chemically hostile environments, reducing the frequency with which they must be replaced. Chemical resistance broadens the scope of dielectric polymers' potential applications. They can be utilized in chemical processing facilities, laboratories, oil and gas industries, and other common chemical exposure environments [49–51].

Chemical resistance enables the combination of dielectric polymers with materials such as metals, ceramics, and other polymers. This property assures the compatibility and effectiveness of multi-material systems, as the dielectric polymer is resistant to any potential chemical interactions or injury from other substances. Chemical resistance ensures that the dielectric properties of the polymer are maintained even when exposed to chemicals. It prevents modifications to electrical insulation properties like resistivity, dielectric constant, and disintegration strength [52, 53]. There is often a correlation between chemical resistance and resistance to environmental factors such as humidity, moisture, and ultraviolet radiation. Chemically resistant polymers are frequently more resistant to these factors, ensuring their durability and performance. It is essential to note that the chemical resistance of dielectric polymers can vary based on the polymer's specific composition, additives, and the types and concentrations of chemicals with which they come into contact. Therefore, it is essential to choose a dielectric polymer that has been specially formulated or designed to withstand the substances present in the intended application environment [54].

2.5 Electrical Properties

Resistivity is a crucial characteristic of dielectric polymers. It refers to their resistance to electric current passing through them. In other words, the low conductivity of dielectric polymers means that electrical charges cannot readily move through their structure. Due to their unique chemical structure and composition, dielectric polymers exhibit high resistivity. Dielectric polymers have a distinct arrangement of atoms and electrons than conductors, such as metals, which have a high concentration of free electrons that can readily carry electric charges [15, 18, 49].

In dielectric polymers, the electrons are securely bound within the individual molecules, and the structure of the material hinders their mobility. This restricted electron mobility makes it challenging for an electric current to travel through the polymer, resulting in a high resistivity [55].

High electrical resistivity indicates that dielectric polymers do not readily conduct electricity. This characteristic is crucial to their use as insulators.

As insulating materials, dielectric polymers are desirable due to their high resistivity. They prevent the leakage of electrical charges when used as insulation, ensuring that the current follows its intended path without interference or loss. In applications where electrical isolation and safety are of the utmost importance, such as in high-voltage cables, capacitors, transformers, and printed circuit boards, this property is especially important.

Notably, the resistivity of dielectric polymers can vary based on variables such as temperature, humidity, and impurities. However, dielectric polymers typically have a much higher resistivity than conductive materials, making them suitable for electrical insulation applications [56–58].

The volume resistivity of a material quantifies its resistance throughout its volume. Resistivity can be calculated by applying a voltage and measuring the current. Typically, the volume resistivities of plastics are on the order of 10^{16} m, indicating their superb insulating properties. Surface resistivity quantifies the resistance of a material to current passage along its surface. It is the relationship between the applied voltage and the resulting current per unit width. Typically, surface resistivity is measured in ohms. Dielectric polymers are distinguished by their high resistivity, which indicates that they present a substantial barrier to the flow of electrical current. This characteristic is essential for their use as insulators, as it prevents leakage currents and ensures effective electrical insulation. Through the meticulous selection of polymer materials and the optimization of their molecular structure, which reduces the mobility of charge carriers within the material, high resistivity can be achieved [59–61].

2.5.1 Dielectric Constant

The dielectric constant, which measures a dielectric material's ability to separate capacitor plates, is defined as the capacitance with the material divided by the capacitance with a vacuum. It measures how well a material can insulate the capacitor's plates. For vacuum, the relative dielectric constant is 1, whereas it is greater than 1 for all other substances [61].

Low dielectric constant materials reduce energy losses caused by dielectric heating or electrical energy dissipation. This characteristic is desirable for energy-efficient applications including electrical power transmission, high-frequency communication systems, and electronic devices. Low dielectric constant polymers provide effective electrical insulation by minimizing capacitive coupling between conducting elements. Printed circuit boards (PCBs) and integrated circuits (ICs) are examples of applications that require electrical isolation and crosstalk suppression [62].

By decreasing the dielectric constant, a low dielectric constant serves to preserve the integrity and quality of electrical signals in high-frequency applications. It reduces signal distortion, propagation delays, and reflections, facilitating signal transmission with greater precision. As the size of electronic devices continues to shrink, it becomes essential to employ materials with a low dielectric constant. These materials permit the creation of smaller, more compact devices without compromising their electrical performance or dependability [15, 63–65]. Polymers with a low dielectric constant are commonly used in high-frequency applications, such as microwave circuits, antennas, and radar systems. The low dielectric constant contributes to the precision and stability of signal propagation at these frequencies [66].

It is important to note that the dielectric constant can vary based on frequency, temperature, and other factors. Consequently, the selection of a dielectric polymer with a specific dielectric constant is determined by the application's requirements and constraints [50, 51, 67, 68]. Polar dielectric constant of the plastics varies with the frequency of the alternating current. At low frequencies, dipoles have sufficient time to align with the field before its direction changes, resulting in a high dielectric constant. At high frequencies, there is insufficient time for the dipoles to align, resulting in a lower dielectric constant. Polar plastics have dielectric constants between 3 and 9 at low frequencies (60 Hz) and 3 to 5 at high frequencies (106 Hz). In contrast, the dielectric constants of non-polar plastics are less than 3, and their values are independent of the alternating current frequency [69, 70].

Low dielectric constant, or relative permittivity, is exhibited by dielectric polymers. The dielectric constant quantifies the ability of a material to store electrical energy when exposed to an electric field. A dielectric material with a low dielectric constant ensures that energy loss and signal attenuation are minimized. It also facilitates the efficient storage of energy in capacitors and the transmission of signals at high speeds in electronic device [71–73].

2.5.2 Dielectric Strength

Dielectric strength is the direct current voltage at which dielectric breakdown occurs. It assesses the insulating properties of a material. The dielectric strength of the majority of plastics falls between 100 and 300 kV/cm [74].

The breakdown strength of a dielectric polymer is its capacity to withstand high electric fields without electrical failure or breakdown. To ensure the safe and dependable operation of electrical systems, a high breakdown strength is crucial. It is affected by variables such as the polymer's molecular structure, chain architecture, and the presence of impurities or defects. Dielectric polymers with a high breakdown strength are resistant to electrical breakdown and dielectric failure at higher voltages [59].

The loss tangent, also known as the dissipation factor, quantifies the energy lost as heat when an electric field is applied to a dielectric material. Low-loss dielectric polymers display a low loss tangent, indicating minimal energy losses resulting from dielectric heating. This property is especially crucial in high-frequency applications, where excessive energy losses can result in decreased efficiency and signal degradation [75–78].

The electrical properties of dielectric polymers are frequency-dependent. The frequency dependence of the dielectric constant, loss tangent, and other electrical parameters is possible. It is crucial to characterize and comprehend the frequency response of dielectric polymers, particularly for applications involving high-frequency signals or variable operating conditions [79].

Temperature, relative humidity, and applied voltage can affect the electrical properties of dielectric polymers. Consequently, when selecting and designing with dielectric polymers, it is vital to consider the operating conditions and environmental factors [75, 77, 78, 80]. Engineers can ensure the effective use of dielectric polymers in applications, such as capacitors, electrical insulation, printed circuit boards, transformers, and other electronic devices by optimizing their electrical properties. Customizing these properties permits improved energy efficiency, decreased power losses, enhanced signal transmission, and increased electrical system dependability.

2.5.3 Power Factor or Power Coefficient (Alternating Current)

The power factor measures the amount of energy absorbed by a material as the direction of the alternating current changes. It is determined by the phase angle, which represents the time gap between dipole alignment and the external field. Indicating the internal friction caused by the alternating current, the power factor determines the heating of a material in an alternating field. The power factor is dependent on the frequency of the alternating current for polar plastics. The power factor and internal heating are low at very low and high frequencies. At intermediate frequencies, the power factor is optimal, resulting in considerable internal friction and heating. Typically, the power factor of non-polar plastics is less than 0.0003, indicating minimal internal friction and heating [47, 76, 81, 82].

2.5.4 Tracking and Arc Resistance (Direct Current)

Arc resistance is the ability of a material to resist the formation of a continuous conduction path in the presence of a high voltage and low current arc. It indicates the length of time a material can withstand an arc without experiencing a significant current flow. The electrical properties of plastics, particularly polar plastics, can be significantly influenced by environmental factors such as humidity and temperature. These materials tend to absorb moisture, which raises their dielectric constant and lowers their volume and surface resistivities. Temperature increases promote polymer chain movement and dipole alignment, resulting in an increase in the dielectric constants of polar plastics, particularly above their glass transition temperature (Tg). Nonpolar plastics, such as fluoropolymers, are less affected by water absorption, and variations in temperature have little effect on their electronic polarization [83].

2.6 Summary

Numerous industries, including electronics, telecommunications, power transmission and distribution, capacitors, energy storage devices, and sensors, rely heavily on dielectric polymers. Their exceptional electrical insulating properties, combined with benefits such as their light weight, flexibility, and simplicity of processing, make them highly desirable for a variety of applications. The structural properties of dielectric polymers, such as their molecular structure, molecular weight, chain architecture, and morphology, have a significant impact on their electrical, mechanical, and thermal properties. Engineers can optimize dielectric polymers for specific applications by manipulating these properties, thereby obtaining desirable characteristics such as high breakdown strength, high resistivity, low moisture absorption and low dielectric constant. Chemical resistance is an essential characteristic of dielectric polymers that ensures their long-lasting performance and durability. In severe environments, their resistance to solvents, acids, bases, and other chemicals contributes to their dependability. In conclusion, the vast array of properties and benefits offered by dielectric polymers renders them indispensable in modern technology. By understanding their preparation methods, structural properties, chemical properties, electrical properties, and other characteristics, engineers are able to customize dielectric polymers to meet the requirements of a variety of applications. As technology advances, the demand for dielectric polymers with improved properties and efficacy will increase. Ongoing research and development efforts seek to enhance dielectric polymers further, enabling their application in emerging technologies and expanding their capabilities. Overall, dielectric polymers serve as a crucial foundation for the efficient and dependable operation of electrical and electronic systems, contributing to the development of numerous industries and enhancing our quality of life.

Acknowledgments

The authors thank the Department of Metallurgical Engineering & Materials Science, Indian Institute of Technology Bombay. Also, the authors like to thank the Faculty SEED Grant (2022) Indian Institute of Technology Bombay (IITB) for funding to initiate research in Polymer nanocomposite.

AI disclosure statement: The author acknowledges the use of AI-based tools for minor language editing or grammatical correction. However, all intellectual contributions, analyses, and conclusions presented in this chapter are entirely the author's own.

References

1. Ozsecen, M.Y., Dielectric electroactive polymer based biomedical devices: control, sensing and interfacing, Northeastern University, 2010.
2. Zhang, Q.M., Bharti, V., Zhao, X., Giant electrostriction and relaxor ferroelectric behavior in electron-irradiated poly (vinylidene fluoride-tri fluoroethylene) copolymer. *Science*, 280, 5372, 2101–2104, 1998.
3. Peltonen, J., Paajanen, M., Lekkala, J., Determination of the actuator sensitivity of electromechanical polypropylene films by atomic force microscopy. *J. Appl. Phys.*, 88, 8, 4789–4793, 2000.
4. Luo, H., *et al.*, Progress on polymer dielectrics for electrostatic capacitors application. *Adv. Sci.*, 9, 29, 2202438, 2022.
5. Mazumder, M.A.J., Sheardown, H., Al-Ahmed, A., *Functional polymers*, Springer, Cham, Switzerland, 2019.

6. Brosseau, C., Modelling and simulation of dielectric heterostructures: a physical survey from an historical perspective. *J. Phys. Appl. Phys.*, 39, 7, 1277, 2006.
7. Gupta, M. and Leong, E.W.W., *Microwaves and metals*, John Wiley & Sons, 2008.
8. Khanbareh, H., Expanding the functionality of piezo-particulate composites, 2016.
9. Pihtili, G., Torğut, G., Biryan, F., Electrical properties of two-armed poly (Ɛ-CL-co-BMA) composites filled with bentonite. *J. Polym. Res.*, 27, 1–11, 2020.
10. Mishra, S. and Mishra, P., Synthesis and characterization of CCTO0. 5-BT0. 5/epoxy hybrid ceramic polymer composite for electronic applications. *Mater. Today Proc.*, 67, 372–376, 2022.
11. Banu, N., Structural, electrical and optical properties of diethanolamine thin films synthesized by plasma polymerization technique, 2018.
12. Yu, A., Ramesh, P., Itkis, M.E., Bekyarova, E., Haddon, R.C., Graphite nanoplatelet–epoxy composite thermal interface materials. *J. Phys. Chem. C*, 111, 21, 7565–7569, 2007.
13. Chen, X., Zheng, Y.P., Kang, F., Shen, W.C., Preparation and structure analysis of carbon/carbon composite made from phenolic resin impregnation into exfoliated graphite. *J. Phys. Chem. Solids*, 67, 5–6, 1141–1144, 2006.
14. Chrissafis, K., Paraskevopoulos, K.M., Stavrev, S.Y., Docoslis, A., Vassiliou, A., Bikiaris, D.N., Characterization and thermal degradation mechanism of isotactic polypropylene/carbon black nanocomposites. *Thermochim. Acta*, 465, 1–2, 6–17, 2007.
15. Hu, H., Zhang, F., Luo, S., Chang, W., Yue, J., Wang, C.-H., Recent advances in rational design of polymer nanocomposite dielectrics for energy storage. *Nano Energy*, 74, 104844, 2020.
16. Gao, F., Zhang, K., Guo, Y., Xu, J., Szafran, M., (Ba, Sr) TiO3/polymer dielectric composites–progress and perspective. *Prog. Mater. Sci.*, 121, 100813, 2021.
17. Ning, N., *et al.*, Largely improved electromechanical properties of thermoplastic polyurethane dielectric elastomers by the synergistic effect of polyethylene glycol and partially reduced graphene oxide. *Compos. Sci. Technol.*, 142, 311–320, 2017.
18. Ahmad, Z., Polymer dielectric materials, in: *Dielectric material*, IntechOpen, 1–26, 2012.
19. Shen, Y., Lin, Y., Zhang, Q.M., Polymer nanocomposites with high energy storage densities. *MRS Bull.*, 40, 9, 753–759, 2015.
20. Jung, H.M., Kang, J.-H., Yang, S.Y., Won, J.C., Kim, Y.S., Barium titanate nanoparticles with diblock copolymer shielding layers for high-energy density nanocomposites. *Chem. Mater.*, 22, 2, 450–456, 2010.

21. Hergenrother, P.M., The use, design, synthesis, and properties of high performance/high temperature polymers: an overview. *High Perform. Polym.*, 15, 1, 3–45, 2003.
22. Wang, Y., *et al.*, Compositional tailoring effect on electric field distribution for significantly enhanced breakdown strength and restrained conductive loss in sandwich-structured ceramic/polymer nanocomposites. *J. Mater. Chem. A*, 5, 9, 4710–4718, 2017.
23. Takahashi, Y., Iijima, M., Fukada, E., Pyroelectricity in poled thin films of aromatic polyurea prepared by vapor deposition polymerization. *Jpn. J. Appl. Phys.*, 28, 12A, L2245, 1989.
24. Wang, D.H., Riley, J.K., Fillery, S.P., Durstock, M.F., Vaia, R.A., Tan, L.-S., Synthesis and characterization of unsymmetrical benzonitrile-containing polyimides: Viscosity-lowering effect and dielectric properties. *J. Polym. Sci. Part Polym. Chem.*, 51, 23, 4998–5011, 2013.
25. Wang, D.H., Kurish, B.A., Treufeld, I., Zhu, L., Tan, L.-S., Synthesis and characterization of high nitrile content polyimides as dielectric films for electrical energy storage. *J. Polym. Sci. Part Polym. Chem.*, 53, 3, 422–436, 2015.
26. Hougham, G., Tesoro, G., Viehbeck, A., Chapple-Sokol, J.D., Polarization effects of fluorine on the relative permittivity in polyimides. *Macromolecules*, 27, 21, 5964–5971, 1994.
27. Alias, A., Ahmad, Z., Ismail, A.B., Preparation of polyimide/Al2O3 composite films as improved solid dielectrics. *Mater. Sci. Eng., B*, 176, 10, 799–804, 2011.
28. Simpson, J.O. and Clair, A.S., Fundamental insight on developing low dielectric constant polyimides. *Thin Solid Films*, 308, 480–485, 1997.
29. Treufeld, I., Wang, D.H., Kurish, B.A., Tan, L.-S., Zhu, L., Enhancing electrical energy storage using polar polyimides with nitrile groups directly attached to the main chain. *J. Mater. Chem. A*, 2, 48, 20683–20696, 2014.
30. Ma, R., *et al.*, Rationally designed polyimides for high-energy density capacitor applications. *ACS Appl. Mater. Interfaces*, 6, 13, 10445–10451, 2014.
31. Pilania, G., *et al.*, New group IV chemical motifs for improved dielectric permittivity of polyethylene. *J. Chem. Inf. Model.*, 53, 4, 879–886, 2013.
32. Baldwin, A.F., *et al.*, Poly (dimethyltin glutarate) as a prospective material for high dielectric applications. *Adv. Mater.*, 27, 2, 346–351, 2015.
33. Baldwin, A.F., *et al.*, Rational design of organotin polyesters. *Macromolecules*, 48, 8, 2422–2428, 2015.
34. Baldwin, A.F., Ma, R., Huan, T.D., Cao, Y., Ramprasad, R., Sotzing, G.A., Effect of incorporating aromatic and chiral groups on the dielectric properties of poly (dimethyltin esters). *Macromol. Rapid Commun.*, 35, 24, 2082–2088, 2014.
35. Mannodi-Kanakkithodi, A., Wang, C.C., Ramprasad, R., Compounds based on Group 14 elements: building blocks for advanced insulator dielectrics design. *J. Mater. Sci.*, 50, 801–807, 2015.

36. Misra, M., *et al.*, Enhanced polymeric dielectrics through incorporation of hydroxyl groups. *Macromolecules*, 47, 3, 3, 2014.
37. Huan, T.D., *et al.*, Advanced polymeric dielectrics for high energy density applications. *Prog. Mater. Sci.*, 83, 236–269, 2016.
38. Zhao, J., *et al.*, Manipulation of interactions at membrane interfaces for energy and environmental applications. *Prog. Polym. Sci.*, 80, 125–152, 2018.
39. Dang, Z.-M., Yuan, J.-K., Zha, J.-W., Zhou, T., Li, S.-T., Hu, G.-H., Fundamentals, processes and applications of high-permittivity polymer-matrix composites. *Prog. Mater. Sci.*, 57, 4, 660–723, 2012.
40. Zhu, Y., Jiang, P., Zhang, Z., Huang, X., Dielectric phenomena and electrical energy storage of poly (vinylidene fluoride) based high-k polymers. *Chin. Chem. Lett.*, 28, 11, 2027–2035, 2017.
41. Qiao, Y., *et al.*, Oligothiophene-containing polymer brushes by ROMP and RAFT: Synthesis, characterization and dielectric properties. *Polymer*, 72, 428–435, 2015.
42. Behera, R. and Elanseralathan, K., Improved electrical properties in three phase poly (vinylidene fluoride) nanocomposites with hybrid fillers of titanium dioxide and MXene. *Mater. Today Commun.*, 34, 105334, 2023.
43. Yin, X., Qiao, Y., Gadinski, M.R., Wang, Q., Tang, C., Flexible thiophene polymers: A concerted macromolecular architecture for dielectrics. *Polym. Chem.*, 7, 17, 2929–2933, 2016.
44. Xu, Y., *et al.*, Synthesis and properties of CO2-based plastics: Environmentally-friendly, energy-saving and biomedical polymeric materials. *Prog. Polym. Sci.*, 80, 163–182, 2018.
45. Qiao, Y., *et al.*, Polymers containing highly polarizable conjugated side chains as high-performance all-organic nanodielectric materials. *Adv. Funct. Mater.*, 23, 45, 5638–5646, 2013.
46. Qiao, Y., Yin, X., Tang, C., Progress in side-chain thiophene-containing polymers: synthesis, properties and applications. *Sci. China Chem.*, 58, 1641–1650, 2015.
47. Blattmann, C.O. and Pratsinis, S.E., Nanoparticle filler content and shape in polymer nanocomposites. *Kona Powder Part. J.*, 36, 3–32, 2019.
48. Paredes, M., Angammana, C.J., Jayaram, S.H., Experimental study of dielectric properties of fumed silica/silicone composites, in: *2016 IEEE Conference on Electrical Insulation and Dielectric Phenomena (CEIDP)*, IEEE, pp. 683–686, 2016.
49. Qiao, Y., Yin, X., Zhu, T., Li, H., Tang, C., Dielectric polymers with novel chemistry, compositions and architectures. *Prog. Polym. Sci.*, 80, 153–162, 2018.
50. Barber, P., *et al.*, Polymer composite and nanocomposite dielectric materials for pulse power energy storage. *Materials*, 2, 4, 1697–1733, 2009.
51. Chen, Q., Shen, Y., Zhang, S., Zhang, Q.M., Polymer-based dielectrics with high energy storage density. *Annu. Rev. Mater. Res.*, 45, 433–458, 2015.

52. Li, Q., *et al.*, Flexible high-temperature dielectric materials from polymer nanocomposites. *Nature*, 523, 7562, 576–579, 2015.
53. Li, Q., *et al.*, Sandwich-structured polymer nanocomposites with high energy density and great charge–discharge efficiency at elevated temperatures. *Proc. Natl. Acad. Sci.*, 113, 36, 9995–10000, 2016.
54. Wang, Y., *et al.*, Ultrahigh energy density and greatly enhanced discharged efficiency of sandwich-structured polymer nanocomposites with optimized spatial organization. *Nano Energy*, 44, 364–370, 2018.
55. Xie, X., Yang, C., Qi, X., Yang, J., Zhou, Z., Wang, Y., Constructing polymeric interlayer with dual effects toward high dielectric constant and low dielectric loss. *Chem. Eng. J.*, 366, 378–389, 2019.
56. Wang, R., *et al.*, High-energy-density ferroelectric polymer nanocomposites utilizing the Coulomb-blockade effect. *Mater. Today Nano*, 20, 100260, 2022.
57. Thakur, Y., *et al.*, Optimizing nanostructure to achieve high dielectric response with low loss in strongly dipolar polymers. *Nano Energy*, 16, 227–234, 2015.
58. Peng, X., *et al.*, Development of high dielectric polyimides containing bipyridine units for polymer film capacitor. *React. Funct. Polym.*, 106, 93–98, 2016.
59. Kim, P., *et al.*, Phosphonic acid-modified barium titanate polymer nanocomposites with high permittivity and dielectric strength. *Adv. Mater.*, 19, 7, 1001–1005, 2007.
60. Tang, H., Zhou, Z., Sodano, H.A., Relationship between BaTiO3 nanowire aspect ratio and the dielectric permittivity of nanocomposites. *ACS Appl. Mater. Interfaces*, 6, 8, 5450–5455, 2014.
61. Zhu, L., Exploring strategies for high dielectric constant and low loss polymer dielectrics. *J. Phys. Chem. Lett.*, 5, 21, 3677–3687, 2014.
62. Zhu, L. and Wang, Q., Novel ferroelectric polymers for high energy density and low loss dielectrics. *Macromolecules*, 45, 7, 2937–2954, 2012.
63. Wang, Q. and Zhu, L., Polymer nanocomposites for electrical energy storage. *J. Polym. Sci. Part B Polym. Phys.*, 49, 20, 1421–1429, 2011.
64. Dang, Z.-M., Yuan, J.-K., Yao, S.-H., Liao, R.-J., Flexible nanodielectric materials with high permittivity for power energy storage. *Adv. Mater.*, 25, 44, 6334–6365, 2013.
65. Wei, J., *et al.*, Achieving high dielectric constant and low loss property in a dipolar glass polymer containing strongly dipolar and small-sized sulfone groups. *ACS Appl. Mater. Interfaces*, 7, 9, 5248–5257, 2015.
66. Huang, C., Zhang, Q.M., Su, J., High-dielectric-constant all-polymer percolative composites. *Appl. Phys. Lett.*, 82, 20, 3502–3504, 2003.
67. Guo, M., Hayakawa, T., Kakimoto, M., Goodson III, T., Organic macromolecular high dielectric constant materials: synthesis, characterization, and applications. *J. Phys. Chem. B*, 115, 46, 13419–13432, 2011.
68. Huang, X. and Jiang, P., Core–shell structured high-k polymer nanocomposites for energy storage and dielectric applications. *Adv. Mater.*, 27, 3, 546–554, 2015.

69. Dang, Z.-M., Lin, Y.-H., Nan, C.-W., Novel ferroelectric polymer composites with high dielectric constants. *Adv. Mater.*, 15, 19, 1625–1629, 2003.
70. Xu, J. and Wong, C.P., Low-loss percolative dielectric composite. *Appl. Phys. Lett.*, 87, 8, 082907, 2005.
71. Qi, L., Lee, B.I., Chen, S., Samuels, W.D., Exarhos, G.J., High-dielectric-constant silver–epoxy composites as embedded dielectrics. *Adv. Mater.*, 17, 14, 1777–1781, 2005.
72. Wang, J.-W., Shen, Q.-D., Yang, C.-Z., Zhang, Q.-M., High dielectric constant composite of P (VDF– TrFE) with grafted copper phthalocyanine oligomer. *Macromolecules*, 37, 6, 2294–2298, 2004.
73. Shannon, R.D., Dielectric polarizabilities of ions in oxides and fluorides. *J. Appl. Phys.*, 73, 1, 348–366, 1993.
74. Popielarz, R., Chiang, C.K., Nozaki, R., Obrzut, J., Dielectric properties of polymer/ferroelectric ceramic composites from 100 Hz to 10 GHz. *Macromolecules*, 34, 17, 5910–5915, 2001.
75. Rajib, M., *et al.*, Enhanced energy storage of dielectric nanocomposites at elevated temperatures. *Int. J. Appl. Ceram. Technol.*, 13, 1, 125–132, 2016.
76. Gultekin, S., Yildirim, S., Celik, E., Arican Alicikus, L.Z., Dielectrical, optical, and structural characterization of TiO_2/PVA nanocomposite films for dielectric applications. *J. Aust. Ceram. Soc.*, 54, 523–531, 2018.
77. Palit, S., Varghese, D., Guo, H., Krishnan, S., Alam, M.A., The role of dielectric heating and effects of ambient humidity in the electrical breakdown of polymer dielectrics. *IEEE Trans. Device Mater. Reliab.*, 15, 3, 308–318, 2015.
78. Abed, M.E.A., Hadi, H., Slama, M.E.A., Belarbi, A.W., Zekri, N., Experimental investigation on dielectric/electric properties of aged polymeric insulator: correlation of permittivity and dielectric loss tangent with surface hydrophobicity. *Electr. Eng.*, 104, 3, 1539–1552, 2021.
79. Sun, L., *et al.*, Layer-structured $BaTiO_3$/P (VDF–HFP) composites with concurrently improved dielectric permittivity and breakdown strength toward capacitive energy-storage applications. *J. Mater. Chem. C*, 8, 30, 10257–10265, 2020.
80. Shojaatalhoseini, M., Conductivity and Relaxation in Polymer Based Solid Electrolytes, 2016.
81. Meena, J.S., *et al.*, Polystyrene-block-poly (methylmethacrylate) composite material film as a gate dielectric for plastic thin-film transistor applications. *RSC Adv.*, 4, 36, 18493–18502, 2014.
82. Banu, N. and Bhuiyan, A.H., Understanding the dielectric and electrical transport behaviors of plasma polymerized diethanolamine thin films under alternating electric field. *Thin Solid Films*, 704, 138018, 2020.
83. Pattanshetti, V.V., Shashidhara, G.M., Veena, M.G., Dielectric and thermal properties of magnesium oxide/poly (aryl ether ketone) nanocomposites. *Sci. Eng. Compos. Mater.*, 25, 5, 915–925, 2018.

3

Biodegradable PLA: Synthesis, Properties and PLA-Based Nanocomposite Applications

Anju Rajan, Raneesh Balakrishnan* and Karthika S.

Department of Physics, Catholicate College, Pathanamthitta, Kerala, India

Abstract

The biodegradability and environmental friendliness of biopolymer composites have drawn much attention due to their excellent thermal and acoustic attributes, improved specific strength and stiffness, low pollution levels, and thermal recycling ability.

Polylactic acid (PLA) is a widely used biopolymer suitable for the development of composite materials. PLA-based materials have garnered significant attention due to their superior mechanical, electrical, and physical characteristics. This book chapter provides an overview of PLA and its composites, as it provides a concise and comprehensive discussion of the properties, applications, and potential future advancements of PLA-based nanocomposite materials. The chapter encourages further research on these materials, thereby supporting innovation in both academic and industrial sectors.

Keywords: Electroactive polymers, piezoelectricity, polylactic acid, nanocomposites, tissue engineering, drug delivery, biosensors, implants

3.1 Introduction

Polymer-based nanotechnology has experienced a significant boom because of its low production costs, lightweight, and increased mechanical strength [1, 2]. In the modern world, the environmental friendliness and biodegradability provided by biopolymers make their products extremely

**Corresponding author*: raneesh.b@catholicatecollege.ac.in; ORCID: 0000-0003-0056-559X

Raneesh Balakrishnan, Sobi K. Chacko and Visakh P.M. (eds.) Electroactive Polymer-Based Smart Materials and Applications, (59–90) © 2026 Scrivener Publishing LLC

important. These biopolymer composites provide numerous benefits, including minimal pollution levels and excellent electrical resistance [1, 3].

As a result of its low density and processing temperature, elastic behaviour, resistance to corrosion, and versatile processing methods, PLA became one of the most popular biomaterials among biopolymers [4, 5]. Because PLA has superior mechanical efficiency, minimal toxicity, bioresorption, biodegradation, and biocompatibility, its use in medicine has grown. This synthetic biodegradable polyester is manufactured by synthesizing lactic acid from renewable resources like corn or sugar cane [2, 4, 6]. Notably, PLA also exhibits piezoelectric activity and it can reach a value up to 10 pC/N [4, 7]. As a result, it is seen as an attractive option for bioapplications utilizing its electroactivity generated by stress, which is beneficial for bone formation, regeneration, or neurological recovery [4].

The complex interplay between physiological reactions and material properties, such as piezoelectricity, is one of the fundamental features of biological systems [4]. In the 1950s, semicrystalline biopolymers like wood and bone were discovered to exhibit piezoelectricity. Since then, researchers have investigated the piezoelectric characteristics of other biological materials, including proteins, deoxyribonucleates, and polysaccharides [8]. Additionally, living tissues such as bone, dentin, and tendon have also been shown to possess inherent piezoelectricity [9–11].

The pyroelectric and piezoelectric properties of bone have been associated with collagen, the crystalline organic matrix of the bone composed of highly aligned polar protein molecules [11, 12]. The organised structure of bone collagen is devoid of a symmetry center. This fulfils the essential need for the development of a piezoelectric effect. Only when the force of shearing is exerted on the collagen fibres in the bone causing them to slide relative to each other does the piezoelectric effect manifest [13, 14].

This chapter provides a comprehensive overview of biodegradable, electroactive polymer PLA and PLA-based nanocomposites. The chapter highlights some basic features of PLA and the prospective applications of PLA-based nanocomposites across several industries. It becomes clear that these revolutionary substances have enormous potential for diverse applications. The synergistic combination of biodegradability and electroactivity of PLA nanocomposites gives new possibilities for solving modern problems.

3.2 Polylactic Acid

PLA is an environment-friendly, bio-based, biocompatible, biodegradable, non-toxic, polymer coming under the family of aliphatic polyesters [15–17]. PLA can be synthesised from natural resources like maize starch, cassava roots, and sugarcane. Since it is non-toxic and non-pollutant, it is a good alternative to the widely used commodity plastics made from petroleum [2]. In 1932, Carothers synthesized PLA for the first time by heating lactic acid under vacuum [16]. Now, PLA has become one of the highly utilized thermoplastic polymers with tunable mechanical properties. Its unique features render it suitable for several biological purposes, including sutures, implants, and drug delivery systems. It can be utilized for tissue engineering as a synthetic polymer that is immunologically inert [2, 18].

Even though PLA has numerous benefits, it also has drawbacks. Its applications are restricted by its poor hardness, slow crystallization, weak gas barrier characteristics, moderate degradation rate, and hydrophobicity. PLA has been extensively mixed with multiple nanofillers to enhance its physical and mechanical properties [16, 19, 20]. The characteristics of PLA are improved by incorporating the nanosized compounds, and as a result, a variety of applications are possible based on the compounds incorporated into the PLA matrix.

3.3 Structure of PLA

In relation to its ability to degrade, biological compatibility, superior mechanical qualities, and good processibility, PLA is a promising material as a substitute for conventional plastics. It is often synthesised through fermentation by bacteria or polymerization of some natural resources [21, 22]. It has two mesoforms: L and D-PLA. The higher crystallinity of L PLA

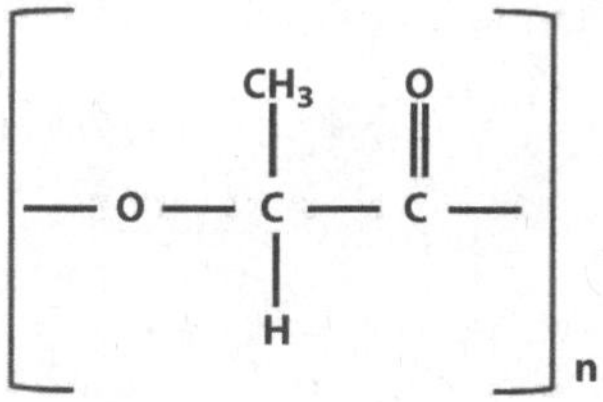

Figure 3.1 Chemical structure of PLA.

results in a higher melting temperature and brittleness. Amorphous PLAs are formed by the copolymerization of L and D mesoforms [23]. The structure of PLA is provided in Figure 3.1.

3.4 Solubility of PLA

The crystalline structure of PLA determines its solubility. Amorphous PLA is soluble in organic solvents such as acetone, acetonitrile. Only at very high temperatures does crystalline PLA dissolve in some solvents like benzene. Generally, PLA compounds are completely dissolved with chloroform, dioxane, acetonitrile, 1,1,2-trichloroethane, dichloroacetic acid, dichloromethane (DCM), N,N-dimethylformamide (DMF), and partially dissolved in toluene, acetone, ethyl benzene, and tetrahydrofuran at ambient condition and readily soluble at boiling temperature [16, 24, 25]. Figure 3.2 shows the solubility of PLA in different solvents.

Health care and biomedical studies have paid substantial attention to nanofibers with distinctive physicochemical and biological features [26]. To establish better solution properties, efficiency, and structure of nanofibers, it is crucial to choose a proper harmless solvent system [27]. Depending on the solvent option made to solubilize the polymer, the procedure produces porous fibers that dry quickly [28].

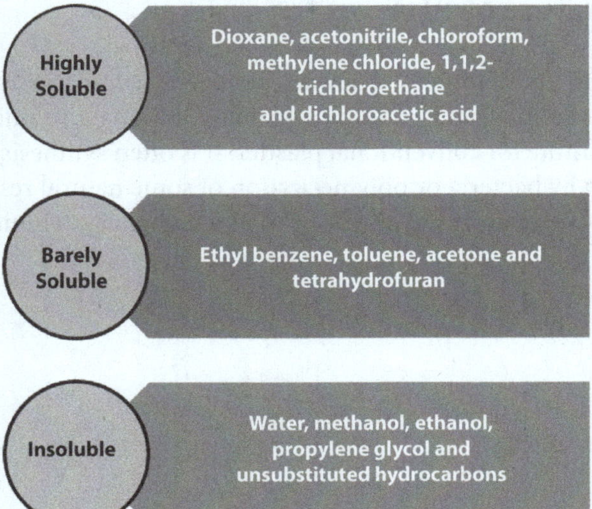

Figure 3.2 Schematic representation of solubility of PLA, adapted from [25].

Casasola *et al.* investigated how the solvent system affected the diameter and shape of the fiber. Both binary and single solvent systems were employed. Acetone (atn) was the only single system that can promote superior fibre morphology. The binary systems they chose were atn/1,4-dioxane (DX), atn/THF, atn/chloroform, atn/dichloromethane, atn/DMF, and atn/Dimethylacetamide (DMAc). Only atn/DMF and acetone/DMAc, among them, produced fibre that was defect-free. They observed that as the boiling point of the second solvent in the binary-solvent system rose, the mean nanofiber diameter decreased. Additionally, an increase in the acetone content in the solvent system may produce defect-free nanofibers, although the mean nanofiber diameter increased [27].

Factors like conductivity, viscosity, and molecular weight of PLA frequently affect how soluble PLA is. Researchers use a variety of solvent options depending on these various factors.

3.5 Degradability of PLA

For several applications, non-toxic degradation products are frequently regarded as a crucial selection criterion. Non-toxic degradation products are present in many aliphatic polyesters. PLA stands out as one of the most attractive and practical biodegradable polyester [24, 29]. The rate of decomposition of PLA is influenced by time, temperature, molecular weight and its arrangement, impurities, catalyst concentration, crystallinity permeability, surface characteristics, microstructure, branching in the molecular chain, solubility, diffusion, hydrogen bonds, and stereoisomeric contents [2, 29–33]. Figure 3.3(a) illustrates the schematic representation of synthesis, decomposition and degradation mechanism of PLA products. Proper disposal of PLA products enables their hydrolysis into nontoxic substances, offering a practical and cost-effective solution for plastic packaging disposal [31, 32, 34]. Even though the degradation products are non-toxic, the gradual breakdown of PLA results in an extended *in vivo* lifespan. Years of *in vivo* life are a serious problem, which limits the application of PLA in many areas [16]. As shown in Figure 3.3(b), several substances, such as water, microorganisms, enzymes, and ultraviolet light (UV), can contribute to the degradation of PLA [35].

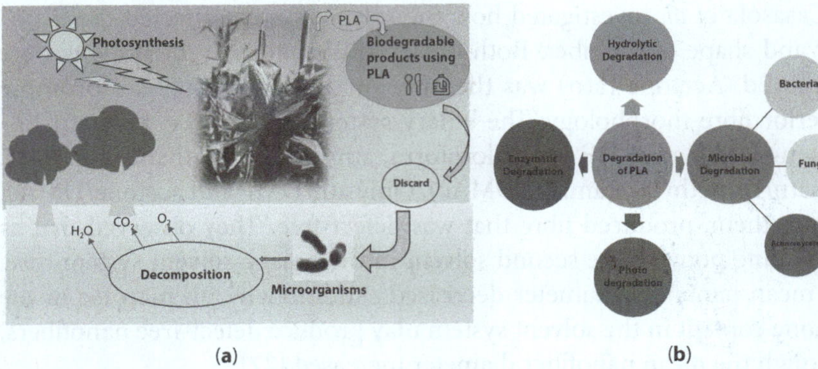

Figure 3.3 (a) Schematic illustration of the production and degradation of PLA compounds. (b) Degradation mechanism of PLA.

3.6 Synthesis Methods of PLA

A biodegradable derivative of lactic acid, PLA is made using naturally occurring crops like corn starch, sugarcane. In developing fields, including orthopedics, drug delivery, sutures, scaffolds, etc., PLA finds many uses. Because its monomers come from renewable resources, they have easier processing and superior mechanical qualities. PLA is conventionally synthesised by sugar fermentation and hydrolysis into monomers. Different synthesis processes were examined for both commercial and industrial applications which are all not simple and easy to operate [24, 36–38]. Generally, four different methods are used for the synthesis of PLA [18, 31] as shown in Figure 3.4.

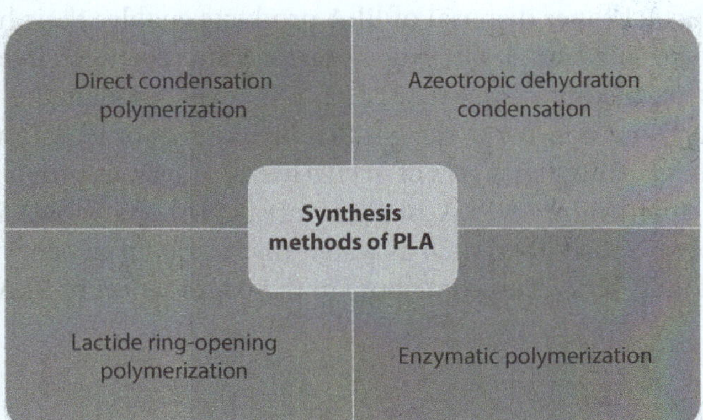

Figure 3.4 Various techniques to synthesize PLA.

1. Direct condensation polymerization
2. Azeotropic dehydration condensation
3. Lactide ring-opening polymerization
4. Enzymatic polymerization

3.6.1 Direct Condensation Polymerization

This is a low-cost synthesis method for PLA. Here, lactic acid undergoes polymerization with the help of a chemical agent. A highly viscous reaction mixture is obtained due to the dehydration condensation of the hydroxyl and carboxy groups. Low molecular weight PLA is obtained from this viscous combination since it is hard to extract water from it. Chain coupling agents can be used to get around this. However, the usability of PLA, particularly in medical applications, is seriously complicated when the chain coupling agent reacts with either the carboxyl or hydroxyl groups in PLA, resulting in impurities [18, 31].

3.6.2 Azeotropic Dehydration Condensation

Using this process, lactic acid is polymerized azeotropically in an azeotropic solvent to create PLA polymer. The remaining amount of water may be eliminated azeotropically, and the solvent is dehydrated and reintroduced into the process. Since this method can be used to synthesise highly pure PLA with a molecular weight of up to 300,000, the presence of any remaining toxic, nonbiodegradable catalyst can seriously harm human tissue, making it unsuitable for use in medical applications [18, 31].

3.6.3 Lactide Ring-Opening Polymerization

This method was initially adopted by Carothers for the synthesis of high-molecular-weight PLA. This approach is frequently employed because of its ability to manipulate the characteristics of the obtained PLA, which broadens its applications. The dominant and effective method for manufacturing PLA is the ring-opening polymerization [39]. According to the reaction mechanism and initiator type, the three main types of ring opening polymerization are anionic and cationic polymerization, and coordination-insertion processes [18, 31].

3.6.4 Enzymatic Polymerization

The process of making PLA by any above mentioned methods in the presence of an enzyme is known as enzymatic polymerization. It was shown that the resultant PLA had a 93% yield and molecular weight between 5428 and 2854 g/mol [35, 40–42].

Table 3.1 Commercial PLA general properties [19, 47, 48].

Sl. no:	Properties	
1.	Glass transition temperature	50–70°C
2.	Flexural strength	106 MPa
3.	Specific tensile strength	16–48 Nm/g
4.	Temperature of decomposition	500–600°C
5.	Tensile modulus	0.35-3.5 GPa
6.	Shear modulus	1287 MPa
7.	Young's modulus	1280 MPa
8.	Poisson's ratio	0.36
9.	Elongation at break (%)	100–180
10.	Percent of elongation	11.3%
11.	Ultimate strain	2.5-6%
12.	Specific heat	1955J/kg°C at 100°C
13.	Melting temperature	130–215°C
14.	Thermal conductivity	0.197W/m°C
15.	Solid Density	1.25 kg/cm^3
16.	Elastic modulus	3600 MPa
17.	Melt density	1.073 kg/cm^3
18.	Degradation rate	50% in 1–2 years
19.	Tensile strength	21-60 MPa
20.	Degree of crystallinity	10–40

3.7 Properties of PLA

Polylactic acid (PLA) is a widely used, eco-friendly, and sustainable linear polyester with significant commercial value and promising market potential. The domains of industry and medicine benefit greatly from using PLA [16, 34]. Numerous variables, including annealing duration, viscosity, affect the properties of PLA [17, 29, 31, 33]. Compared to other oil-based polymers, PLA exhibits favorable optical, mechanical, physical, and barrier properties [2, 43]. Table 3.1 describes some general properties of commercial PLA.

PLA has sufficient thermal stability below 200°C; hence, it can retard degradation and maintain performance to some extent. At temperatures higher than 200°C, PLA undergo the following processes: hydrolysis, oxidative main chain scission, lactide recombination, and intra- and intermolecular transesterification [24, 44–46].

A high degree of mechanical and thermal properties is necessary for several possible applications; therefore, these qualities must be enhanced. This can be accomplished by incoperating with modified nanofillers during synthesis or by blending directly with PLA matrix [2, 24, 49–51].

3.8 Advantages and Disadvantages of PLA

One of the most appealing and practical recyclable aliphatic polyesters is PLA. It is synthesisable from low-cost biomaterials. It can be applied to many different fields due to its high strength and modulus. It can be handled using straightforward, traditional processing techniques [31]. Although PLA has many more advantages than conventional polymers, it also has some drawbacks. These shortcomings restrain its use [52]. The key advantages and disadvantages of PLA are summarized in Table 3.2.

Table 3.2 Major advantages and disadvantages of PLA [16, 30, 32, 46, 53–58].

Sl. no:	Advantages	Disadvantages
1.	Transparency: Because PLA is so transparent, it may be utilised for various applications like radio-controlled toys, drones, aircraft, automobiles, prototyping, and many more [16, 54]	Poor toughness: PLA has very poor toughness, with less than 10% elongation at break. Its inadequate toughness restricts its use to scenarios requiring plastic deformation under elevated stress conditions [16, 30, 57, 58]

(*Continued*)

Table 3.2 Major advantages and disadvantages of PLA [16, 30, 32, 46, 53–58]. (*Continued*)

Sl. no:	Advantages	Disadvantages
2.	Environmentally friendly: PLA is a synthetic biopolymer that is biodegradable, compostable, recyclable, and bioderived. It uses less carbon dioxide during its manufacture. Its decomposition by-products are harmless. PLA is a promising biomaterial due to these sustainable behaviour [53, 54]	Slow degradation rate: Although the degradation products are non-toxic, the slow degradation rate of PLA yields an extended *in-vivo* lifespan. Years of *in-vivo* life are a serious problem, which limits the applications of PLA various fields [32]
3.	Biocompatibility: An important characteristic of PLA is its biocompatibility, making it suitable for implantation in living systems, including humans. Due to this property PLA has significant potential for use in tissue culturing, medicinal applications, and food packaging because of their biocompatibility [30, 46]	Hydrophobicity: PLA exhibits inherent hydrophobicity, characterized by a static water contact angle of approximately 80° [16, 30]
4.	Processability: Various methods can be employed to process PLA. Extrusion, injection stretch blow molding, thermoforming, injection molding, blown film, blending, fiber spinning, foaming, and compounding are a few typical techniques [30, 55, 59]	Lack of reactive side-chain groups: Due to the absence of side-chain groups, the surface and bulk modifications of PLA are challenging task [16, 30]
5.	Energy savings: Compared to conventional polymers, manufacturing of PLA requires 25–55% less energy, and shortly, this requirement can be reduced even further to less than 10% [30, 56]	Hydrolysis rate and thermal resistance: Because PLA has a high hydrolysis rate and lower thermal resistance, PLA-based food packaging is not recommended for hot liquid [30]

3.9 Applications of PLA

One significant advancement in the polymer sector for solving environmental challenges is the development of biodegradable polymers [60]. Biodegradable polymers can be employed in various sectors, including biomedicine, additive manufacturing, film, textiles, packaging, automotive, and agriculture [61]. The "polymer of the 21st century" is PLA, the ideal polymers employed in such applications. It is notable for being a biocompatible, biodegradable, and biobased polymers and has been manufactured on a large scale [62]. With ever-increasing applications and needs for greater manufacturing, it appears to be the most environmentally friendly polymer today. The demand for PLA is greatly increasing because it emits fewer carbon emissions than traditional polymers [63]. Because PLA has high biocompatibility, is non-toxic, and is frequently employed in the biomedical industry [24]. The non-biomedical applications of PLA are briefly discussed and schematically represented in Figure 3.5, whereas the biomedical applications are elaborated in detail in the following section.

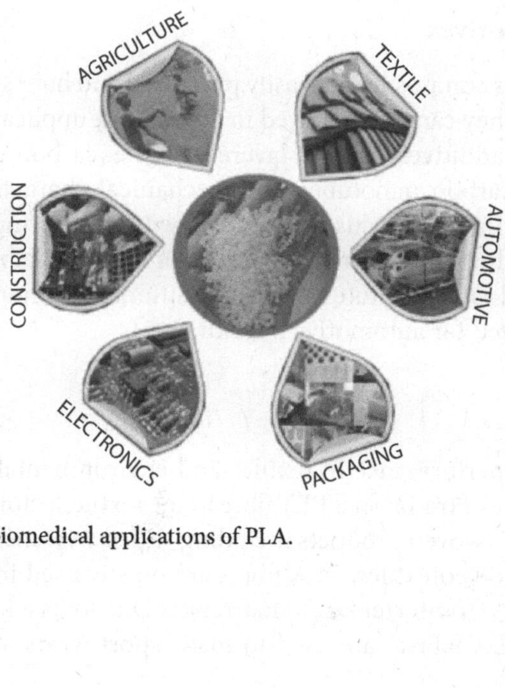

Figure 3.5 Non-biomedical applications of PLA.

3.9.1 Packaging

PLA is used frequently in food packaging due to its excellent biocompatibility, non-toxicity, and physical characteristics. The production of different kinds of containers for fresh food packaging makes extensive use of PLA. Still, the application of PLA is constrained by its low barrier properties, low versatility, and poor crystallization behavior. Therefore, a variety of substances are frequently combined with the proper additives to boost its performance [63–66].

3.9.2 Single-Use Products

There is a need to change our delivery strategies to offer more environmentally sustainable items because of the adverse impacts of plastics. Growing efforts are being made to provide affordable, single-use goods that are eco-friendly. Many single-use products, such as straws, trays, containers, plates, bowls, hot and cold cups, lids, cutlery sets, and have entered the market in recent years. PLA polymer can be used to make any of these. There are also gift boxes, animal waste bags, bags of various kinds, and gloves made completely with PLA available in the market [63].

3.9.3 Automotives

Since PLA and its composites are easily processed and have strong mechanical properties, they can be employed in automobile applications. By using the appropriate additives, such as layered silicate, carbon derivatives like graphene, and carbon nanotubes, the mechanical characteristics can be enhanced [67]. The material is well recognized for lowering carbon emissions and providing other benefits because of its remarkable bio-content. PLA can be used as a substitute for most traditional thermoplastics that are commonly utilized for automotive interiors [65].

3.9.4 Textile

Due to its high performance in textiles and environmental benefits, PLA can take the place of traditional PET fibre in the textile sector. In addition to being used in nonwoven products including wipes, hygiene items, agricultural textiles, and geotextiles, PLA fibres are mostly used in bedding such as duvets, pillows, comforters, and mattresses. Due to its effective moisture management, PLA fabrics are used to make sport wears and fashionable clothes. However, the production of PLA fibre or fabrics is hampered by heat deterioration and low hydrolytic resistance to high acidity. In the textile business, pairing PLA with other biobased or synthetic polymers is quite frequent [68–71].

3.9.5 Agricultural Use

The worldwide environmental conditions and quantity of food produced have a significant impact on the agricultural methods used. Intensive agriculture or pastoral use already occupies around half of the world's usable land. Agriculture changes natural ecosystems and enriches terrestrial ecosystems with large-scale, environmentally harmful nitrogen and phosphorus concentrations. Global food security is under severe risk as a result of decades of agricultural intensification to boost crop yields. Major obstacles to feeding a growing human population include soil erosion, different types of pollution as well as climate change [72]. Renewable resource-based polymers were proposed to lessen fossil fuel consumption and their environmental impact. Among these, PLA is a promising bio-based polymer with wide applicability in agriculture. Still, certain modifications are required, mostly to reduce brittleness. On various agricultural uses, chemical procedures are used to improve the hydrophobicity of PLA. In fact, by lowering the humidity in the crop environment, fungal infections can develop less, further decreasing the need for fungicides [73].

3.9.6 Electronics

Modern society must move away from excessive reliance on materials and energy derived from fossil fuels and toward technology built on secure and sustainable resources. One of the main strategies for reducing the release of carbon dioxide is the responsible utilization of natural assets and the widespread application of green power. Printed and hybrid integrated circuits constructed of recycled and renewable materials can reduce energy consumption and storage while also slowing the exhaustion of finite natural resources. Over the last twenty years, the field of organic electronics has undergone significant expansion as a result of extraordinary advancements in both materials and equipment. Recycled polymers, in contrast to biodegradability and disposability, offer an environmentally friendly alternative to conventional polymers by conserving virgin raw material resources,

Table 3.3 Recent developments in the applications of PLA-based nanocomposite materials.

Serial number	Year	PLA composite	Application	References
1.	2020	Surface treated PLA	Agriculture	[73]
2.	2020	PLA/TiO_2	Contaminated water treatment	[78]
3.	2018	PBAT/PLA/$CaCO_3$	Agriculture	[79]
4.	2019	Abamectin/ PLA/-Tannin	Agriculture	[80]
5.	2022	Polyamide/PLA/ Alumina	3D printing	[81]
6.	2021	PLA/poly(3-hydroxy butyrate-co-3-hydroxyvalerate)/ Carbon nanotubes	Electrical and electromagnetic applications	[82]
7.	2022	PLA/PCL/Graphene nanoparticles	Electrically conductive circuit	[83]
8.	2019	PLA/MgO	Food Packaging	[84, 85]
9.	2021	PLA/ZnO	Food Packaging	[86]
10.		PLA/Nano MgO	Food Packaging	
11.	2018	PLA/TiO_2/Oleic acid	Food Packaging	[87]
12.	2020	PLA/Biochar	Food Packaging	[88]
13.	2019	PLA/TiO_2	Food packaging	[89]
14.	2022	PLA/ZnO/aloe barbadensis leaves extract	Food Packaging	[90]
15.	2018	PLA/TiO_2	Food Packaging	[15]
16.	2021	PLA/lignocellulose materials	Single-use products	[91]

helping to promote a circular economy and weaning society off of its reliance on fossil fuels. PLA is a high-quality, favorable, recyclable polymers [74, 75].

3.9.7 Construction

The construction industry consumes significant global resources and severely impacts the environment. Polylactic acid (PLA) may be a possible replacement for present commercial polymers in developing new polymers with a strong environmental focus. PLA is an environmentally friendly and compostable thermoplastic polymer and degrades enzymatically. Over a temperature range from ambient temperature to 343 K, PLA has high dielectric breakdown strength, high electrical resistivity, and good processability. However, as previously mentioned, the PLA has a number of disadvantages that limits its potential in diverse industrial applications. To partially overcome these restrictions, PLA has been strengthened using affordable natural fibers or other nanofillers [76, 77]. Recent advancements in PLA-based nanocomposite applications are summarized in Table 3.3.

3.10 Biomedical Applications of PLA

Natural biodegradable polymers are extensively utilized in biological applications. However, they are impractical to supply in large quantities and difficult to modify structurally. Biodegradable synthetic polymers can help solve these problems. They are highly adaptable and can be chemically modified to suit specific biomedical needs [92]. Polyhydroxy acids, including polyglycolic acid, PLA, and polydioxanone, are the most often used synthetic biodegradable polymers in medical applications [93]. A viable contender for the clinical field could be an anti-bacterial metal/alloy-loaded PLA nanocomposite scaffold [94]. PLA is utilized in tissue engineering, delivery systems, membranes, bioabsorbable medical implants, and cosmetic sutures due to its versatility and ability to adapt to various applications [95]. To obtain the desired behavior, it may be commonly blended or copolymerized with polymeric or non-polymeric constituents [31].

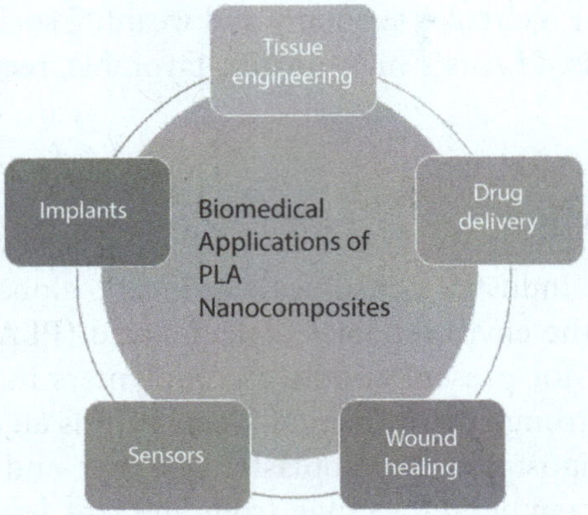

Figure 3.6 Biomedical applications of PLA.

One of the key strategies for addressing the issues with PLA in healthcare is the formulation of composite substances based on PLA. The combination of PLA with alternative materials may yield products exhibiting optimized physical and biological characteristics. Incorporating nanoparticles into PLA has been explored recently to achieve improved performance. Due to their distinctive characteristics, nanocomposites have received a lot of interest [96, 97]. Due to their capacity to encapsulate drugs, deliver sustained drug release, and have biocompatibility and bioavailability, polymer nanocarriers are currently active area of research for controlled local drug delivery applications [92]. The following section reviews the current status of PLA biocomposites in tissue engineering, drug delivery, wound therapy, implants, and biosensors for potential use in medical applications. Figure 3.6 shows some biomedical uses for PLA composites in a schematic manner.

3.10.1 Tissue Engineering

The multidisciplinary field of tissue engineering harnesses advancements in chemistry, physics, engineering, and biological sciences to address critical medical challenges, including loss of tissue and organ failure [98]. A potentially effective method for regenerative medicine is tissue engineering. Tissue-engineered biomaterials should have three-dimensional (3D) structures with the appropriate pore volume, pore size, and surface topography for cell migration and the formation of new tissue. It should be simple for nutrients, soluble components, and by-products to diffuse *via* the 3D pores. To facilitate the installation process and patient movements following surgery, biological substitutes should have mechanical qualities that are comparable to or even superior to the tissues to be recovered [92]. Linear aliphatic polyesters (PGA, PLA, etc.), manufactured into biomedical research scaffolds, are degradable polymers employed for tissue-engineering applications [98]. Blending PLA with other polymers enhances its properties, enabling the development of materials suited for diverse biomedical applications, including tissue engineering, where PLA and its copolymers are widely used due to their excellent biocompatibility and mechanical properties [98].

PLA composite scaffolds made of nano-hydroxyapatite (n-HA) were created by Wang *et al.* (2021) using the Fused Deposition Modelling method. The mechanical properties test and material characterization were used to optimize the composite scaffold. Finally, a rabbit model has been developed for assessing the *in vivo* bone regenerating potential of the fabricated material. The findings indicated that the resultant materials exhibit excellent printability and adjustable mechanical strength corresponding to the amount of n-HA. It was demonstrated that the biological suitability and osteogenic induction capabilities of the material made from PLA/n-HA were superior to those from pure PLA. These scaffolds made of PLA and n-HA can repair large bone defects [99].

Figure 3.7 illustrates the preparation and application of the PLA–nHA scaffold. It also summarizes *in vivo* and *in vitro* evaluations.

Grémare *et al.* (2018) created PLA scaffolds with varying pore sizes using the FDM technique. They discovered that the semicrystalline nature of the polymer was unaffected by 3D printing. The mechanical characteristics of the scaffold are independent of the pore- size. The scaffolds showed no cytotoxicity towards human bone marrow stromal cells. The scaffolds showed good cell viability and uniform distribution [100].

Hassanajili *et al.*, 2019 employed an indirect method of three-dimensional printing for developing bone substitutes from PLA, PCL, and HA composites. To develop simultaneously macro- and micro-porous materials, a soluble 3D-printed negative mold was filled with composite suspensions by means of freeze-drying. The porosity, microstructure, mechanical strength, and pore size of scaffolds were studied, and the compressive modulus was also calculated. The cytotoxicity evaluation indicates the developing cells could potentially proliferate and endure on the scaffolds. The findings showed the scaffolds with mass ratios of 70:30 for PLA and PCL have better biocompatibility, viability, and osteo-induction properties [101].

To achieve good filler dispersion, Silva *et al.* (2021) created composite filaments of PLA that were chemically functionalized, enhanced with

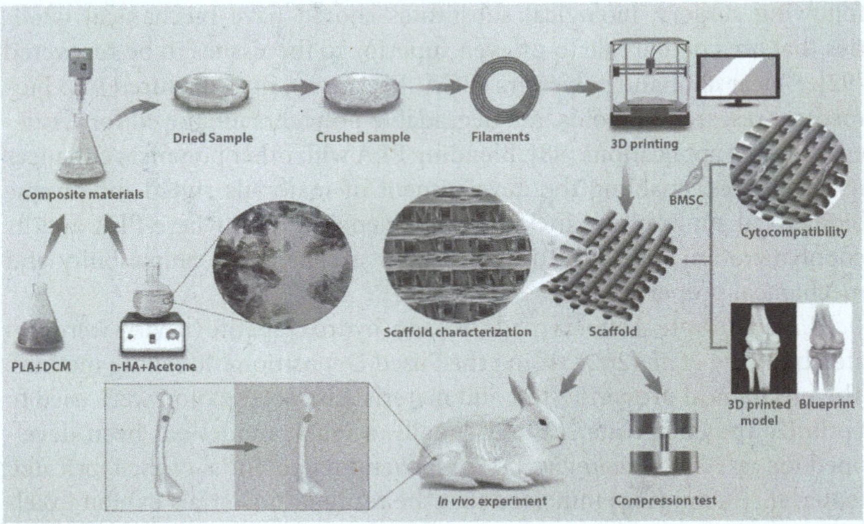

Figure 3.7 Diagram showing the steps involved in preparing and using PLA/n-HA scaffolds for osteogenesis reconstruction. Reproduced with permission [99], Copyright Elsevier.

graphite nanoplatelets, and embellished with silver nanoparticles. The resultant nanocomposite tubes exhibited thermal stability, and the addition of graphite makes them stiffer and less electrically resistive than the materials exclusively made of PLA. After 27 days, none of the filaments experienced significant deterioration. Composite filaments were processed into 3D scaffolds with controlled size and porosity, making them suitable for ligament tissue engineering [102].

3.10.2 Wound Healing

A wound is a disruption of the skin resulting from trauma or medical disorders, categorized as superficial, partial thickness, or full thickness according to the depth and extent involved. An optimal wound healing product must maintain skin moisture, inhibit infection, and absorb exudates and toxins [103]. The process of healing a wound is extremely complicated and influenced by numerous variables. One of the elements that significantly affects the healing process is bacterial infection. The application of broad-spectrum antibacterial agents is one approach to managing wound infections. The PLA-based hybrid nanocomposites are promising for wound-healing applications [104].

Bi *et al.* (2020), compared electrospun PLA and PLA/Polyvinyl alcohol (PVA)/Sodium alginate (SDA) composite fiber for wound closure *in vitro* and *in vivo*. They created porous nanocomposite fiber with an extensive specific surface area using electrospinning. *In vitro* tests showed that both PLA and PLA/PVA/SDA composite fiber effectively supported rat fibroblast growth, demonstrating their potential for wound therapy. Additionally, rat fibroblasts showed marginally superior adherence and growth on the developed film compared to the PLA film. The study showed no discernible changes in wound closure between pure and composite membranes [105].

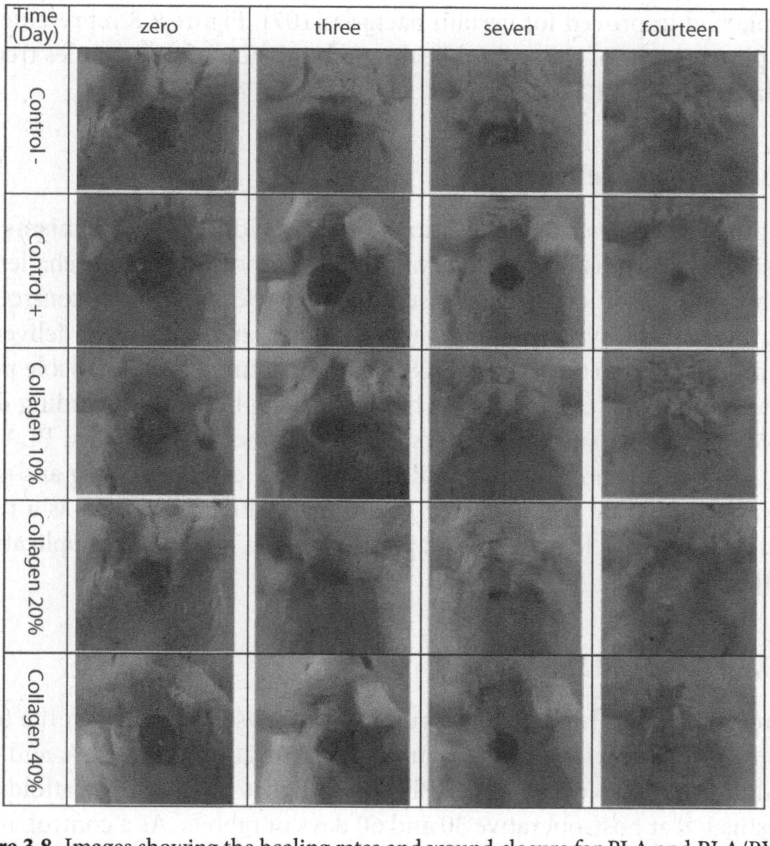

Figure 3.8 Images showing the healing rates and wound closure for PLA and PLA/PVA/SA fiber-treated groups on days 1, 7, 10, and 16. Reproduced with permission [107], Copyright Elsevier.

Clarithromycin drug release from PLA nanofibers combined with graphene-oxide (GO) and selenium (Se) was shown by Ciftci *et al.* (2021). They found that the fabricated fiber has high antibacterial properties and reveals its potential for wound healing. The generated nanofibers showed significantly appropriate mechanical qualities and controlled release. The results suggested that the composite made of Se-integrated PLA-GO-drug is a potential material for wound healing [106].

To specifically aim to release the enclosed compounds, Hajikhani *et al.* (2021) created Polyvinylpyrrolidone /PLA/Polyethylene oxide composite nanofibers containing collagen and cefazolin scaffold. The speed of healing was seen to differ significantly from the control sample at dosages of collagen 10 and 20%, while at 40%, the rate of wound closure in mice was seen to slow down. The outcomes showed that the antibacterial activity of the sample had improved for certain bacteria [107]. Figure 3.8 represents the real photographs of the wound healing progression of the samples treated on days 0, 3, 7, and 14.

3.10.3 Drug Delivery

The process of delivering medications and xenobiotics to an organism's target site is known as drug delivery. Drug delivery systems face challenges in ensuring sustained drug release, targeting specific tissues, controlling drug bioavailability, improving cell specificity, and effectively delivering macromolecules to target cells [108]. Nanostructured biodegradable polymers are potential candidates for biomedical applications, including drug delivery, imaging, and cell carriers [92, 109]. Since its discovery, PLA has been used as a promising material for scaffolds, catheters, drug and gene delivery, and tissue regeneration. Additionally, PLA is utilized as a regulated delivery vehicle in oncology, plastic surgery, and cardiac applications [103].

3.10.4 Medical Implants

A prototype of an Implantable Drug Delivery System (IDDS) in the form of a disc was developed by Fayzullin *et al.* in 2021 using PLA and pirfenidone (PF)loaded PLA(PLA/PF). The impact of PLA/PF scaffold was investigated at post-operative 30 and 60 days in rabbits. As a control, identical single-dose PF injections were administered without PLA implants. By modulating the foreign body response with progressively biodegradable PF-loaded IDDS, they demonstrated the viability of local regulation of the fibrotic response to implants [110].

Using a bio 3D printer and fused deposition modeling, Saniei *et al.* (2020) fabricated PLA screw implants that were then surface modified to improve their bioactivity and cell affinity. PVA-nHA nanofibers with varying nHA concentrations were electrospun onto the screws for coating. By submerging the coated screws in a simulated body fluid, bioactivity was assessed, and the cytotoxicity was measured using MTT assay. The outcomes demonstrated that the PVA-nHA nanocomposite fibers could considerably improve both the attachment and proliferation of the MC3T3-E1 cells [111].

By hot melt extrusion method, Bode *et al.*, (2019) developed various polylactic-co-glycolic acid (PLGA) and PLA derived systems for regulated dexamethasone release. Both the drug loading (1–15%) and the PLA: PLGA ratio were altered (1:1, 3:1, and 1:0). The implants were examined before and after being placed in a phosphate buffer at 37°C [112].

A poly(L-lactide-co-D, L-lactide) 70:30 nasal implant's biocompatibility and absorption profile were investigated by Marian *et al.* (2018). In 11 nasal dorsa, 66 rod-shaped absorbable implants were inserted. Histological and gross examinations were performed. The findings demonstrated that there were no surgical problems, indications of infection, or tissue rejection during the follow-up periods. The results of gross pathological investigations revealed no irregularities [113].

3.10.5 Biosensors

Since additive manufacturing facilitates the customised and decentralised generation of inexpensive sensors and actuators, it has started to gain popularity for prototyping at biosensing avenues. Electrochemical and optical devices, among other areas, might greatly benefit from the additive manufacture of bioanalytical platforms; nevertheless, some crucial issues should be resolved to produce active and reliable biosensing systems [114].

The antithesis of subtractive manufacturing, which creates three-dimensional items using material-removal operations, including drilling, milling, sawing, and broaching, is additive manufacturing, or 3D printing. Both additive and subtractive manufacturing are acceptable processes for rapid prototyping. The decision depends on considering a variety of elements. Plastic-based products are typically made using additive manufacturing, whereas metal and timber products are typically made using subtractive manufacturing. Along with these distinctions, it is important to note that additive manufacturing has the potential to drastically minimise material waste and energy costs by using machines with lower energy requirements [115].

By using the Carreau-Yasuda model, Zare *et al.* (2019) examined the complex viscosity of PLA/polyethylene oxide (PEO) and PLA/PEO/CNT biosensors with respect to frequency. The model correlates the complex viscosity at zero, relaxation time, power-law index, shear stress, and the width of the transition region between Newtonian and power-law behavior. At many frequency ranges, the Carreau-Yasuda model's computations successfully match the experimental results. With the help of the prepared samples, nano-biosensors for the early detection of diseases like cancer can be generated [116].

Using electrospun PLA/polyethylene glycol (PEG) and PLA/PEG/Biotin fibres as examples, Shepherd *et al.* (2016) discuss the influence of solvent and co-polymer block lengths on the stability of both fibres in water. Customizing the PEG block length improved the water stability of the electrospun fibers compared to previous designs. They showed that solvent choice also affected fiber stability and hydrophilicity [117].

A composite polymer based on activated biochar that is suitable for biosensors in smart food packaging was created and studied by Sobhan *et al.* in 2021. Corn stover was used to create biochar, which was then activated using steam. Activated charcoal (AC) was created using a solvent casting process using PLA. The electrical conductivity and mechanical strength of the developed composite fibers were measured and it was concluded that as the AC content increased, the current increased, while as the PLA content increased, the mechanical properties improved [118].

Zare *et al.* (2020) proposed a straightforward model that takes into account CNT concentration and diameters, network size, and interphase thickness. The effective filler concentration and the percolation threshold, which control the proportion of networked CNT in nanocomposite biosensors, are expressed by the CNT concentration and size, and interphase thickness [119].

3.11 Conclusions and Future Perspectives

This chapter highlights the potential of electroactive, biodegradable PLA-based nanocomposites as multifunctional materials for various applications. They hold promise for biomedical applications, such as drug delivery and tissue engineering, and in advancing environmental sustainability. Their emerging roles in electronics and sensing technologies point to a future of eco-conscious innovations. To fully realize their capabilities, research should prioritize nanofiller optimization, interface engineering, and scalable manufacturing. With sustained progress in these areas,

PLA-based nanocomposites could bridge high-performance technology and environmental stewardship. This synergy paves the way for an innovative and sustainable future.

AI disclosure statement: The author acknowledges the use of AI-based tools for minor language editing or grammatical correction. However, all intellectual contributions, analyses, and conclusions presented in this chapter are entirely the author's own.

References

1. Ponnamma, D., Ogunleye, G.J., Sharma, P., AlMaadeed, M.A., Piezo- and Thermoelectric Materials from Biopolymer Composites, in: *Biopolymer Composites in Electronics*, pp. 333–352, Elsevier Inc., 2017, doi:10.1016/B978-0-12-809261-3.00012-7.
2. Ameer Ali, M. and Shanavas, A., Polylactic Acid-Based Nanocomposites: An Important Class of Biodegradable Composites, pp. 221–231, 2019, doi:10.1007/978-981-13-8063-1_9.
3. Sadasivuni, K.K., Abdullahil K., Lindong Z., Hyun-U. K., Seongcheol M., Jaehwan K., Transparent and flexible cellulose nanocrystal/reduced graphene oxide film for proximity sensing. *Small*, 11, 8, 994–1002, 2015. https://doi.org/10.1002/smll.201402109.
4. Sencadas, V., *et al.*, Local piezoelectric activity of single poly(L-lactic acid) (PLLA) microfibers. *Appl. Phys. A Mater. Sci. Process.*, 109, 51–55, 2012. https://link.springer.com/article/10.1007/s00339-012-7095-z.
5. Jordan, J., Jacob, K.I., Tannenbaum, R., Sharaf, M.A., Jasiuk, I., Experimental trends in polymer nanocomposites - A review. *Mater. Sci. Eng., A*, 393, 1–11, 2005, Preprint at https://doi.org/10.1016/j.msea.2004.09.044.
6. Tsuji, H., Daimon, H., Fujie, K., A new strategy for recycling and preparation of poly(L-lactic acid): Hydrolysis in the melt. *Biomacromolecules*, 4, 835–840, 2003. https://pubs.acs.org/doi/abs/10.1021/bm034060j.
7. Hattori, T., Iwauchi, S., Nagano, K., Tanaka, T., Effect of heat treatment on the interface characteristics in reactively sputtered Al2O3-Si structures. *Jpn. J. Appl. Phys.*, 10, 2, 1971. http://iopscience.iop.org/1347-4065/10/.
8. Fukada, E., Piezoelectricity of biopolymers. *Biorheology*, 32, 6, 593–609, 1995. https://doi.org/10.1016/0006-355X(95)00039-C.
9. Ahn, A.C. and Grodzinsky, A.J., Relevance of collagen piezoelectricity to 'Wolff's Law': A critical review. *Med. Eng. Phys.*, 31, 733–741, 2009, Preprint at https://doi.org/10.1016/j.medengphy.2009.02.006.
10. Wang, T., Feng, Z., Song, Y., Chen, X., Piezoelectric properties of human dentin and some influencing factors. *Dent. Mater.*, 23, 450–453, 2007. https://doi.org/10.1016/j.dental.2006.02.016.

11. Halperin, C., Mutchnik, S., Agronin, A., Molotskii, M., Urenski, P., Salai, M., Rosenman, G., Piezoelectric effect in human bones studied in nanometer scale. *Nano Lett.*, 4, 7, 1253–1256, 2004. https://pubs.acs.org/doi/abs/10.1021/nl049453i.
12. Athenstaedt, H., Pyroelectric and piezoelectric properties of vertebrates. *Ann. N. Y. Acad. Sci.*, 238, 1, 68–94, 1974. https://nyaspubs.onlinelibrary.wiley.com/doi/abs/10.1111/j.1749-6632.1974.tb26780.x
13. Kwon, J. and Hanna Cho. Collagen piezoelectricity in osteogenesis imperfecta and its role in intrafibrillar mineralization. *Commun. Biol.*, 5, 1, 1229, 2022. https://www.nature.com/articles/s42003-022-04204-z.
14. Hastings, G.W. and Mahmud, F.A., Electrical effects in bone. *J. Biomed. Eng.*, 10, 6, 515–521, 1988. https://www.sciencedirect.com/science/article/pii/0141542588901094.
15. Segura González, E.A., Olmos, D., Lorente, M.Á., Vélaz, I., González-Benito, J., Preparation and characterization of polymer composite materials based on PLA/TiO2 for antibacterial packaging. *Polymers*, 10, 12, 1365, 2018. https://www.mdpi.com/2073-4360/10/12/1365.
16. Farah, S., Anderson, D.G., Langer, R., Physical and mechanical properties of PLA, and their functions in widespread applications — A comprehensive review. *Adv. Drug Deliv. Rev.*, 107, 367–392, 2016, Preprint at https://doi.org/10.1016/j.addr.2016.06.012.
17. Murariu, M. and Dubois, P., PLA composites: From production to properties. *Adv. Drug Deliv. Rev.*, 107, 17–46, 2016, Preprint at https://doi.org/10.1016/j.addr.2016.04.003.
18. Hu, Y., *et al.*, Preparation and drug release behaviors of nimodipine-loaded poly(caprolactone)-poly(ethylene oxide)-polylactide amphiphilic copolymer nanoparticles. *Biomaterials*, 24, 2395–2404, 2003. https://www.sciencedirect.com/science/article/pii/S0142961203000218
19. Tarani, E., *et al.*, Cold crystallization kinetics and thermal degradation of pla composites with metal oxide nanofillers. *Appl. Sci. (Switzerland)*, 11, 7, 3004, 2021. https://www.mdpi.com/2076-3417/11/7/3004
20. Ngo, T.D., Kashani, A., Imbalzano, G., Nguyen, K.T.Q., Hui, D., Additive manufacturing (3D printing): A review of materials, methods, applications and challenges. *Composites, Part B*, 143, 172–196, 2018, Preprint at https://doi.org/10.1016/j.compositesb.2018.02.012.
21. McKeen, L.W., Environmentally Friendly Polymers, in: *Permeability Properties of Plastics and Elastomers*, pp. 287–304, Elsevier, 2012, doi:10.1016/b978-1-4377-3469-0.10013-x.
22. McKeen, L., Renewable Resource and Biodegradable Polymers, in: *The Effect of Sterilization on Plastics and Elastomers*, pp. 305–317, Elsevier, 2012, doi:10.1016/b978-1-4557-2598-4.00012-5.
23. Deshmukh, K., *et al.*, Biopolymer Composites with High Dielectric Performance: Interface Engineering, in: *Biopolymer Composites in Electronics*, pp. 27–128, Elsevier Inc., 2017, doi:10.1016/B978-0-12-809261-3.00003-6.

24. Li, G., *et al.*, Synthesis and Biological Application of Polylactic Acid. *Molecules*, 25, 21, 5023, 2020, Preprint at https://doi.org/10.3390/molecules25215023.
25. Konta, A.A., García-Piña, M., Serrano, D.R., Personalised 3D printed medicines: Which techniques and polymers are more successful? *Bioengineering*, 4, 4, 79, 2017, Preprint at https://doi.org/10.3390/bioengineering4040079.
26. Shahriar, S.M.S., *et al.*, Electrospinning nanofibers for therapeutics delivery. *Nanomaterials*, 9, 4, 532, 2019, Preprint at https://doi.org/10.3390/nano9040532.
27. Casasola, R., Thomas, N.L., Trybala, A., Georgiadou, S., Electrospun poly lactic acid (PLA) fibres: Effect of different solvent systems on fibre morphology and diameter. *Polymer (Guildf)*, 55, 4728–4737, 2014. https://www.sciencedirect.com/science/article/pii/S0032386114005266
28. Bhattarai, R.S., Bachu, R.D., Boddu, S.H.S., Bhaduri, S., Biomedical applications of electrospun nanofibers: Drug and nanoparticle delivery. *Pharmaceutics*, 11, 1, 5, 2019, Preprint at https://doi.org/10.3390/pharmaceutics11010005.
29. Pantani, R., De Santis, F., Sorrentino, A., De Maio, F., Titomanlio, G., Crystallization kinetics of virgin and processed poly(lactic acid). *Polym. Degrad. Stab.*, 95, 1148–1159, 2010, https://www.sciencedirect.com/science/article/pii/S0141391010001825.
30. Sharif, A., Mondal, S., Hoque, M.E., Polylactic acid (PLA)-based nanocomposites: Processing and properties, in: *Bio-based Polymers and Nanocomposites: Preparation, Processing, Properties & Performance*, pp. 233–254, Springer International Publishing, 2019, doi:10.1007/978-3-030-05825-8_11.
31. Gupta, A.P. and Kumar, V., New emerging trends in synthetic biodegradable polymers - Polylactide: A critique. *Eur. Polym. J.*, 43, 4053–4074, 2007, Preprint at https://doi.org/10.1016/j.eurpolymj.2007.06.045.
32. Jamshidi, K., Hyon, S.H., Ikada, Y., Thermal characterization of polylactides. *Polymer,* 29, 12, 2229–2234, 1988. https://www.sciencedirect.com/science/article/pii/0032386188901164
33. Bergsma, J.E., De Bruijn, W.C., Rozema, F.R., Bos, R.R.M., Boering, G., Late degradation tissue response to poly (L-lactide) bone plates and screws. *Biomaterials,* 16, 1995. https://www.sciencedirect.com/science/article/pii/014296129591092D
34. Garlotta, D., A literature review of poly (lactic acid). *J. Polym. Environ.*, 9, 2, 63–84, 2001. https://link.springer.com/article/10.1023/A:1020200822435
35. Swetha, T.A., Ananthi, V., Bora, A., Sengottuvelan, N., Ponnuchamy, K., Muthusamy, G., Arun, A., A review on biodegradable polylactic acid (PLA) production from fermentative food waste-Its applications and degradation. *Int. J. Biol. Macromol.*, 234, 123703, 2023, https://www.sciencedirect.com/science/article/pii/S0141813023005962.
36. Vert, M., Schwarch, G., Coudane, J., Present and Future of PLA Polymers. *J. Macromol. Sci. Part A*, 32, 787–796, 1995, https://www.tandfonline.com/doi/abs/10.1080/10601329508010289.

37. Mainil-Varlet, P., Rahn, B., Gogolewski, S., Long-Term *in vivo* Degradation and Bone Reaction to Various Polylactides 1. One-Year Results. *Biomaterials*, 18, 3, 257–266, 1997, https://www.sciencedirect.com/science/article/pii/S0142961296001263.
38. Zhang, X., MacDonald, D.A., Goosen, M.F., McAuley, K.B., Mechanism of lactide polymerization in the presence of stannous octoate: the effect of hydroxy and carboxylic acid substances. *J. Polym. Sci. A: Polym. Chem.*, 32, 15, 2965–2970, 1994, https://onlinelibrary.wiley.com/doi/abs/10.1002/pola.1994.080321519
39. Dubey, S.P., *et al.*, Progress in environmental-friendly polymer nanocomposite material from PLA: Synthesis, processing and applications. *Vacuum*, 146, 655–663, 2017, https://www.sciencedirect.com/science/article/pii/S0042207X17305547.
40. Hu, Y., Daoud, W.A., Cheuk, K.K.L., Lin, C.S.K., Newly developed techniques on polycondensation, ring-opening polymerization and polymer modification: Focus on poly(lactic acid). *Materials*, 9, 3, 133, 2016, Preprint at https://doi.org/10.3390/ma9030133.
41. García-Arrazola, R., López-Guerrero, D.A., Gimeno, M., Bárzana, E., Lipase-catalyzed synthesis of poly-l-lactide using supercritical carbon dioxide. *J. Supercrit. Fluids*, 51, 197–201, 2009, https://www.sciencedirect.com/science/article/pii/S0896844609002873.
42. Utsunomia, C., *et al.*, Synthesis of lactate (LA)-based poly(ester-urethane) using hydroxyl-terminated LA-based oligomers from a microbial secretion system. *J. Polym. Res.*, 24, 10, 167, 2017, https://link.springer.com/article/10.1007/s10965-017-1325-1.
43. Lim, L.T., Auras, R., Rubino, M., Processing technologies for poly(lactic acid). *Prog. Polym. Sci. (Oxford)*, 33, 820–852, 2008, Preprint at https://doi.org/10.1016/j.progpolymsci.2008.05.004.
44. Agarwal, M., Koelling, K.W., Chalmers, J.J., Characterization of the Degradation of Polylactic Acid Polymer in a Solid Substrate Environment. *Biotechnol. Prog.*, 14, 3, 517–526, 1998, https://aiche.onlinelibrary.wiley.com/doi/abs/10.1021/bp980015p
45. Abdel-Rahman, M.A., Tashiro, Y., Sonomoto, K., Recent advances in lactic acid production by microbial fermentation processes. *Biotechnol. Adv.*, 31, 877–902, 2013, Preprint at https://doi.org/10.1016/j.biotechadv.2013.04.002.
46. Athanasiou, K.A., Niederauer, G.G., Agrawal, C.M., Sterilization, Toxicity, Biocompatibility and Clinical Applications of Polylactic Acid/ Polyglycolic Acid Copolymers. *Biomaterials*, 17, 2, 93–102, 1996.
47. Ranakoti, L., Gangil, B., Mishra, S.K., Singh, T., Sharma, S., Ilyas, R.A., El-Khatib, S., Critical review on polylactic acid: properties, structure, processing, biocomposites, and nanocomposites. *Materials*, 15, 12, 4312, 2022.
48. Ilyas, R.A., *et al.*, Natural Fiber-Reinforced Polylactic Acid, Polylactic Acid Blends and Their Composites for Advanced Applications. *Polymers*, 14, 1, 202, 2022, Preprint at https://doi.org/10.3390/polym14010202.

49. Agrawal, C.M. and Athanasiou, K.A., Technique to control pH in vicinity of biodegrading PLA-PGA implants. *J. Biomed. Mater. Res.*, 38, 105–114, 1997 https://onlinelibrary.wiley.com/doi/abs/10.1002/(SICI)1097-4636(199722)38:2%3C105::AID-JBM4%3E3.0.CO;2-U.
50. Zalusky, A.S., Olayo-Valles, R., Wolf, J.H., Hillmyer, M.A., Ordered nanoporous polymers from polystyrene-polylactide block copolymers. *J. Am. Chem. Soc.*, 124, 12761–12773, 2002 ,https://pubs.acs.org/doi/abs/10.1021/ja0278584.
51. Auras, R.A., Harte, B., Selke, S., Hernandez, R., Mechanical, physical, and barrier properties of poly(lactide) films. *J. Plast. Film Sheeting*, 19, 123–135, 2003, https://journals.sagepub.com/doi/abs/10.1177/8756087903039702.
52. Casalini, T., Rossi, F., Castrovinci, A., Perale, G., A Perspective on Polylactic Acid-Based Polymers Use for Nanoparticles Synthesis and Applications. *Front. Bioeng. Biotechnol.*, 7, 5, 787421, 2019, Preprint at https://doi.org/10.3389/fbioe.2019.00259.
53. Sawyer, D.J., Bioprocessing - No Longer a Field of Dreams. *Macromol. Symp.*, 201, 1, 271–282, 2003, Octobere, Weinheim: WILEY-VCH, Verlag. https://journals.sagepub.com/doi/abs/10.1177/8756087903039702
54. Rajendran, D.S., Venkataraman, S., Jha, S.K., Chakrabarty, D., Vaidyanathan, V.K., A review on bio-based polymer polylactic acid potential on sustainable food packaging. *Food Sci. Biotechnol.*, 33, 8, 1759-1788, 2024, ttps://doi.org/10.1007/s10068-024-01543-x.
55. Eling, B., Gogolewski, S., Pennings, A.J., Biodegradable materials of poly (l-lactic acid): 1. Melt-spun and solution-spun fibres. *Polymer*, 23, 11, 1587–1593, 1982. https://www.sciencedirect.com/science/article/pii/0032386182901768
56. Vink, E.T.H., Rábago, K.R., Glassner, D.A., Gruber, P.R., Applications of life cycle assessment to NatureWorks™ polylactide (PLA) production. *Polym. Degrad. Stab.*, 80, 403–419, 2003, https://www.sciencedirect.com/science/article/pii/S0141391002003725.
57. Raquez, J.M., Habibi, Y., Murariu, M., Dubois, P., Polylactide (PLA)-based nanocomposites. *Prog. Polym. Sci.*, 38, 1504–1542, 2013, Preprint at https://doi.org/10.1016/j.progpolymsci.2013.05.014.
58. Hiljanen-Vainio, M., Varpomaa, P., Seppälä, J., Törmälä, P., Modification of poly (L-lactides) by blending: mechanical and hydrolytic behavior. *Macromol. Chem. Phys.*, 197, 4, 1503–1523, 1996. https://onlinelibrary.wiley.com/doi/abs/10.1002/macp.1996.021970427.
59. Okada, M., Chemical syntheses of biodegradable polymers. *Prog. Polym. Sci.*, 27, 1, 87–133, 2002. https://www.sciencedirect.com/science/article/pii/S0079670001000399
60. Nakajima, H., Dijkstra, P., Loos, K., The recent developments in biobased polymers toward general and engineering applications: Polymers that are upgraded from biodegradable polymers, analogous to petroleum-derived polymers, and newly developed. *Polymers*, 9, 10, 523, 2017, Preprint at https://doi.org/10.3390/polym9100523.

61. Höhnemann, T., *et al.*, Poly(Ethylene furanoate) along its life-cycle from a polycondensation approach to high-performance yarn and its recyclate. *Materials*, 14, 1–15, 2021, https://www.mdpi.com/1996-1944/14/4/1044.
62. Pretula, J., Slomkowski, S., Penczek, S., Polylactides—Methods of synthesis and characterization. *Adv. Drug Deliv. Rev.*, 107, 3–16, 2016, Preprint at https://doi.org/10.1016/j.addr.2016.05.002.
63. Balla, E., *et al.*, Poly(lactic acid): A versatile biobased polymer for the future with multifunctional properties-from monomer synthesis, polymerization techniques and molecular weight increase to PLA applications. *Polymers*, 13, 11, 1822, 2021, Preprint at https://doi.org/10.3390/polym13111822.
64. Fredi, G., Rigotti, D., Bikiaris, D.N., Dorigato, A., Tuning thermo-mechanical properties of poly(lactic acid) films through blending with bioderived poly(alkylene furanoate)s with different alkyl chain length for sustainable packaging. *Polymer (Guildf)*, 218, 123527, 2021, https://www.sciencedirect.com/science/article/pii/S0032386121001506.
65. Barage, S., *et al.*, Nanomaterial in Food Packaging: A Comprehensive Review. *J. Nanomater.*, 2022, 1, 6053922, Preprint at https://doi.org/10.1155/2022/6053922.
66. Lagarón, J.M., Polylactic acid (PLA) nanocomposites for food packaging applications, in: *Multifunctional and Nanoreinforced Polymers for Food Packaging*, pp. 485–497, Elsevier, 2011, doi:10.1533/9780857092786.4.485.
67. Arjmandi, R., Hassan, A., Zakaria, Z., Polylactic acid green nanocomposites for automotive applications, in: *Green Energy and Technology*, pp. 193–208, 2017, https://link.springer.com/chapter/10.1007/978-3-319-49382-4_9.
68. Delgado-Aguilar, M., Puig, R., Sazdovski, I., Fullana-i-Palmer, P., Polylactic acid/polycaprolactone blends: On the path to circular economy, substituting single-use commodity plastic products. *Materials*, 13, 11, 2655, 2020.
69. Yang, Y., *et al.*, Poly(lactic acid) fibers, yarns and fabrics: Manufacturing, properties and applications. *Text. Res. J.*, 91, 1641–1669, 2021, Preprint at https://doi.org/10.1177/0040517520984101.
70. Avinc, O. and Khoddami, A., Overview of poly(lactic acid) (PLA) Fibre Part I: Production, Properties, Performance, Environmental Impact, and End-Use Applications of Poly(Lactic Acid) Fibres. *Fibre Chem.*, 41, 6, 391–401, 2009, https://link.springer.com/article/10.1007/s10692-010-9213-z.
71. Baji, A., Agarwal, K., Oopath, S.V., Emerging developments in the use of electrospun fibers and membranes for protective clothing applications. *Polymers*, 12, 2, 492, 2020, Preprint at https://doi.org/10.3390/polym12020492.
72. Tilman, K., Cassman, G., Matson, P.A., Naylor, R., Polasky, S., Agricultural sustainability and intensive production practices, insight review articles. *Nature*, 418, 671–677, 2002, www.nature.com/nature, 2002.
73. Knoch, S., *et al.*, Surface modification of PLA nets intended for agricultural applications. *Colloids Surf., A*, 598, 124787, 2020, https://www.sciencedirect.com/science/article/pii/S0927775720303800.

74. Mattana, G., Briand, D., Marette, A., Vásquez Quintero, A., De Rooij, N.F., Polylactic acid as a biodegradable material for all-solution-processed organic electronic devices. *Org. Electron.*, 17, 77–86, 2015, https://www.sciencedirect.com/science/article/pii/S1566119914005138.
75. Välimäki, M.K., *et al.*, Printed and hybrid integrated electronics using bio-based and recycled materials—increasing sustainability with greener materials and technologies. *Int. J. Adv. Manuf. Technol.*, 111, 325–339, 2020, https://link.springer.com/article/10.1007/s00170-020-06029-8.
76. Barkhad, M.S., Abu-Jdayil, B., Iqbal, M.Z., Mourad, A.H.I., Thermal insulation using biodegradable poly(lactic acid)/date pit composites. *Constr. Build. Mater.*, 261, 120533, 2020, https://www.sciencedirect.com/science/article/pii/S0950061820325381.
77. Abu Hassan, N.A., Ahmad, S., Chen, R.S., Shahdan, D., Cells analyses, mechanical and thermal stability of extruded polylactic acid/kenaf bio-composite foams. *Constr. Build. Mater.*, 240, 117884, 2020, https://www.sciencedirect.com/science/article/pii/S0950061819333379.
78. Gupta, N., *et al.*, Photocatalytic Nanocomposite Microsponges of Polylactide-Titania for Chemical Remediation in Water. *ACS Appl. Polym. Mater.*, 2, 5188–5197, 2020, https://pubs.acs.org/doi/abs/10.1021/acsapm.0c00937.
79. Rocha, D.B., Souza de Carvalho, J., de Oliveira, S.A., dos Santos Rosa, D., A new approach for flexible PBAT/PLA/CaCO3 films into agriculture. *J. Appl. Polym. Sci.*, 135, 35, 46660, 2018, https://onlinelibrary.wiley.com/doi/abs/10.1002/app.46660.
80. Yu, M., *et al.*, Tannic acid-based nanopesticides coating with highly improved foliage adhesion to enhance foliar retention. *RSC Adv.*, 9, 27096–27104, 2019, https://pubs.rsc.org/en/content/articlehtml/2019/ra/c9ra05843e.
81. Petousis, M., Vidakis, N., Mountakis, N., Papadakis, V., Tzounis, L., Three-Dimensional Printed Polyamide 12 (PA12) and Polylactic Acid (PLA) Alumina (Al2O3) Nanocomposites with Significantly Enhanced Tensile, Flexural, and Impact Properties. *Nanomaterials*, 12, 23, 4292, 2022, https://www.mdpi.com/2079-4991/12/23/4292.
82. Silva, A.P.B., Montagna, L.S., Passador, F.R., Rezende, M.C., Lemes, A.P., Biodegradable nanocomposites based on PLA/PHBV blend reinforced with carbon nanotubes with potential for electrical and electromagnetic applications. *eXPRESS Polym. Lett.*, 15, 987–1003, 2021, https://search.proquest.com/openview/9782c8f258d59dfb0b0a97bf9490e9f1/1?pq-origsite=gscholar&cbl=2044900.
83. Masarra, N.A., *et al.*, Fabrication of PLA/PCL/Graphene Nanoplatelet (GNP) Electrically Conductive Circuit Using the Fused Filament Fabrication (FFF) 3D Printing Technique. *Materials*, 15, 3, 762, 2022, https://www.mdpi.com/1996-1944/15/3/762.

84. Swaroop, C. and Shukla, M., Development of blown polylactic acid-MgO nanocomposite films for food packaging. *Compos. Part A Appl. Sci. Manuf.*, 124, 105482, 2019, https://www.sciencedirect.com/science/article/pii/S1359835X19302313.
85. Swaroop, C. and Shukla, M., Nano-magnesium oxide reinforced polylactic acid biofilms for food packaging applications. *Int. J. Biol. Macromol.*, 113, 729–736, 2018, https://www.sciencedirect.com/science/article/pii/S0141813018303039.
86. Zhang, R., *et al.*, Development of polylactic acid/ZnO composite membranes prepared by ultrasonication and electrospinning for food packaging. *LWT*, 135, 110072, 2021, https://www.sciencedirect.com/science/article/pii/S0023643820310616.
87. Baek, N., Kim, Y.T., Marcy, J.E., Duncan, S.E., O'Keefe, S.F., Physical properties of nanocomposite polylactic acid films prepared with oleic acid modified titanium dioxide. *Food Packag. Shelf Life*, 17, 30–38, 2018, https://www.sciencedirect.com/science/article/pii/S2214289417303848.
88. Sobhan, A., Muthukumarappan, K., Wei, L., Development of bionanocomposite films by combination of PLA and biochar for smart food packaging, in: *ASABE 2020 Annual International Meeting*, American Society of Agricultural and Biological Engineers, 2020, doi:10.13031/aim.202000566.
89. Feng, S., Zhang, F., Ahmed, S., Liu, Y., Physico-mechanical and antibacterial properties of PLA/TiO2 composite materials synthesized *via* electrospinning and solution casting processes. *Coatings*, 9, 8, 525, 2019, https://www.mdpi.com/2079-6412/9/8/525.
90. Batool, M., *et al.*, Quality Control of Nano-food Packing Material for Grapes (Vitis vinifera) Based on ZnO and Polylactic Acid (PLA) biofilm. *Arabian J. Sci. Eng.*, 47, 319–331, 2022, https://link.springer.com/article/10.1007/s13369-021-05361-9.
91. Scaffaro, R., Maio, A., Gulino, E.F., Alaimo, G., Morreale, M., Green composites based on pla and agricultural or marine waste prepared by fdm. *Polymers (Basel)*, 13, 9, 1361, 2021, https://www.mdpi.com/2073-4360/13/9/1361.
92. Liu, S., *et al.*, Current applications of poly(lactic acid) composites in tissue engineering and drug delivery. *Composites, Part B*, 199, 108238, 2020, Preprint at https://doi.org/10.1016/j.compositesb.2020.108238.
93. Middleton, J.C. and Tipton, A.J., Synthetic biodegradable polymers as orthopedic devices. *Biomaterials*, 21, 23, 2335–2346, 2000. https://www.sciencedirect.com/science/article/pii/S0142961200001010
94. Alam, F., Shukla, V.R., Varadarajan, K.M., Kumar, S., Microarchitected 3D printed polylactic acid (PLA) nanocomposite scaffolds for biomedical applications. *J. Mech. Behav. Biomed. Mater.*, 103, 103576, 2020, https://www.sciencedirect.com/science/article/pii/S1751616119312779.
95. Gupta, N.S., *et al.*, Experimental evidence for the formation of geomacromolecules from plant leaf lipids. *Org. Geochem.*, 38, 28–36, 2007, https://www.sciencedirect.com/science/article/pii/S0146638006002476.

96. Liu, S., Chevali, V.S., Xu, Z., Hui, D., Wang, H., A review of extending performance of epoxy resins using carbon nanomaterials. *Composites, Part B*, 136, 197–214, 2018, Preprint at https://doi.org/10.1016/j.compositesb.2017.08.020.
97. Buzarovska, A., Dinescu, S., Chitoiu, L., Costache, M., Porous poly(l-lactic acid) nanocomposite scaffolds with functionalized TiO2 nanoparticles: properties, cytocompatibility and drug release capability. *J. Mater. Sci.*, 53, 11151–11166, 2018, https://link.springer.com/article/10.1007/s10853-018-2415-0.
98. Singhvi, M.S., Zinjarde, S.S., Gokhale, D.V., Polylactic acid: synthesis and biomedical applications. *J. Appl. Microbiol.*, 127, 1612–1626, 2019, Preprint at https://doi.org/10.1111/jam.14290.
99. Wang, W., *et al.*, 3D printing of PLA/n-HA composite scaffolds with customized mechanical properties and biological functions for bone tissue engineering. *Compos. B Eng.*, 224, 109192, 2021, https://www.sciencedirect.com/science/article/pii/S1359836821005709.
100. Grémare, A., *et al.*, Characterization of printed PLA scaffolds for bone tissue engineering. *J. Biomed. Mater. Res. A*, 106, 887–894, 2018, https://onlinelibrary.wiley.com/doi/abs/10.1002/jbm.a.36289.
101. Hassanajili, S., Karami-Pour, A., Oryan, A., Talaei-Khozani, T., Preparation and characterization of PLA/PCL/HA composite scaffolds using indirect 3D printing for bone tissue engineering. *Mater. Sci. Eng., C*, 104, 109960, 2019, https://www.sciencedirect.com/science/article/pii/S0928493119308756.
102. Silva, M., *et al.*, Poly(Lactic acid)/graphite nanoplatelet nanocomposite filaments for ligament scaffolds. *Nanomaterials*, 11, 11, 2796, 2021, https://www.mdpi.com/2079-4991/11/11/2796.
103. Munim, S.A. and Raza, Z.A., Poly(lactic acid) based hydrogels: formation, characteristics and biomedical applications. *J. Porous Mater.*, 26, 881–901, 2019, https://link.springer.com/article/10.1007/s10934-018-0687-z.
104. Ignatova, M., Manolova, N., Markova, N., Rashkov, I., Electrospun nonwoven nanofibrous hybrid mats based on chitosan and PLA for wound-dressing applications. *Macromol. Biosci.*, 9, 102–111, 2009, https://onlinelibrary.wiley.com/doi/abs/10.1002/mabi.200800189.
105. Bi, H., Feng, T., Li, B., Han, Y., *In vitro* and *in vivo* comparison study of electrospun PLA and PLA/PVA/SA fiber membranes for wound healing. *Polymers (Basel)*, 12, 4, 839, 2020, https://www.mdpi.com/2073-4360/12/4/839.
106. Ciftci, F., *et al.*, Selenium and clarithromycin loaded PLA-GO composite wound dressings by electrospinning method. *Int. J. Polym. Mater. Polym. Biomater.*, 71, 898–909, 2022, https://www.tandfonline.com/doi/abs/10.1080/00914037.2021.1925276.
107. Hajikhani, M., Emam-Djomeh, Z., Askari, G., Fabrication and characterization of mucoadhesive bioplastic patch *via* coaxial polylactic acid (PLA) based electrospun nanofibers with antimicrobial and wound healing application. *Int. J. Biol. Macromol.*, 172, 143–153, 2021, https://www.sciencedirect.com/science/article/pii/S0141813021000726.

108. Allen, T.M. and Cullis, P.R., 27. Drug Delivery Systems: Entering the Mainstream, v. 38. M. S. Lesney. *Todays Chem. Work*, 16, 1818–1822, 2002, www.sciencemag.org.
109. Campardelli, R., *et al.*, Au-PLA nanocomposites for photothermally controlled drug delivery. *J. Mater. Chem. B*, 2, 409–417, 2014, https://pubs.rsc.org/en/content/articlehtml/2014/tb/c3tb21099e.
110. Fayzullin, A., *et al.*, Local delivery of pirfenidone by pla implants modifies foreign body reaction and prevents fibrosis. *Biomedicines*, 9, 8, 853, 2021, https://www.mdpi.com/2227-9059/9/8/853.
111. Saniei, H. and Mousavi, S., Surface modification of PLA 3D-printed implants by electrospinning with enhanced bioactivity and cell affinity. *Polymer (Guildf)*, 196, 122467, 2020, https://www.sciencedirect.com/science/article/pii/S0032386120303025.
112. Bode, C., Kranz, H., Fivez, A., Siepmann, F., Siepmann, J., Often neglected: PLGA/PLA swelling orchestrates drug release: HME implants. *J. Control. Release*, 306, 97–107, 2019. https://www.sciencedirect.com/science/article/pii/S0168365919303037
113. Rippy, M.K., Baron, S., Rosenthal, M., Senior, B.A., Evaluation of absorbable PLA nasal implants in an ovine model. *Laryngoscope Investig. Otolaryngol.*, 3, 156–161, 2018, https://onlinelibrary.wiley.com/doi/abs/10.1002/lio2.166.
114. Muñoz, J. and Pumera, M., 3D-printed biosensors for electrochemical and optical applications. *TrAC, Trends Anal. Chem.*, 128, 115933, 2020, Preprint at https://doi.org/10.1016/j.trac.2020.115933.
115. Ambrosi, A. and Pumera, M.3D-printing technologies for electrochemical applications. *Chem. Soc. Rev.*, 45, 10, 2740–2755, 2016, https://pubs.rsc.org/en/content/articlehtml/2016/cs/c5cs00714c
116. Zare, Y., Park, S.P., Rhee, K.Y., Analysis of complex viscosity and shear thinning behavior in poly (lactic acid)/poly (ethylene oxide)/carbon nanotubes biosensor based on Carreau–Yasuda model. *Results Phys.*, 13, 102245, 2019, https://www.sciencedirect.com/science/article/pii/S2211379718334946.
117. Shepherd, L.M., Gonzalez, E., Chen, E.X., Frey, M.W., Increasing stability of biotin functionalized electrospun fibers for biosensor applications. *ACS Appl. Mater. Interfaces*, 9, 1968–1974, 2017, https://pubs.acs.org/doi/abs/10.1021/acsami.6b14348.
118. Sobhan, A., *et al.*, Development and characterization of a novel activated biochar-based polymer composite for biosensors. *Int. J. Polym. Anal. Charact.*, 26, 544–560, 2021, https://www.tandfonline.com/doi/abs/10.1080/1023666X.2021.1921497.
119. Zare, Y. and Rhee, K.Y., Effects of interphase regions and filler networks on the viscosity of PLA/PEO/carbon nanotubes biosensor. *Polym. Compos.*, 40, 4135–4141, 2019, https://4spepublications.onlinelibrary.wiley.com/doi/abs/10.1002/pc.25274.

4

Optical, Dielectric, and Electrical Properties of Poly(2-Ethyl-2-Oxazoline)-Polyvinylpyrrolidone-Graphene Nanocomposites

Shubha A.[1,2] and S. R. Manohara[1]*

[1]*Nano-Composites and Materials Research Lab, Department of Physics, Siddaganga Institute of Technology, Tumakuru, Karnataka, India*
[2]*Department of Physics, B.M.S. College of Engineering, Bengaluru, Karnataka, India*

Abstract

This chapter provides an overview of polymer blends, their various types, and their methods of preparation, and the processing techniques used for polymer nanocomposites. Blends of poly(2-ethyl-2-oxazoline) [PEOX] and polyvinylpyrrolidone [PVP] with compositions of 20, 40, 60, and 80 wt% PVP, as well as PEOX-PVP (80:20 wt%) nanocomposites with graphene nanoplatelets (GNPs) at concentrations of 1, 5, 10, 15, and 20 wt%, were synthesised using the solution casting technique.

The absorption band edge, optical energy band gap (E_g^{opt}), and the width of the localized states' tail in the band gap (ΔE) were observed to vary based on the PVP content in the PEOX-PVP blends. The dielectric loss, dielectric constant, and loss tangent were found to decrease as the frequency of the electric field increased, while AC electrical conductivity exhibited an upward trend with increasing frequency. The PEOX-PVP (80:20 wt%) blend emerged as an optimized material within the PEOX-PVP system, demonstrating superior optical, dielectric, and electrical characteristics.

**Corresponding author*: manoharasr@sit.ac.in; ORCID: https://orcid.org/0000-0003-3093-5416

Shubha A.: ORCID: https://orcid.org/0000-0002-8028-5629

Raneesh Balakrishnan, Sobi K. Chacko and Visakh P.M. (eds.) Electroactive Polymer-Based Smart Materials and Applications, (91–124) © 2026 Scrivener Publishing LLC

In PEOX-PVP-graphene nanocomposites, a reduction in the optical energy band gap and an increase in Urbach energy were noted with higher GNP concentrations. The dielectric properties, including the dielectric loss, dielectric constant, and tangent loss, decreased with increasing frequency. Conversely, the AC electrical conductivity of these nanocomposites remained constant at lower frequencies but increased with higher frequencies. The PEOX-PVP-10 wt% graphene nanocomposite exhibited the most favorable electrical, dielectric, and optical properties among the nanocomposites studied. Due to these enhanced properties, this new class of nanocomposites holds potential for industrial applications.

***Keywords*:** Poly(2-ethyl-2-oxazoline), polyvinylpyrrolidone, graphene nanocomposites, optical properties, dielectric behavior, electrical conductivity, polymer composites, solution casting

4.1 Introduction

The term "polymer" originates from the Greek words "poly" meaning "many," and "mer" meaning "parts." In simple terms, a polymer is a long-chain molecule composed of numerous repeating units known as monomers.

These chains are formed as carbon atoms bond together in sequence, either through a single σ bond or a combination of σ and π bonds (as in double or triple bonds), creating what is referred to as the polymer's backbone.

Some polymers occur naturally, such as cellulose, proteins, and silk, while others, like polystyrene, nylon, and polyethylene, are produced synthetically through chemical processes.

In metals, only one type of bond holds all the atoms together. But, in case of polymers, various types of bonds hold the atoms together. They are primary covalent bonds, hydrogen bond, dipole interaction, van der Walls and ionic bonds. Except covalent bond others are called as secondary bonds. Polymers possess several fundamental properties:

(i) They are extremely lightweight while maintaining significant strength.
(ii) Polymers come with inherent properties that can be further modified and enhanced.
(iii) They can exhibit strong resistance to chemicals and weathering, though they are reactive with specific solvents.
(iv) Polymers can function as both thermal and electrical insulators.
(v) They have excellent transport properties, such as high diffusivity.
(vi) They have good mechanical and other physical properties, and hence can replace many materials.

4.1.1 Polymer Blends and Its Types

A polymer blend is a material composed of a mixture of at least two different polymers, which may or may not have chemical bonds between the constituent polymers. The polymer blend will have different physical properties compared to its constituent polymers. The blends are analogous to metal alloys. The blend with unique properties, which is homogeneous and miscible, can be obtained by using polymers having proper molar mass, entropy, interfacial surface tension, interface adhesion and specific interactions.

Based on miscibility, the polymers can be classified into three types,

4.1.1.1 *Completely Miscible Blend*

The term miscible is used to depict a single phase polymer blend. The miscibility refers to the scale of mixing of polymer constituents which yields the properties of a single phase system. The change in Gibb's free energy of mixing, ΔG_m, for a miscible polymer blends should be negative and is given by eq. (4.1) [1],

$$\Delta G_{\mathrm{m}} = \Delta H_{\mathrm{m}} - T\Delta S_{\mathrm{m}} \tag{4.1}$$

where T is the temperature, and ΔS_m and ΔH_m are the changes in entropy and enthalpy of mixing, respectively. This type of blend will have single thermal transition (e.g. T_g) and a single phase for all the blending ratios. The miscibility of this blends are credited to the specific interaction that exists between the constituent polymers [1]. The important factors which are essential for the development of miscible blends are entropy and intermolecular interactions, for example, the formation of hydrogen bonds between the constituent polymers can upset the interaction between constituents.

4.1.1.2 *Partially Miscible Blend*

Some polymer blends are neither completely miscible nor immiscible. The interactions are limited in these types of blends, but a small fraction of one polymer will be miscible in other polymer. Most of the polymers are miscible when there is more of one of the constituent polymer compared to other. These blends will have good phase morphology. Thermodynamically, these blends show the miscibility upto the binodal composition of the blend.

4.1.1.3 *Completely Immiscible Blend*

The blends which have positive value of Gibb's free energy of mixing, ΔG_m, are called immiscible blends. In common language, the blend is termed as immiscible if it is composed of two separate phases comprising constituent polymers. Majority of polymer blends are immiscible [1] and are characterized by multiple thermal transitions, opacity, etc. In this type of blends, each constituent polymer forms its own domain in the blend, due to lack of specific interactions, which results in poor interfacial properties. As a consequence, there will be less adhesion strength at the interface which leads to poor mechanical properties and limit their practical applications.

4.1.2 Methods of Preparation of Polymer Blends

Some of the most significant and widely used methods for preparing polymer blends are as follows.

4.1.2.1 *Mechanical–Melt Mixing*

This is the most widely used method in industry for preparing polymer blends. The process involves blending two or more polymers, typically thermoplastics, under molten conditions using appropriate equipment such as extruders, rollers, or kneading machines, all under controlled reaction conditions.

4.1.2.2 *Solution Cast Technique*

This simple method is employed to create thin-layer blend films. It involves mixing two or more polymer solutions and then evaporating the solvent. Once the solvent has evaporated, the resulting film is removed from the substrate. The advantages of the solution casting technique over the melt process include: (i) the production of high-quality films with uniform thickness, and (ii) the creation of highly pure, clear films that are free from residuals and pinholes.

4.1.2.3 *Fine Powder Mixing*

In this method, polymer powders are mixed through diffusion at elevated temperatures, typically above the glass transition temperature of the constituent polymers.

4.1.2.4 In Situ Polymerization

In this method, the polymerization of one polymer occurs in the presence of another polymer. Interpenetrating polymer networks are formed by polymerizing different polymer networks together. This technique is commonly used to create polymer electrolytes.

4.1.3 Processing of Polymer Nanocomposites

There are several methods for preparing polymer-based nanocomposites, including solution blending, melt blending, *in situ* polymerization, layer-by-layer assembly, surfactant-assisted mixing, and dry powder mixing. Each of these approaches yields different outcomes concerning the dispersion of nanostructures, interfacial interactions between components, the resulting properties of the nanocomposites, and their potential applications. The following paragraphs will provide a brief overview of the three most widely used and significant techniques for producing polymer nanocomposites.

4.1.3.1 Solution Casting Method

The solution casting or solution blending technique is widely used for creating polymer nanocomposites due to its simplicity. This process involves mixing nanostructures with a polymer in a suitable solvent. The chosen solvent must be capable of fully dissolving the polymer and promoting the effective dispersion of the nanofiller. The method typically consists of three steps: dispersing the nanostructures in the solvent, blending them with the polymer (either at room or elevated temperature), and then carefully evaporating the solvent to form a nanocomposite film [2]. As the solvent evaporates, the nanofiller remains evenly dispersed within the polymer matrix. Both organic and aqueous solvents can be utilized for producing polymer nanocomposites.

In the solution blending technique, thorough mixing of nanostructures with the polymer in the solvent helps to prevent aggregation and ensures good dispersion. This can be achieved through shear mixing, magnetic stirring, reflux, or most commonly, ultrasonication [3]. Ultrasonication can be done either in a mild manner using a water bath or with high power using a probe-type sonicator. However, prolonged use of high-power ultrasonication can shorten the length of nanostructures, particularly carbon nanotubes (CNTs), reducing their aspect ratio, which negatively impacts the properties of the nanocomposite. Surfactants may also be used in this method to facilitate the dispersion of higher concentrations of nanostructures.

The advantages of this method include ease of processing, applicability to a wide range of nanofillers, suitability for both thermoplastic and thermoset polymers, and the ability to reduce gas permeability. However, its main drawbacks include the potential for nanofiller agglomeration [4] and its limitation to polymers that are soluble in water [5].

4.1.3.2 Melt Blending or Melt Intercalation Method

While solution blending is an effective method for dispersing nanofillers and forming nanocomposites, it is less suitable for large-scale industrial production. Melt processing, on the other hand, is favored in industrial settings due to its simplicity and cost-effectiveness, making it ideal for commercial applications. This method is particularly suitable for polymers that are insoluble in solvents, typically thermoplastic polymers.

Melt blending generally involves heating the polymer until it melts into a viscous liquid, into which nanofillers (such as graphene, carbon nanotubes, nanoclay, nanoparticles, or nanometal oxides) are added. High shear forces are applied to disperse the nanofillers throughout the polymer matrix. In this process, twin-screw extruders are commonly used to blend the polymer with the nanofillers.

Though melt blending is a straightforward technique, the application of high shear force and high temperatures can sometimes degrade the properties of the resulting nanocomposite. Additionally, excessive shear force may cause nano-fragmentation of the fillers, so it is important to apply optimal shear stress to achieve the desired dispersion while minimizing damage to the nanofillers.

The advantages of melt blending include improvements in mechanical properties [6–9], enhanced thermal stability [10], effective dispersion of nanofillers [11], and its cost-effective and eco-friendly nature. However, the method has some limitations, such as the difficulty in controlling the interaction between the polymer and nanofillers, as well as processing conditions like residence time and temperature [12]. Achieving uniform distribution of nanofillers in the polymer matrix can also be challenging in some cases.

4.1.3.3 In-Situ Polymerization

In-situ polymerization, another technique, involves dispersing the nanofillers in a monomer solution before polymerization. Polymerization is then initiated using stimuli such as a suitable initiator or heat, allowing the polymer to form on the surface of the nanomaterial [2]. This method is advantageous for achieving high nanofiller loading without agglomeration [3] and is often used for polymers that are insoluble or thermally unstable and cannot be processed using solution or melt blending [13]. Key benefits of *in-situ* polymerization include producing polymer nanocomposites with high transparency [14, 15], good interfacial adhesion of nanofillers [16], and controllable nanofiller morphology [17].

4.2 Poly(2-Ethyl-2-Oxazoline)

Poly(2-ethyl-2-oxazoline) (PEOX) is a non-ionic, water-soluble, amorphous, and thermoplastic polymer. It falls under the category of polyoxazolines and is classified as a tertiary amide polymer. Polyoxazolines have various applications, such as in biocompatible materials for drug delivery and as steric stabilizers [18–20]. PEOX exhibits favorable Newtonian properties and stable shear melt flow, making it a promising alternative to polyvinyl alcohol and polyvinylpyrrolidone, with notable thermal stability. PEOX may be used as an indirect food additive, according to US Food and Drug Administration (FDA) approval. Studies have shown that liposomes grafted with PEOX circulate in the bloodstream for extended periods and demonstrate reduced clearance rates in the liver and spleen [21]. By employing boron trifluoride diethyl etherate (BF_3Et_2O) as an initiator, PEOX can be produced *via* cationic ring-opening polymerization of 2-ethyl-2-oxazoline [22].

Properties of PEOX:

- **Molecular formula:** $(C_6H_9NO)_n$
- **Chemical structure:**

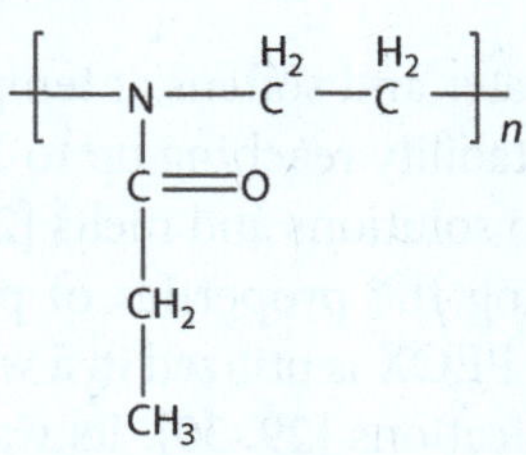

- **Glass transition temperature** (T_g): 70 °C
- **Appearance (form and colour):** Solid and off-white to light yellow

Table 4.1 Solubility of poly(2-ethyl-2-oxazoline) [23].

Solubility parameter, δ,	Solvent*
9.3	Chloroform
9.3	Methyl ethyl ketone
9.5	Ethylhexol
9.6	Methyl acetate
9.7	Methylene chloride
9.9	Acetone
10.0	Dioxane
11.9	Acetonitrile
12.7	Nitromethane
12.7	Ethanol
---	Dimethylformamide (DMF)
14.5	Methanol
23.4	Water

*The solubility is greater than 25 % by weight.

- **Solubility:** The way PEOX dissolves in different solvents is determined by their solubility parameter, δ, and is given in the Table 4.1. PEOX is soluble in a wide range of solvents, including water and organic polar solvents.

Applications of PEOX

PEOX dissolves easily in water and softens at temperatures between 110 °C and 120 °C, with thermal stability reaching up to 380 °C, making it suitable for use as adhesives, both in solutions and melts [24, 25]. It blends well with various polymers, enhancing the properties of polymer systems [25–28]. Due to its biocompatibility, PEOX is utilized in a wide range of personal care products and medical applications [29, 30]. Its water solubility and dispersant properties make it an effective compatibilizing agent in personal care products [31]. Additionally, PEOX is used in photoresists, inks, and coatings [32, 33]. Being both hydrophilic and pH-responsive, PEOX allows for precise control of drug delivery and release, triggered by pH changes at the target site [34–36].

4.3 Polyvinylpyrrolidone

Polyvinylpyrrolidone (PVP), also known as povidone or polyvidone, is a non-ionic polymer that dissolves in water. It is produced through the free radical polymerization of N-vinylpyrrolidone, either in water or isopropanol [37]. PVP is highly hygroscopic and absorbs atmospheric water up to 40 % of its weight. In solid state, it has excellent wetting property and can readily form films.

Properties of PVP:

- **Molecular formula:** $(C_6H_9NO)_n$
- **Chemical structure:**

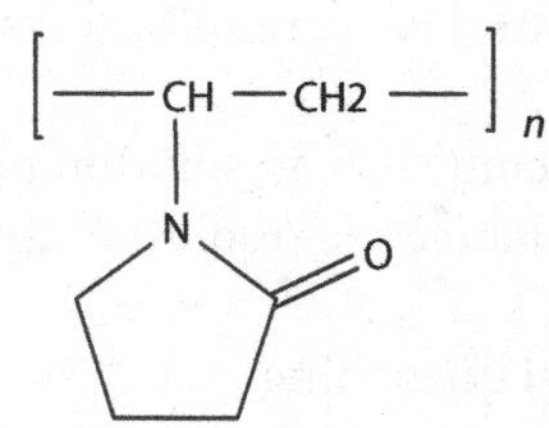

- **Melting point (T_m) and Glass transition temperature (T_g):** 150 to 180 °C
- **Appearance (form and colour):** Solid and white to light yellow and amorphous powder.
- **Solubility:** Water and other polar solvents like chloroform and alcohols (e.g. methanol and ethanol).

Applications of PVP:
PVP is known for its biocompatibility, cost-effectiveness, and versatility. It is widely used in numerous industries, including pharmaceuticals, cosmetics, inks, textiles, and coatings, with a particular emphasis on its role as an adhesive [38].

4.3.1 Medical Applications

PVP has many medical applications. However, long-term effects on the human body are unknown.

- Due to its biocompatibility and biodegradability, PVP is employed as a binder in various pharmaceutical tablets [39].
- Povidone-iodine, a complex formed when PVP is added to iodine, is antiseptic, and hence, this product is employed in variety of goods such as soaps, lotion, and surgical scrubs [40].
- PVP lowers friction, making it suitable for use as a lubricant in certain contact lenses [41] and their packaging solutions [42].
- PVP is utilized as a lubricant in various eye drops, such as Soothe by Bausch and Lomb [43].

4.3.2 Technological Applications

Polyvidone has many technological applications:

- Adhesive in hot-melt adhesives and glue stick.
- Additives for batteries, ceramics, inks, fiberglass inkjet paper, etc.
- Disintegrant and emulsifier in solution polymerization.
- Photoresists to enhance the resolution for cathode ray tubes (CRT) [44].
- Water-based metal quenching.
- Membranes, e.g. water purification filters and dialysis.
- Used in agriculture for crop protection, seed treatment, and coating, serving as a complexation agent and binder.
- Serves as a thickening agent in tooth whitening gels [45].
- Synthesis of nanoparticles and their self-assembly as a reducing agent, surfactant, dispersant, and shape controlling agent [38].
- Functions as a stabilizing agent in inorganic solar cells [46].

4.3.3 Other Applications

Owing to its polar properties, PVP exhibits strong affinity for polar molecules. This characteristic enables its use in ink formulations for inkjet printers and as a coating material for photo papers and transparencies. PVP is also incorporated into personal care products like toothpaste and shampoo, as well as in paints and adhesives. Additionally, it serves as an additive in the production of beer and white wine within the wine industry.

4.4 Poly(2-Ethyl-2-Oxazoline)-Polyvinylpyrrolidone Systems

4.4.1 Poly(2-Ethyl-2-Oxazoline)-Polyvinylpyrrolidone Blends

Poly(2-ethyl-2-oxazoline) [PEOX] with a molecular weight of approximately 200,000 and polyvinylpyrrolidone [PVP] with a molecular weight of around 58,000 were obtained from Alfa Aesar, USA, and used without further purification. Solid polymer blend films with varying weight percentages of PEOX and PVP were formed using a straightforward solution blending technique. Specifically, PEOX was dissolved in Milli-Q water at 50 °C with stirring (e.g., for an 80:20 wt% PEOX-PVP blend, 4.8 g of PEOX and 1.2 g of PVP were used). Concurrently, PVP was dissolved in Milli-Q water at room temperature. The two solutions were then combined at 50 °C using a magnetic stirrer to form homogeneous blends in ratios of 80:20, 60:40, 40:60, and 20:80 wt%. The resulting homogeneous solution was poured into polypropylene Petri dishes and placed in a vacuum oven at 50 °C for 5 days to allow for water evaporation. Once dried, the films were removed from the Petri dishes and heated in a vacuum oven at 100 °C for three hours to eliminate any remaining water and promote cross-linking. Pure PEOX and PVP films were also prepared for comparison. The thickness of both the pure and blended films, measured with a digital vernier caliper with a resolution of 0.01 mm, was approximately 0.50 mm.

4.4.2 PEOX-PVP-Graphene Nanocomposites

The graphene nanoplatelets used in this study were sourced from J. K. Impex, Mumbai, India, and used without additional purification. To prepare the PEOX-PVP (80:20 wt%) blend solution, 4.8 g of PEOX was dissolved in de-ionized water at 50 °C with magnetic stirring, while 1.2 g of PVP was dissolved at room temperature. The solutions were combined at 50 °C to achieve a homogeneous blend. A film of this PEOX-PVP blend was then formed.

For the PEOX-PVP-graphene nanocomposites, a specific amount of PVP was dissolved in de-ionized water at room temperature with magnetic stirring. Graphene slurry was prepared with varying amounts of graphene nanoplatelets [1 wt% (60.6 mg), 5 wt% (315.8 mg), 10 wt% (666.7 mg), 15 wt% (1058.8 mg), and 20 wt% (1500 mg) relative to PEOX-PVP weight] using about 1 ml of the PVP solution. This slurry was mixed with the PVP solution, followed by stirring using magnetic and ultrasonic treatment using ultrasonicator for 10 and 20 minutes, respectively, to ensure good dispersion. The resulting PVP-graphene solution was then added to the PEOX solution and stirred for 15 minutes. An additional 30 minutes of ultrasonic treatment was applied to the PEOX-PVP-graphene solution to achieve a homogeneous dispersion and reduce graphene agglomeration.

Finally, the PEOX-PVP-graphene nanocomposite solution was poured into polypropylene Petri dishes and placed in a vacuum oven at 60 °C for 5 days to form thin films. The thickness of both pure and nanocomposite films, measured with a digital vernier caliper with a 0.01 mm resolution, ranged from 0.40 mm to 0.60 mm.

4.5 Optical, Electrical, and Dielectric Properties

4.5.1 PEOX-PVP Blends

UV-Vis measurements were conducted at room temperature using a UV-Vis spectrophotometer (Double-beam Shimadzu 1800) with a 1 nm resolution across the 200–800 nm wavelength range. From the absorbance data, the absorption coefficient, absorption band edge, optical energy band gap, and Urbach energy were determined. Figure 4.1(a) shows the optical absorption spectra for pure PEOX and PEOX-PVP blends over the 200 to 800 nm range. Pure PEOX exhibits a $\pi \rightarrow \pi^*$ transition with an absorption band at 210 nm, while the PEOX-PVP blends display absorption peaks at 237, 235, 218, and 217 nm for the 80:20, 60:40, 40:60, and 20:80 wt% ratios, respectively. The absorbance for the PEOX-PVP blends is greater than that of pure PEOX, with the PEOX-PVP (80:20 wt%) blend showing the highest absorbance. This increased absorption coefficient for the PEOX-PVP (80:20 wt%) blend may be attributed to the even distribution of PVP within the PEOX matrix. Conversely, blends with higher PVP content (40, 60, and 80 wt%) exhibit lower absorption due to the resistance of PVP aggregation to light absorption.

The absorption edge, direct optical band gap (E_g^{opt}), and Urbach energy (ΔE) were derived from plots of the absorption coefficient (α) versus photon energy (hv), $(\alpha hv)^2$ versus hv, and $\ln(\alpha)$ versus hv, respectively, as illustrated in Figure 4.1(b) and Figures 4.2(a) & (b). The absorption edge and E_g^{opt} values were obtained by extrapolating the linear regions of the curves to zero absorption in the high absorption region, while ΔE was determined from the slopes of the linear portions in the $\ln(\alpha)$ versus hv plots. It was found that as the PVP content increased, the absorbance and absorption edges marginally shifted towards lower wavelength.

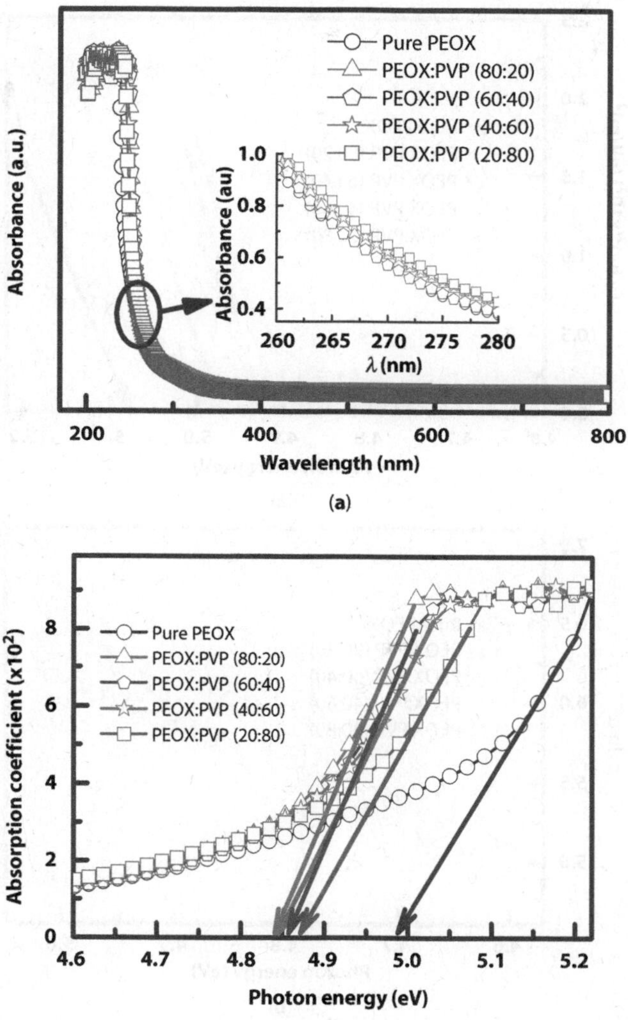

Figure 4.1 Graph of (a) absorbance as a function of wavelength and (b) absorption coefficient against photon energy for pure PEOX and PEOX-PVP blends [50]. Copyright 2024. Reproduced with permission from Elsevier.

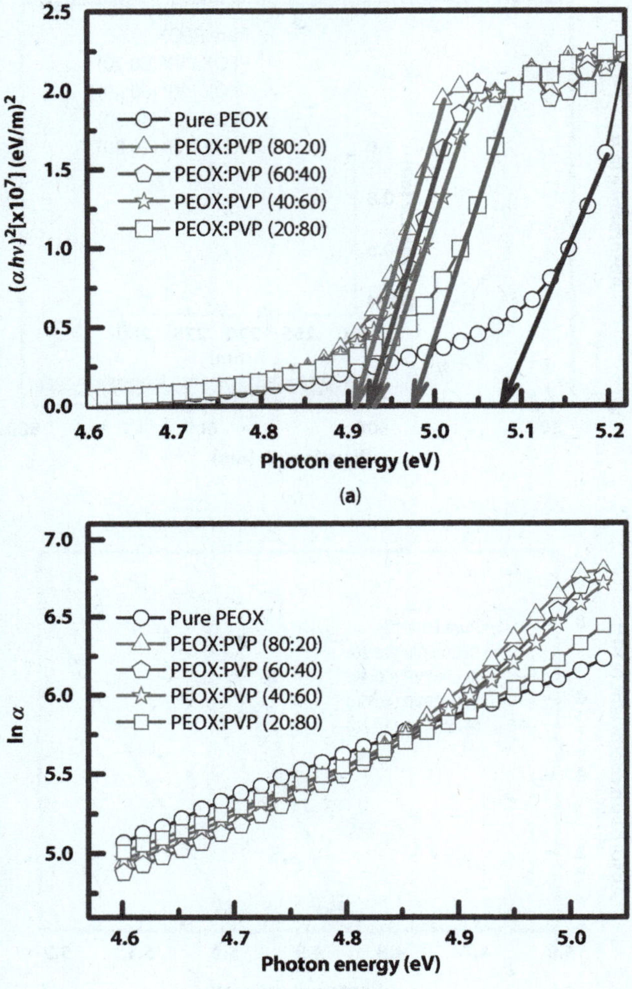

Figure 4.2 Graph of (a) $(\alpha hv)^2$ as a function of photon energy and (b) lnα against photon energy for pure PEOX and PEOX-PVP blends [50]. Copyright 2024. Reproduced with permission from Elsevier.

Table 4.2 summarizes the optical parameters. The absorption edge and direct band gaps decrease slightly as PVP concentration increases, with the minimum values observed for the PEOX (80:20 wt%) blend. This reduction may be due to the introduction of localized energy states in the band gap caused by compositional disorder or changes in the number of

Table 4.2 Absorption edge, directly allowed optical band gap, E_g^{opt}, and Urbach energy, ΔE, values of pristine PEOX and PEOX-PVP blends [50]. Copyright 2024. Reproduced with permission from Elsevier.

Sample name	Absorption edge (eV)	E_g^{opt} (eV)	ΔE (eV)
Pure PEOX	4.98	5.08	0.38
PEOX-PVP (80:20 wt%)	4.84	4.91	0.22
PEOX-PVP (60:40 wt%)	4.86	4.92	0.23
PEOX-PVP (40:60 wt%)	4.84	4.93	0.26
PEOX-PVP (20:80 wt%)	4.87	4.98	0.31

available final states [47, 48]. Additionally, the increased density of unsaturated defects and final states in the band structure likely contributes to the observed decrease in the optical band gap [49].

The dielectric constant, ε', and dielectric loss, ε'', as a function of frequency of electric field for pristine PEOX and PEOX-PVP blends is graphically shown in Figures 4.3(a, b) at room temperature. The error in measurement of ε' and ε'' are found to be less than 1%. In the low frequency region (5 Hz-1 kHz), both ε' and ε'' decreases sharply with increasing frequency for PEOX-PVP (40:60 wt%) and PEOX-PVP (80:20 wt%) blends, and become almost constant in the high frequencies (> 1 kHz). It can also be seen that ε' and ε'' are almost independent of frequency for other PEOX-PVP blends, the highest values being seen for PEOX-PVP (80:20 wt%) at all frequencies. Higher values of ε' and ε'' at lower frequencies may be attributed to the Maxwell-Wagner effect, where AC current aligns with the applied potential, or to increased DC conductivity due to greater ion mobility, or a combination of both factors [51, 52].

The dielectric constant, ε', of the PEOX-PVP blends decreases as the frequency increases. This behavior is typical of most dielectric materials and is attributed to dielectric relaxation, which involves the reorientation of molecules within the material. The dielectric constant is influenced by polarizability, which increases with greater polarizability. ε' in materials is primarily derived from electronic or atomic, ionic, orientation, and interfacial or space charge polarizations. When exposed to an electric field, dipoles (or polar dielectric molecules) in the material align with the field. In the case of an alternating electric field, these dipoles continuously adjust their alignment direction.

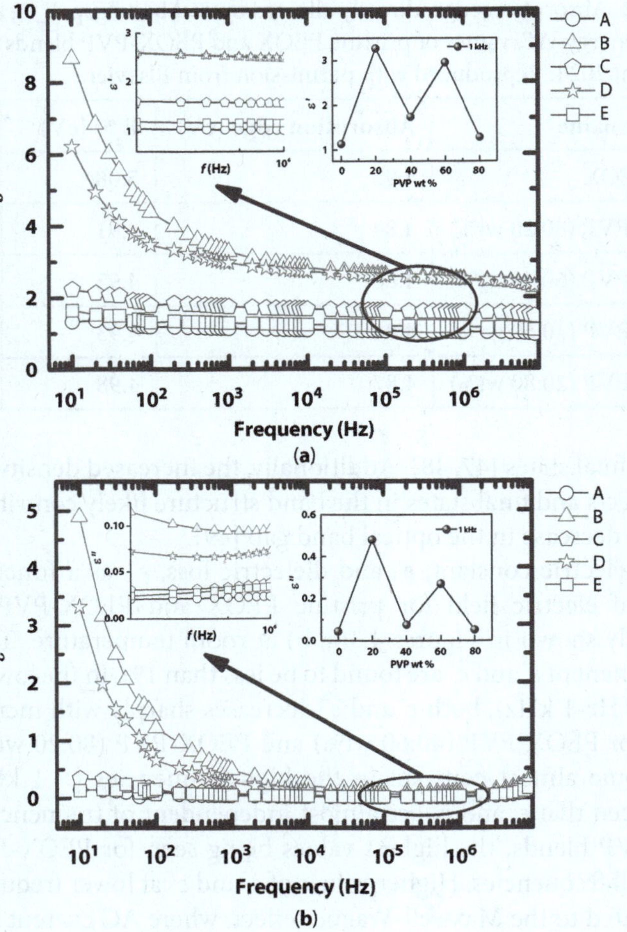

Figure 4.3 Variation of (a) real, ε′, and (b) imaginary, ε″, part of relative permittivity with frequency for: (A) pure PEOX, (B) PEOX-PVP (80:20 wt%), (C) PEOX-PVP (60:40 wt%), (D) PEOX-PVP (40:60 wt%), and (E) PEOX-PVP (20:80 wt%) at room temperature of 300 K. Inset graph (on right) shows the dependence of ε′ on weight percentage of PVP at 1 kHz [50]. Copyright 2024. Reproduced with permission from Elsevier.

However, dipoles require time to reach an equilibrium state, known as the *relaxation time*, with its reciprocal termed the *relaxation frequency*. This relaxation time varies depending on the type of polarization. Orientation polarization generally has a longer relaxation time compared to atomic or electronic polarization. Materials typically show higher dielectric constants at lower frequencies due to increased dipole alignment. At higher frequencies, the dipoles cannot align quickly enough with the alternating electric field, leading to reduced ε′ values.

The dielectric loss, ε″, decreases as the frequency increases, primarily due to the frictional effects between the dipoles. When subjected to an AC electric field, dipoles continuously change their alignment direction, creating frictional effects that result in energy dissipation as heat within the dielectric material leading to energy loss, called *dielectric loss*. At lower frequencies, dipoles have sufficient time to align fully with the field, leading to greater internal friction and higher dielectric losses. In contrast, at higher frequencies, the dipoles align less effectively, resulting in reduced internal friction and lower dielectric losses. Electronic and ionic polarizations dominate at higher frequencies, contributing to these lower losses, while orientation and interfacial polarizations, which are more effective at lower frequencies, lead to higher dielectric losses.

The inset graph in Figures 4.3(a, b) shows that as the percentage of PVP increases in the PEOX-PVP blends, both ε′ and ε″ exhibit anomalous behavior but remain higher compared to pure PEOX. The ε′ and ε″ values for the PEOX-PVP (80:20 wt%) blend are notably higher than those for pure PEOX and other blends. Specifically, for PEOX-PVP (80:20 wt%), ε′ = 3.27 and ε″ = 0.48 at 1 kHz, which are significantly higher than the values for pure PEOX (ε′ = 1.14 and ε″ = 0.42). The increased ε′ in this blend is largely due to enhanced dipole orientation. The interaction between the –CH3 groups of PEOX and the carbonyl groups of PVP reduces the cohesive

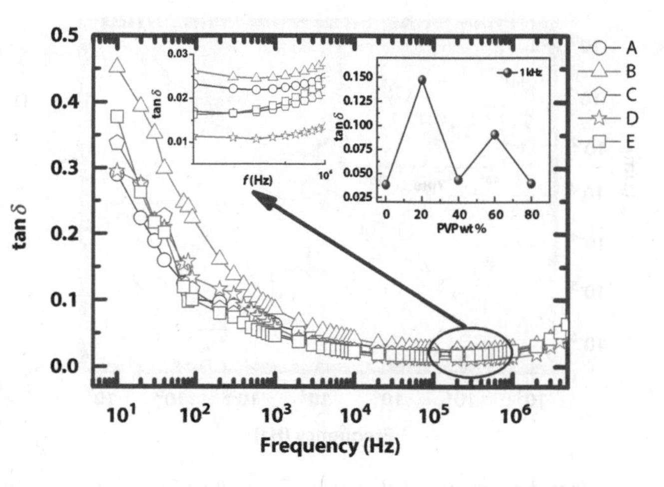

Figure 4.4 Variation of loss tangent, tanδ, with frequency for: (A) pure PEOX, (B) PEOX-PVP (80:20 wt%), (C) PEOX-PVP (60:40 wt%), (D) PEOX-PVP (40:60 wt%), and (E) PEOX-PVP (20:80 wt%) at room temperature of 300 K. Inset graph (on right) shows the dependence of tanδ on weight percentage of PVP at 1 kHz [50]. Copyright 2024. Reproduced with permission from Elsevier.

forces between the PEOX and PVP polymer chains, thereby improving segmental mobility and increasing the number of dipoles. As the PVP concentration rises, the blend experiences greater agglomeration and reduced segmental distances, leading to decreased ε′ and ε″ with higher PVP content.

Figure 4.4 illustrates how the loss tangent, tan δ, varies with frequency for pure PEOX and PEOX-PVP blends. The data show that tan δ is higher at lower frequencies, decreases with increase in frequency, and stabilizes at higher frequencies. The elevated loss tangent values at lower frequencies result from the combined effects of ionic and interfacial polarization in the blends [53]. At higher frequencies, tan δ is reduced due to the difficulty in orienting the dipoles within the blend. The inset graph in Figure 4.4 indicates that tan δ is greater for the blends compared to pure PEOX, with the larger value observed in the 80:20 wt% PEOX-PVP blend. The increased tan δ at this ratio is attributed to stronger intermolecular interactions between PEOX and PVP, which lead to more self-associated bonds. However, tan δ decreases at PVP concentrations above 20 wt%, likely due to the formation of discrete aggregates at higher PVP content, which reduces tan δ.

Figure 4.5 displays the AC electrical conductivity, σ_{AC}, of pure PEOX and PEOX-PVP blends as a function of frequency at room temperature. The plot reveals a strong frequency dependence of AC conductivity across

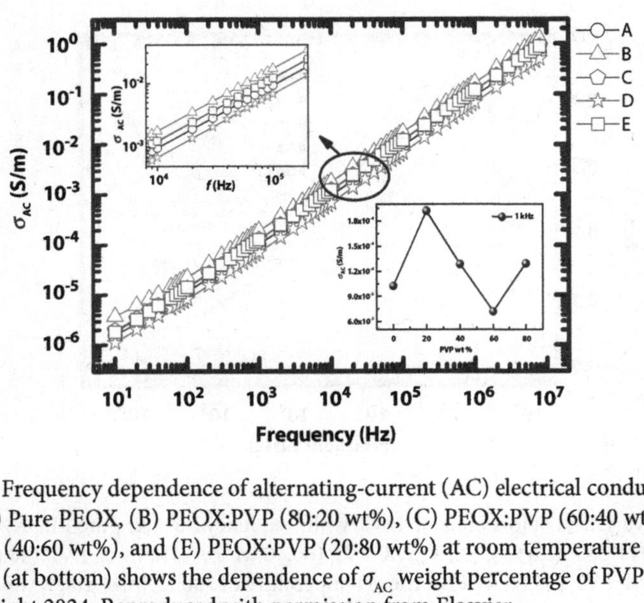

Figure 4.5 Frequency dependence of alternating-current (AC) electrical conductivity, σ_{AC}, for: (A) Pure PEOX, (B) PEOX:PVP (80:20 wt%), (C) PEOX:PVP (60:40 wt%), (D) PEOX:PVP (40:60 wt%), and (E) PEOX:PVP (20:80 wt%) at room temperature of 300 K. Inset graph (at bottom) shows the dependence of σ_{AC} weight percentage of PVP at 1 kHz [50]. Copyright 2024. Reproduced with permission from Elsevier.

all samples, reflecting their insulating characteristics. Similar frequency-dependent behaviors have been observed in other polymer blends [54, 55].

This frequency-dependent behavior of σ_{AC} is accredited to a combination of interface charge polarization (Maxwell-Wagner-Sillar effect) and intrinsic electric polarization [56]. According to the Maxwell-Wagner-Sillar interfacial model, the interfaces between PEOX and PVP segments act as potential barriers, causing charge carriers within the segments to behave like charges in a potential well. While charge carriers can move freely within individual polymer segments, migration between segments is hindered by these resistive interfaces. At low frequencies, a small number of charge carriers may tunnel between segments, allowing for limited electrical conduction. As the frequency increases, more charge carriers tunnel between segments, leading to an increase in σ_{AC}. At high frequencies, the charge carriers gain enough energy to overcome the potential barriers, resulting in higher σ_{AC}.

The inset graph at the bottom of Figure 4.5 shows that the AC conductivity, σ_{AC}, is higher for the PEOX-PVP (80:20 wt%) blend compared to pristine PEOX and other blends. However, σ_{AC} does not increase with additional PVP beyond the 20 wt% level. This composition represents the *percolation threshold* or *limit* of the blend [57, 58]. The conductivity of the polymer blend is affected by the amount of filler polymer used, its propensity to form aggregates, and the morphology of the matrix polymer. For the PEOX-PVP (80:20 wt%) blend, the transition from insulating to conductive behavior occurs, at the percolation threshold, due to the formation of a conductive network. At higher PVP concentrations, PVP and PEOX form strong aggregates, which result in reduced conductivity due to loose clustering and the creation of internal voids and gaps within the blend. Additionally, the conductivity of the PEOX-PVP (40:60 wt%) blend is lower than that of pure PEOX, likely due to weaker interfacial interactions between PEOX and PVP macromolecules, which diminishes electronic and dipole polarization.

4.5.2 PEOX-PVP-Graphene Nanocomposites

The ultraviolet diffusive reflectance spectra (UV-Vis DRS) of the PEOX-PVP blend and PEOX-PVP-graphene nanocomposites were utilized to determine their energy band gap and Urbach energy. Reflectance data from the samples were converted to relative absorption data using the Kubelka-Munk formula [59], $F(R) = (1-R)^2/(2R)$, where R represents the absolute reflectance value

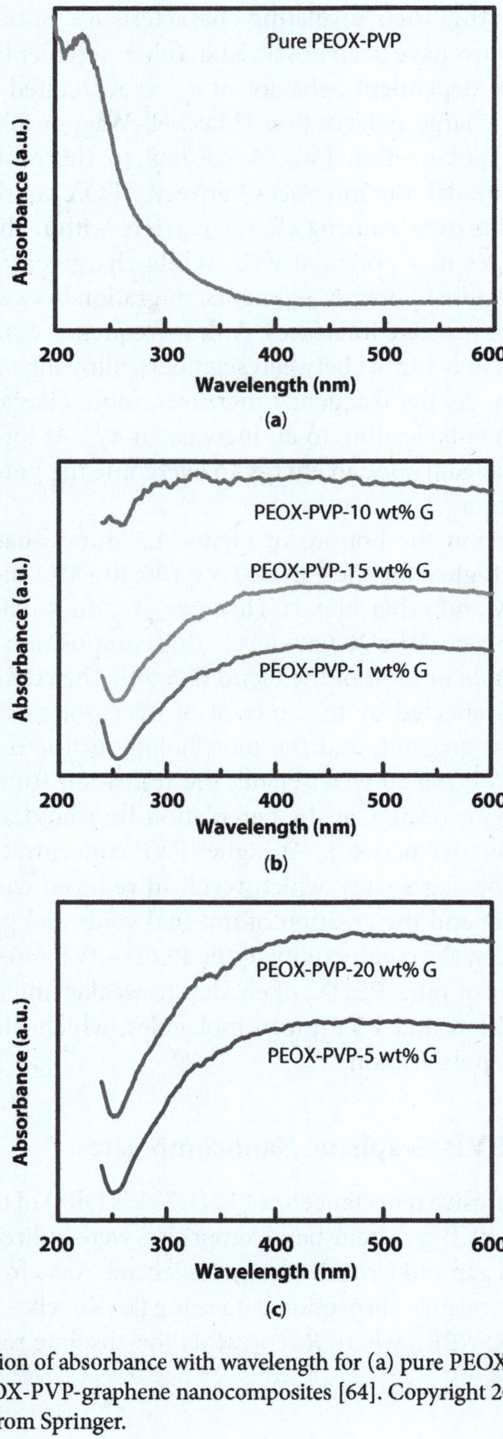

Figure 4.6 Variation of absorbance with wavelength for (a) pure PEOX-PVP (80:20 wt%) blend and (b) PEOX-PVP-graphene nanocomposites [64]. Copyright 2024. Reproduced with permission from Springer.

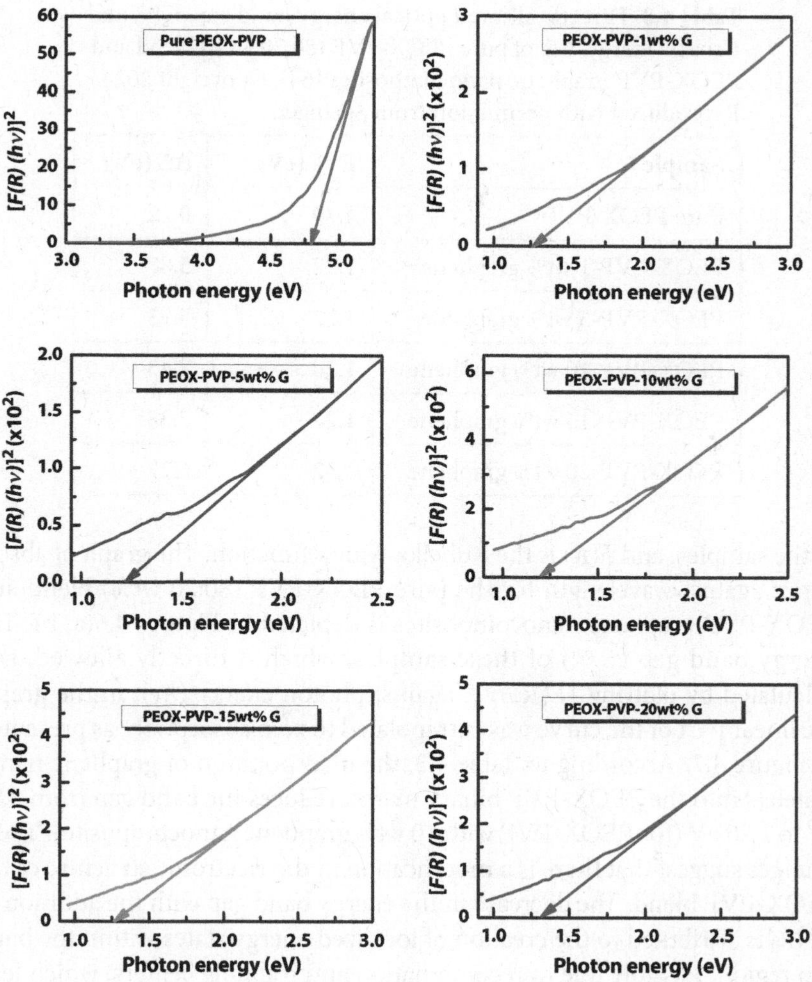

Figure 4.7 Graph of $[F(R)\,(h\nu)]^2$ against photon energy for pure PEOX-PVP (80:20 wt%) blend and PEOX-PVP-graphene nanocomposites [64]. Copyright 2024. Reproduced with permission from Springer.

Table 4.3 Directly allowed optical energy band gap, E_g^{opt}, and Urbach energy, ΔE, of pure PEOX-PVP (80:20 wt%) blend and PEOX-PVP-graphene nanocomposites [64]. Copyright 2024. Reproduced with permission from Springer.

Sample	E_g^{opt} (eV)	ΔE (eV)
Pure PEOX-PVP	4.79	0.22
PEOX-PVP-1 wt% graphene	1.27	5.87
PEOX-PVP-5 wt% graphene	1.22	5.93
PEOX-PVP-10 wt% graphene	1.20	7.43
PEOX-PVP-15 wt% graphene	1.27	7.38
PEOX-PVP-20 wt% graphene	1.32	6.77

of the samples, and $F(R)$ is the Kubelka-Munk function. The graph of absorbance against wavelength for the pure PEOX-PVP (80:20 wt%) blend and PEOX-PVP-graphene nanocomposites is depicted in Figures 4.6(a, b). The energy band gap (E_g^{opt}) of these samples, which is directly allowed, was calculated by plotting $[F(R)h\nu]^2$ against photon energy ($h\nu$). In the graph, the linear part of the curve was extrapolated to zero absorption, as presented in Figure 4.7. According to Table 4.3, the incorporation of graphene nanoplatelets into the PEOX-PVP binary matrix reduces the band gap from 4.79 eV to 1.20 eV (for PEOX-PVP with 10 wt% graphene nanocomposite). These changes suggest that there is a modification in the electronic structure of the PEOX-PVP blend. The decrease in the energy band gap with the addition of GNPs is attributed to the creation of localized energy states within the band gap region, possibly due to recombination and trapping centers, which lead to the observed reduction in the energy band gap [60].

The Urbach energy (ΔE) of the samples was determined by taking the reciprocal of the slopes of the linear parts of the graph plotting $\ln\alpha$ ($= \ln[F(R)]$) against photon energy ($h\nu$) as depicted in Figure 4.8, where α is absorption coefficient. The corresponding Urbach energy values (ΔE) are presented in Table 4.3. An increase in ΔE values is observed with the incorporation of GNPs into the PEOX-PVP binary matrix. Table 4.3

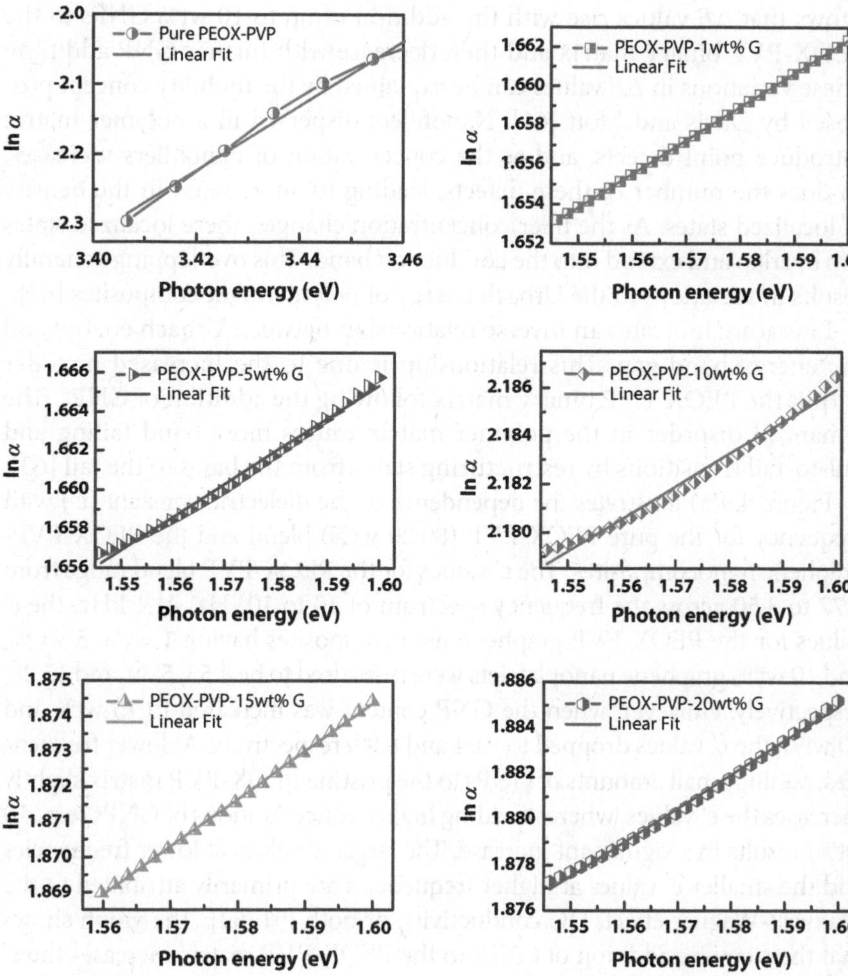

Figure 4.8 Graph of lnα against photon energy for pure PEOX-PVP (80:20 wt%) blend and PEOX-PVP-graphene nanocomposites [64]. Copyright 2024. Reproduced with permission from Springer.

shows that ΔE values rise with the addition of up to 10 wt% GNPs to the PEOX-PVP binary matrix and then decrease with further GNP addition. These variations in ΔE values can be explained by the mobility concept proposed by Davis and Mott [61]. Nanofillers dispersed in a polymer matrix introduce point defects, and as the concentration of nanofillers increases, so does the number of these defects, leading to an increase in the density of localized states. As the filler concentration changes, these localized states can overlap and extend into the conduction band. This overlapping generally results in variations in the Urbach energy of polymer nanocomposites [62].

Literature indicates an inverse relationship between Urbach energy and the energy band gap. This relationship is due to the increased disorder within the PEOX-PVP binary matrix following the addition of GNPs. The enhanced disorder in the polymer matrix causes more band tailing and tail-to-tail transitions by restructuring states from the band to the tail [63].

Figure 4.9(a) illustrates the dependence of the dielectric constant (ε') with frequency for the pure PEOX-PVP (80:20 wt%) blend and the PEOX-PVP-graphene nanocomposites. The ε' values for the PEOX-PVP blend range from 3.77 to 3.50 across the frequency spectrum of 10 to 10^6 Hz. At 1 kHz, the ε' values for the PEOX-PVP-graphene nanocomposites having 1 wt%, 5 wt %, and 10 wt % graphene nanoplatelets were measured to be 4.53, 5.79, and 15.28, respectively. However, when the GNP content was increased to 15 wt% and 20 wt%, the ε' values dropped to 9.04 and 8.83, respectively. At lower frequencies, adding small amounts of GNPs to the pristine PEOX-PVP matrix slightly increases the ε' values, whereas adding higher concentrations of GNPs (e.g., 10 wt%) results in a significant increase. The larger ε' values at lower frequencies and the smaller ε' values at higher frequencies are primarily attributed to the Maxwell-Wagner effect, DC conductivity, or both [51, 52]. The graph shows that the stepwise addition of GNPs to the PEOX-PVP matrix increases the ε' values, which can be explained by the micro capacitor network model [65].

In the PEOX-PVP-graphene nanocomposites, the alternating GNPs function as micro capacitor electrodes, separated by a thin dielectric layer of the PEOX-PVP polymer matrix, which are connected in series or parallel, forming numerous micro capacitors. As the GNP concentration increases, the number of micro capacitors also rises, while there is a decrease in the thickness of the polymer dielectric layer, leading to higher capacitance and consequently higher ε' values.

However, when GNPs are loaded at higher concentrations (i.e., 15 wt% and 20 wt%) in the PEOX-PVP-graphene nanocomposite, aggregation occurs within the PEOX-PVP matrix. According to the micro capacitor model, such aggregation reduces the number of micro capacitors, resulting

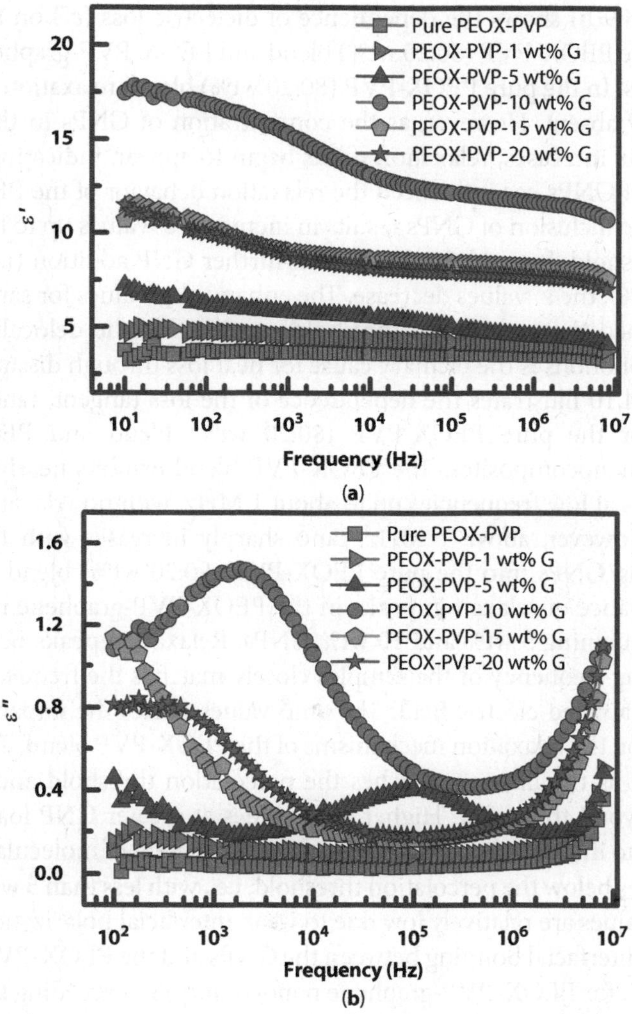

Figure 4.9 Dependence of (a) dielectric constant, ε′, and (b) dielectric loss, ε″, on frequency for pure PEOX-PVP (80:20 wt%) blend and PEOX-PVP graphene nanocomposites [64]. Copyright 2024. Reproduced with permission from Springer.

in lower ε′ values. This phenomenon can also be explained by percolation theory, which suggests that for materials with a high dielectric constant, there is an optimal ratio of polymer to filler content, known as the *percolation threshold*. In this case, the percolation limit is reached at a loading of 5 wt% GNP. Exceeding this limit results in a sharp increase in both permittivity and conductivity, but further addition of filler leads to a decrease in both permittivity and conductivity [66].

Figure 4.9(b) shows the dependence of dielectric loss (ε'') on frequency for the pure PEOX-PVP (80:20 wt%) blend and PEOX-PVP-graphene nanocomposites. In the pure PEOX-PVP (80:20 wt%) blend, relaxation peaks are completely absent. However, as the concentration of GNPs in the PEOX-PVP matrix increases, relaxation peaks begin to appear, indicating that the presence of GNPs has influenced the relaxation behavior of the PEOX-PVP matrix. The inclusion of GNPs results in increased ε'' values up to the percolation threshold. Beyond this limit, with further GNP addition (i.e., at 15% and 20 wt%), the ε'' values decrease. The enhanced ε'' values for samples that have reached the percolation threshold suggest that the delocalization of localized phonons is the primary cause for heat loss through dissipation.

Figure 4.10 illustrates the dependence of the loss tangent, tanδ, on frequency for the pure PEOX-PVP (80:20 wt%) blend and PEOX-PVP-graphene nanocomposites. The PEOX-PVP blend exhibits nearly constant tanδ values at low frequencies up to about 1 MHz, with no relaxation peaks present. However, above 1 MHz, tanδ sharply increases with frequency. Introducing GNPs into the pure PEOX-PVP (80:20 wt%) blend results in the appearance of relaxation peaks in the PEOX-PVP-graphene nanocomposites containing 5 wt% and 10 wt% GNPs. Relaxation peaks occur when the hopping frequency of the samples closely matches the frequency of the externally applied electric field. The tanδ values reflect the impact of GNP inclusion on the relaxation mechanisms of the PEOX-PVP blend. These tanδ values rise as the filler approaches the percolation threshold and increase rapidly beyond this point. Higher tanδ values at greater GNP loadings are attributed to interfacial polarization and the movement of molecular dipoles. For samples below the percolation threshold, i.e., with less than 5 wt% GNPs, the tanδ values are relatively low due to weak interfacial polarization caused by strong interfacial bonding between the GNPs and the PEOX-PVP matrix. In contrast, for PEOX-PVP-graphene nanocomposites exceeding the percolation limit (i.e., more than 5 wt% GNPs), good dispersion between GNPs and the PEOX-PVP matrix leads to the formation of conductive networks, with the dominant conduction process contributing to higher tanδ values. Additionally, adding GNPs to the PEOX-PVP blend matrix at 15 wt% and 20 wt% results in relatively poor dispersion compared to the nanocomposite with 10 wt% GNP, which suppresses the movement of molecular dipoles and leads to lower loss tangent values [67].

The frequency dependence of AC electrical conductivity (σ_{AC}) for the pure PEOX-PVP (80:20 wt%) blend and PEOX-PVP-graphene nanocomposites is depicted in Figure 4.11. The figure reveals that (i) in the lower frequencies, σ_{AC} remains nearly constant and is not dependent on frequency, and (ii) in the higher frequencies, σ_{AC} increases with the concentration of

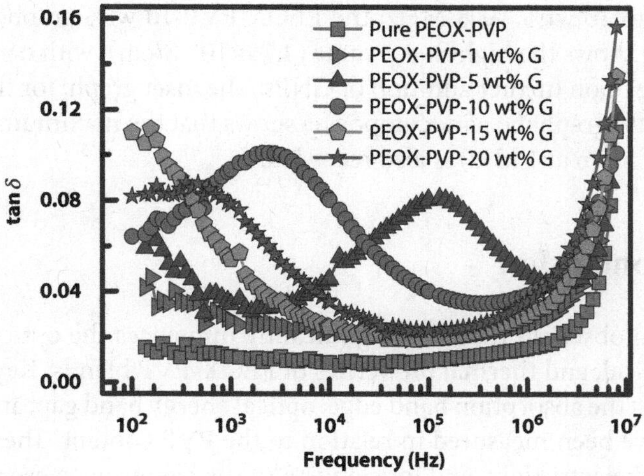

Figure 4.10 Frequency dependence of loss tangent (tan δ) for pure PEOX-PVP (80:20 wt%) blend and PEOX-PVP-graphene nanocomposites [64]. Copyright 2024. Reproduced with permission from Springer.

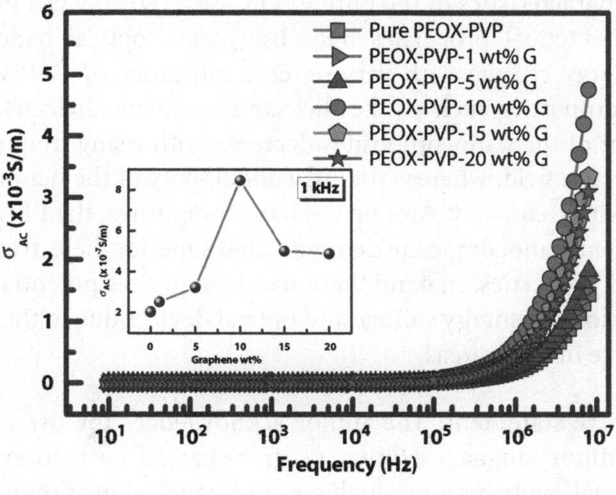

Figure 4.11 Frequency dependence of AC electrical conductivity (σ_{AC}) for pure PEOX-PVP (80:20 wt%) blend and PEOX-PVP-graphene nanocomposites [64]. Copyright 2024. Reproduced with permission from Springer.

GNPs up to 10 wt%. At 8 MHz, the PEOX-PVP-10 wt% graphene nanocomposite shows the highest σ_{AC} value (4.75×10^{-5} S/cm), with conductivity decreasing upon further addition of GNPs. The inset graph, for the PEOX-PVP-10 wt% graphene nanocomposite, shows that the maximum σ_{AC} value (8.47×10^{-9} S/cm at 1 kHz) is also recorded.

4.6 Conclusion

It has been observed that PVP significantly influences the optical, dielectric, electrical, and thermal properties of PEOX-PVP blends. Key parameters such as the absorption band edge, optical energy band gap, and Urbach energy have been measured in relation to the PVP content. The dielectric constant, dielectric loss, and loss tangent of the blends decrease as the frequency of the applied electric field increases, while the AC conductivity shows an upward trend with increasing frequency. The PEOX-PVP (80:20 wt%) blend exhibits superior optical, dielectric, and electrical properties.

The incorporation of GNPs notably affects the optical, dielectric, and electrical characteristics of the pure PEOX-PVP (80:20 wt%) binary polymer blend. Optical properties like absorbance, optical band gap, and Urbach energy change with varying concentrations of GNPs. Similarly, dielectric properties such as the dielectric constant, dielectric loss, and loss tangent of the nanocomposites decrease with rising frequency of the applied electric field, whereas the AC conductivity of the nanocomposites increases with frequency. Among the nanocomposites, the PEOX-PVP-10 wt% graphene nanocomposite demonstrated superior dielectric, electrical, and optical properties, making these nanocomposites potentially suitable for applications in energy storage and optical devices due to their excellent performance in these areas.

AI disclosure statement: The author acknowledges the use of AI-based tools for minor language editing or grammatical correction. However, all intellectual contributions, analyses, and conclusions presented in this chapter are entirely the author's own.

References

1. Utracki, L.A., *Polymer blends handbook*, Kluwer Academic Publishers, USA, 2003.

2. Passador, F.R., Ruvolo-Filho, A., Pessan, L.A., 7 - Nanocomposites of Polymer Matrices and Lamellar Clays, in: *Nanostructures*, A.L. Da Róz, M. Ferreira, F. de Lima Leite, O.N. Oliveira (Eds.), pp. 187–207, William Andrew Publishing, 2017.
3. Naz, A., Kausar, A., Siddiq, M., Choudhary, M.A., Comparative Review on Structure, Properties, Fabrication Techniques, and Relevance of Polymer Nanocomposites Reinforced with Carbon Nanotube and Graphite Fillers. *Polym. Plast. Technol. Eng.*, 55, 171–198, 2016.
4. Cong, H., Radosz, M., Towler, B.F., Shen, Y., Polymer–inorganic nanocomposite membranes for gas separation. *Sep. Purif. Technol.*, 55, 281–291, 2007.
5. Bhattacharya, M., Polymer Nanocomposites—A Comparison between Carbon Nanotubes, Graphene, and Clay as Nanofillers. *Materials*, 9, 262–329, 2016.
6. Garcia, M., van Vliet, G., Jahin, S., Schrauwen, B.A.G., Sarkissov, A., v. Z., W.E., Boukamp, B., Polypropylene/SiO2 nanocomposites with improved mechanical properties. *Rev. Adv. Mater. Sci.*, 6, 169–175, 2004.
7. Etienne, S., Becker, C., Ruch, D., Grignard, B., Cartigny, G., Detrembleur, C., Calberg, C., Jerome, R., Effects of incorporation of modified silica nanoparticles on the mechanical and thermal properties of PMMA. *J. Therm. Anal. Calorim.*, 87, 101–104, 2007.
8. Lin, O.H., Akil, H.M., Mohd Ishak, Z.A., Surface-activated nanosilica treated with silane coupling agents/polypropylene composites: Mechanical, morphological, and thermal studies. *Polym. Compos.*, 32, 1568–1583, 2011.
9. Grala, M., Bartczak, Z., Różański, A., Morphology, thermal and mechanical properties of polypropylene/SiO2 nanocomposites obtained by reactive blending. *J. Polym. Res.*, 23, 25, 2016.
10. Dong, Q., Ding, Y., Wen, B., Wang, F., Dong, H., Zhang, S., Wang, T., Yang, M., Improvement of thermal stability of polypropylene using DOPO-immobilized silica nanoparticles. *Colloid. Polym. Sci.*, 290, 1371–1380, 2012.
11. Mallakpour, S. and Marefatpour, F., Novel chiral poly(amide-imide)/surface modified SiO2 nanocomposites based on N-trimellitylimido-l-methionine: Synthesis and a morphological study. *Prog. Org. Coat.*, 77, 1271–1276, 2014.
12. Fischer, H., Polymer nanocomposites: from fundamental research to specific applications. *Mater. Sci. Eng., C*, 23, 763–772, 2003.
13. Lawal, G.I., Balogun, S.A., Akpan, E.I., Review of green polymer nanocomposites. *J. Miner. Mater. Charact. Eng.*, 11, 385–416, 2012.
14. Yang, F., Yang, W., Zhu, L., Chen, Y., Ye, Z., Preparation and investigation of waterborne fluorinated polyacrylate/silica nanocomposite coatings. *Prog. Org. Coat.*, 95, 1–7, 2016.
15. Jo, C.I., Ko, J., Yin, Z., Kim, Y.-J., Kim, Y.S., Solvent-Free and Highly Transparent SiO2 Nanoparticle–Polymer Composite with an Enhanced Moisture Barrier Property. *Ind. Eng. Chem. Res.*, 55, 9433–9439, 2016.

16. Wang, X., Wang, L., Su, Q., Zheng, J., Use of unmodified SiO2 as nanofiller to improve mechanical properties of polymer-based nanocomposites. *Compos. Sci. Technol.*, 89, 52–60, 2013.
17. Qi, D., Liu, C., Chen, Z., Dong, G., Cao, Z., *In situ* emulsion copolymerization of methyl methacrylate and butyl acrylate in the presence of SiO2 with various surface coupling densities. *Colloid. Polym. Sci.*, 293, 463–471, 2015.
18. Cheon Lee, S., Kim, C., Chan Kwon, I., Chung, H., Young Jeong, S., Polymeric micelles of poly(2-ethyl-2-oxazoline)-block-poly(ε-caprolactone) copolymer as a carrier for paclitaxel. *J. Control. Release*, 89, 437–446, 2003.
19. Gaertner, F.C., Luxenhofer, R., Blechert, B., Jordan, R., Essler, M., Synthesis, biodistribution and excretion of radiolabeled poly(2-alkyl-2-oxazoline)s. *J. Control. Release*, 119, 291–300, 2007.
20. Mero, A., Pasut, G., Via, L.D., Fijten, M.W.M., Schubert, U.S., Hoogenboom, R., Veronese, F.M., Synthesis and characterization of poly(2-ethyl 2-oxazoline)-conjugates with proteins and drugs: Suitable alternatives to PEG-conjugates? *J. Control. Release*, 125, 87–95, 2008.
21. Zalipsky, S., Hansen, C.B., Oaks, J.M., Allen, T.M., Evaluation of Blood Clearance Rates and Biodistribution of Poly(2-oxazoline)-Grafted Liposomes§. *J. Pharm. Sci.*, 85, 133–137, 1996.
22. Rivas, B.L. and Ananías, S.I., Ring opening polymerization of 2-ethyl-2-oxazoline. *Polym. Bull.*, 18, 189–194, 1987.
23. Boubaker, K., A physical explanation to the controversial Urbach tailing universality. *Eur. Phys. J. Plus*, 126, 10, 2011.
24. Chiu, T.T., Thill, B.P., Fairchok, W.J., Poly(2-ethyl-2-oxazoline): A New Water- and Organic-Soluble Adhesive, in: *Water-Soluble Polymers*, pp. 425–433, American Chemical Society, 1986.
25. Keskkula, H. and Paul, D.R., Miscibility of polyethyloxazoline with thermoplastic polymers. *J. Appl. Polym. Sci.*, 31, 1189–1197, 1986.
26. Christova, D., Velichkova, R., Loos, W., Goethals, E.J., Prez, F.D., New thermoresponsive polymer materials based on poly(2-ethyl-2-oxazoline) segments. *Polymer*, 44, 2255–2261, 2003.
27. Keskkula, H. and Paul, D.R., Thermal behavior of polyethyloxazoline. *J. Appl. Polym. Sci.*, 31, 941–950, 1986.
28. Keskkula, H., Paul, D.R., Young, P., Stein, R.S., Diffusion of miscible polymers in multilayer films. *J. Appl. Polym. Sci.*, 34, 1861–1877, 1987.
29. Trieu, T.C.H.H., Lim, R., Lange, E.C., Anderson, K.M., Bruneau, A., Interosteotic implant comprising a metallic or polymeric material for treatment of spinal disorders, 2007.
30. Jensen, S.D., Peroxide/poly(2-ethyl-2-oxazoline) gel compositions of good stability for bleaching of teeth, hair, laundry, 2007.
31. Torizuka, H.S.M., Fujiwara, K., Oda, T., Tanaka, N., Rindo, K., Emulsified, water-in-oil type composition and skin cosmetic preparation, 1997.
32. Hood, D.K., Coating composition for forming an inkjet-printable coating on a substrate, 2007.

33. Kuwahara, K., Nanbu, H., Ibi, A., Fukuyama, Y., Polymer dispersant for pigment and nonaqueous pigment dispersion containing it, 2007.
34. Wang, C.-H., Wang, C.-H., Hsiue, G.-H., Polymeric micelles with a pH-responsive structure as intracellular drug carriers. *J. Control. Release*, 108, 140–149, 2005.
35. Hsiue, G.-H., Wang, C.-H., Lo, C.-L., Wang, C.-H., Li, J.-P., Yang, J.-L., Environmental-sensitive micelles based on poly(2-ethyl-2-oxazoline)-b-poly(l-lactide) diblock copolymer for application in drug delivery. *Int. J. Pharm.*, 317, 69–75, 2006.
36. Liaw, J.J., Integrated semiconductor structure for static random access memory (sram) cells, 2007.
37. Haaf, F., Sanner, A., Straub, F., Polymers of N-Vinylpyrrolidone: Synthesis, Characterization and Uses. *Polym. J.*, 17, 143–152, 1985.
38. Koczkur, K.M., Mourdikoudis, S., Polavarapu, L., Skrabalak, S.E., Polyvinylpyrrolidone (PVP) in nanoparticle synthesis. *Dalton Trans.*, 44, 17883–17905, 2015.
39. Bühler, V., *Polyvinylpyrrolidone Excipients for Pharmaceuticals: Povidone, Crospovidone and Copovidone*, Springer, Berlin, Heidelberg, New York, 2005.
40. Wajid, A.S., Das, S., Irin, F., Ahmed, H.T., Shelburne, J.L., Parviz, D., Fullerton, R.J., Jankowski, A.F., Hedden, R.C., Green, M.J., Polymer-stabilized graphene dispersions at high concentrations in organic solvents for composite production. *Carbon*, 50, 526–534, 2012.
41. Bourlinos, A.B., Georgakilas, V., Zboril, R., Steriotis, T.A., Stubos, A.K., Trapalis, C., Aqueous-phase exfoliation of graphite in the presence of polyvinylpyrrolidone for the production of water-soluble graphenes. *Solid State Commun.*, 149, 2172–2176, 2009.
42. Guo, S., Dong, S., Wang, E., Three-dimensional Pt-on-Pd bimetallic nanodendrites supported on graphene nanosheet: facile synthesis and used as an advanced nanoelectrocatalyst for methanol oxidation. *ACS Nano*, 4, 547–555, 2009.
43. Liu, J., Wang, Z., Li, S., Sun, X., A novel graphene oxide/trans-1,4-polyisoprene (GO/TPI) shape memory polymer nanocomposite and its multifunctional properties. *Nanotechnology*, 30, 255706, 2019.
44. Swei, J. and Talbot, J.B., Development of high-definition aqueous polyvinylpyrrolidone photoresists for cathode ray tubes. *J. Appl. Polym. Sci.*, 102, 1637–1644, 2006.
45. Zhu, Q., Wang, X., Miller, J.D., Advanced Nanoclay-Based Nanocomposite Solid Polymer Electrolyte for Lithium Iron Phosphate Batteries. *ACS Appl. Mater. Interfaces*, 11, 8954–8960, 2019.
46. Li, B., Zhang, Y., Fu, L., Yu, T., Zhou, S., Zhang, L., Yin, L., Surface passivation engineering strategy to fully-inorganic cubic CsPbI3 perovskites for high-performance solar cells. *Nat. Commun.*, 9, 1076, 2018.
47. Mott, N.F. and Davis, E.A., *Electronic processes in non-crystalline materials*, Oxford university press, 2012.

48. El-Samanoudy, M. and Ammar, A., Some Physical and Optical Properties of the a-Se85SbxS15–x Film System. *Phys. Status Solidi A*, 187, 611–621, 2001.
49. Cholakis, C.H., Zingg, W., Sefton, M.V., Effect of heparin–pva hydrogel on platelets in a chronic canine arterio-venous shunt. *J. Biomed. Mater. Res.*, 23, 417–441, 1989.
50. Shubha, A., Manohara, S.R., Gerward, L., Influence of polyvinylpyrrolidone on optical, electrical, and dielectric properties of poly(2-ethyl-2-oxazoline)-polyvinylpyrrolidone blends. *J. Mol. Liq.*, 247, 328–336, 2017.
51. Abd-El-Messieh, S. and Naguib, H., Electrical conductivity and dielectric behavior of some poly (alkyl methacrylate) s/polyvinylpyrrolidone blends. *Polym. Plast. Technol. Eng.*, 44, 1591–1606, 2005.
52. Saad, A., Hassan, A., Youssif, M., Ahmed, M., Studies of electrical properties of some fire-retarding poly (vinyl chloride) compositions. *J. Appl. Polym. Sci.*, 65, 27–35, 1997.
53. Jayakrishnan, P. and Ramesan, M.T., Synthesis, characterization and electrical properties of Fe3O4/poly(vinyl alcohol-co-acrylic acid) nanocomposites. *AIP Conf. Proc.*, 1620, 165–172, 2014.
54. Ramesan, M., Athira, V., Jayakrishnan, P., Gopinathan, C., Preparation, characterization, electrical and antibacterial properties of sericin/poly (vinyl alcohol)/poly (vinyl pyrrolidone) composites. *J. Appl. Polym. Sci.*, 133, 43535, 2016.
55. Bishai, A., Gamil, F., Awni, F., Al-Khayat, B., Dielectric and mechanical properties of poly (vinyl chloride)–dioctylphthalate systems. *J. Appl. Polym. Sci.*, 30, 2009–2020, 1985.
56. Ramesan, M.T., Processing characteristics and mechanical and electrical properties of chlorinated styrene–butadiene rubber/fly ash composites. *J. Thermoplast. Compos. Mater.*, 28, 1286–1300, 2015.
57. Petzelt, J., Nuzhnyy, D., Bovtun, V., Savinov, M., Kempa, M., Rychetsky, I., Broadband dielectric and conductivity spectroscopy of inhomogeneous and composite conductors. *Phys. Status Solidi A*, 210, 2259–2271, 2013.
58. Dang, Z.-M., Yuan, J.-K., Zha, J.-W., Zhou, T., Li, S.-T., Hu, G.-H., Fundamentals, processes and applications of high-permittivity polymer–matrix composites. *Prog. Mater Sci.*, 57, 660–723, 2012.
59. Kubelka, P., New Contributions to the Optics of Intensely Light-Scattering Materials. Part I. *J. Opt. Soc. Am.*, 38, 448–457, 1948.
60. El-Nahass, M., Dongol, M., Abou-Zied, M., El-Denglawey, A., The compositional dependence of the structural and optical properties of amorphous As20Se80– xTlx films. *Physica B*, 368, 179–187, 2005.
61. Mott, N. and Davis, E., *Electronic process in non-crystalline materials*, Oxford University Press, 1971.
62. NB, R.K., Crasta, V., Praveen, B., Kumar, M., Studies on structural, optical and mechanical properties of MWCNTs and ZnO nanoparticles doped PVA nanocomposites. *Nanotechnol. Rev.*, 4, 457–467, 2015.

63. O'Leary, S.K., Zukotynski, S., Perz, J.M., Disorder and optical absorption in amorphous silicon and amorphous germanium. *J. Non-Cryst. Solids*, 210, 249–253, 1997.
64. Shubha, A. and Manohara, S.R., Effect of graphene nanoplatelets concentration on optical, dielectric and electrical properties of poly(2-ethyl-2-oxazoline)–polyvinylpyrrolidone–graphene nanocomposites. *J. Mater. Sci.: Mater. Electron.*, 31, 16498–16510, 2020.
65. Begam, N., Das, A.N., Chandran, S., Ibrahim, M., Padmanabhan, V., Sprung, M., Basu, J.K., Nanoparticle-polymer interfacial layer properties tune fragility and dynamic heterogeneity of athermal polymer nanocomposite films. *Soft Matter*, 14, 8853–8859, 2018.
66. Yang, K., Huang, X., Fang, L., He, J., Jiang, P., Fluoro-polymer functionalized graphene for flexible ferroelectric polymer-based high-k nanocomposites with suppressed dielectric loss and low percolation threshold. *Nanoscale*, 6, 14740–14753, 2014.
67. Roy, M., Nelson, J., MacCrone, R., Schadler, L.S., Reed, C., Keefe, R., Polymer nanocomposite dielectrics-the role of the interface. *IEEE Trans. Dielectr. Electr. Insul.*, 12, 629–643, 2005.

5

Piezoelectric Polymer Composites for Tribo and Piezoelectric Energy Harvesting Applications

Asrar Rafiq Bhat, Akash M. Chandran and Prasanna Kumar Mural*

Department of Metallurgical Engineering and Materials Science, Indian Institute of Technology Bombay (IITB), Powai, Mumbai, Maharashtra, India

Abstract
Electroactive polymers (EAPs) with piezoelectric properties alter their electrical characteristics when mechanically deformed. These electroactive piezoelectric polymers are extensively used in actuators, sensors, energy harvesting, and wearable electronics. This chapter focuses on characterizing a piezoelectric and triboelectric EAP for energy harvesting applications. It provides an overview of EAP piezoelectric polymers, discusses the mechanisms through which their composites produce piezoelectricity, and explores various methods for tailoring these properties. It also summarizes piezoelectric-based EAPs in energy harvesting applications and discusses triboelectric energy harvesting using electroactive polymer composites. Also, the various underlying mechanisms and methods of tailoring it are briefed.

Keywords: Electroactive polymers, piezoelectric polymers, triboelectric energy harvesting, energy harvesters, nanogenerators, sustainable energy, wearable electronics, Internet of Things (IoT)

5.1 Introduction

The environmental impact and depletion of conventional fossil fuel energies are driving the push for research into clean energy production. Thus, the United Nations General Assembly (UN-GA) established clean energy

**Corresponding author*: prasannamural@iitb.ac.in

Raneesh Balakrishnan, Sobi K. Chacko and Visakh P.M. (eds.) Electroactive Polymer-Based Smart Materials and Applications, (125–158) © 2026 Scrivener Publishing LLC

as the Sustainable Development Goals aiming for achievement by 2030 [1]. Owing to long-term sustainability issues from carbon sources of energy (fossil fuel) researchers across the globe are trying to replace them with new green, eco-friendly technologies. Renewable energies are replacing the conventional energy sources. However, the energy-harvesting technique devices are drawing the limelight nowadays owing to their long-lasting and maintenance-free source [2]. Energy harvesters capture the widely distributed irregular and random energies in the ambient conditions and have high energy entropy arising from the mechanical energy that is unexploited in our routine like vibrations, wind, human motion, ocean waves, raindrops and mechanically rotating tires, moving parts, and so on [48]. To capture random energies, energy harvesting devices should have a simple structure, be easy to fabricate, have low production cost, have flexibility in design, and be easy to recover [3].

Energy harvesters provide substitute energy for portable devices such as the Internet of Things (IoT), portable electronics, automatic security systems, mobile phones, and defense technologies. At present for these storage devices batteries or capacitors provide a source of energy that unavoidably has inadequate capacity thus needing regular recharging or replacement which has possible environmental implications (owing to the release of toxic elements like cobalt, lithium, etc.) [4]. With further booming of IoT the problems might escalate, thus nanogenerator (NG) based point of source energy would be an energy harvesting for sustainable energy, eliminating all energy loss and environmental impact. The polymer-based PENG and TENG have simple structures, are easy to fabricate, have low production cost, have flexibility in design, and should be easy to recover thus making it an ideal choice of material over inorganic piezoelectric materials. Applications including energy harvesting, artificial muscles, and smart systems hold tremendous potential for electroactive polymers (EAPs), which alter shape when subjected the electric field.

This chapter provides an in-depth exploration into the characterization of piezo-triboelectric EAPs for energy applications. It provides an overview of EAP piezoelectric polymers, explains the mechanisms through which their composites produce piezoelectricity, and explores various methods for tailoring these properties. Further, it briefs about the piezoelectric-based EAPs in energy harvesting applications. Then, triboelectric energy harvesting using electroactive polymer-based composites along with the various underlying mechanisms and methods of tailoring it are also briefed.

5.2 Piezoelectric Polymers and Their Properties

Electroactive polymers with piezoelectric properties change electrical properties when subjected to mechanical deformation. These electroactive piezoelectric polymers are widely employed in actuators, sensors, energy harvesting, and wearable electronics technologies [5]. Compared to their inorganic counterparts, piezoelectric polymers exhibit significantly higher piezoelectric stress constants and provide greater processing flexibility, allowing them to be formed into complex shapes. These polymers have high strength, impact resistance, ensuring durability and robustness. Despite these advantages, they maintain low elastic stiffness, dielectric constant, and density, making them lightweight and efficient for various applications [5]. For a polymer to exhibit piezoelectricity, it must possess critical elements such as molecular dipoles and the capability to undergo strain when mechanically stressed. Additionally, it is essential that these molecular dipoles can align and maintain their alignment during the poling process. The manifestation of piezoelectricity varies between semi-crystalline and amorphous polymers. In semi-crystalline polymers, piezoelectricity is driven by the presence of electroactive crystalline phase. Conversely, in amorphous region, piezoelectricity emerges from the preserved molecular dipoles in a quasi-stable state during the freezing process. This phenomenon occurs when the molecular structure lacks a regular, crystalline arrangement but still retains dipoles that align in response to mechanical stress, generating an electrical potential. This unique property allows amorphous polymers to exhibit piezoelectric behavior despite their non-crystalline nature. This difference in behavior underscores the distinct mechanisms at play in these two types of polymer structures [5].

5.2.1 Semi-Crystalline Polymers

The semi-crystalline polymers have polymorphic phases that can be polar. The post-processing operations (mechanical stretching, thermal annealing, and voltage treatment) facilitate the rotation of crystallites to induce polar phase. In addition, post-processing operations align the amorphous phase which assists the crystal orientation that maintains stability until Curie temperature. PVDF and its family, polyureas, polyamides, biopolymers, liquid crystalline polymers, and a combination of blends and composites, are semi-crystalline polymers that exhibit piezoelectric activity.

5.2.1.1 *Polyvinylidene Fluoride Polymer*

Carbon (C) and fluorine (F) atoms make up the synthetic semi-crystalline polymer known as PVDF, which belongs to a high-performance and specialized class of thermoplastic polymers [6]. PVDF stands out among partially fluorinated fluoropolymers, attracting substantial research attention owing to its widespread applications across scientific and technological fields. The monomer vinylidene fluoride (VDF, $CF_2 = CH_2$) was used to create PVDF polymer by Thomas A. Ford and Hanford William Edward in 1948 at EI Du Pont de Nemours and Co., also known as DuPont [7]. PVDF polymer, synthesized using vinylidene fluoride (VDF) as a monomer, employs aqueous emulsion or suspension polymerization processes. The CAS number for PVDF is 24937-79-9 [8]. When VDF monomers are added in a head-to-tail fashion, PVDF becomes a semi-crystalline homopolymer with a typical crystallinity of 30% to 40%. However, introducing VDF in a different arrangement can lower its crystallinity [9]. The properties of PVDF, such as molecular weight distribution, production method, and chain irregularities, significantly impact its characteristics. Fluorine-19 nuclear magnetic resonance spectroscopy (19F NMR) is employed for detecting structural defects in PVDF, facilitated by the pronounced interactions between $-CF_2$ and $-CH_2$ groups within the polymer's molecular structure. This technique leverages the sensitivity of fluorine nuclei to their chemical environment, providing insights into how these interactions influence the overall integrity and composition of PVDF. By analyzing the NMR spectra, researchers can pinpoint deviations or anomalies in the CF_2-CH_2 interactions, thereby identifying defects or variations in the polymer structure with high precision. This capability makes 19F NMR a valuable tool in characterizing and understanding the properties of PVDF and similar fluorinated polymers in various scientific and industrial applications. Compared to polytetrafluoroethylene (PTFE), PVDF exhibits smaller amorphous molecular dimensions, contributing to its highest flexural modulus among fluoropolymers [10]. In the market, PVDF is commercially available under brands like Hylar (Solvay), Kynar (Arkema), and Solef (Solvay), each tailored for specific applications in various industries.

5.2.1.1.1 Piezoelectricity in PVDF

Professor Heiji Kawai discovered the piezoelectric properties of PVDF around 1969. He demonstrated that the stretched and poled PVDF film exhibits a piezoelectric constant $5\,pC/N$ [11, 12]. Additionally, Oshiki and Fukada have highlighted the significance of the interaction between residual polarization and the electrostriction effect as the primary factor influencing the piezoelectric capabilities of PVDF polymer films [13]. Later, researchers looked into PVDF's characteristics and found that its piezoelectricity is influenced by the Poisson ratio, the crystal structure, the electrostriction constant, and other factors as well [14]. The piezoelectric effect in PVDF is due to non-centrosymmetric and large dipole moments in the unit cell. PVDF's molecular structure includes a carbon chain with alternating -CF_2 and -CH_2 groups, composed of hydrogen and fluorine atoms. During crystallization, PVDF forms four distinct crystal phases known as alpha (α), beta (β), delta (δ), and gamma (γ) phases, making it a polycrystalline material [15–17]. The most common phase monoclinic unit cell (non-polar α-phase) with TGTG' (trans (*T*) and gauche (*G*) linkages, in conformation exhibiting high thermodynamic stability. The non-polar α-phase (crystal form II) is a non-centrosymmetric crystal with antiparallel packing of the dipoles, and it exhibits a P21/c space group [18].

PVDF, categorized by the electroactive orthorhombic unit cell with a TTTT (all-trans) conformation, demonstrates a wide array of exceptional properties. In its polar phase (as illustrated in Figure 5.1), PVDF exhibits excellent dielectric properties, making it suitable for electrical insulation and capacitor applications. Its piezoelectric nature enables the production of electric charge in response to mechanical tension, essential for sensors and actuators. PVDF also displays pyroelectric behavior, where it generates electric charge during the temperature change, useful in infrared detectors. Moreover, its ferroelectric properties allow for the reversible polarization of electric dipoles under an applied electric field, contributing to memory devices and non-volatile storage technologies. The significant spontaneous polarization observed in PVDF further enhances its utility

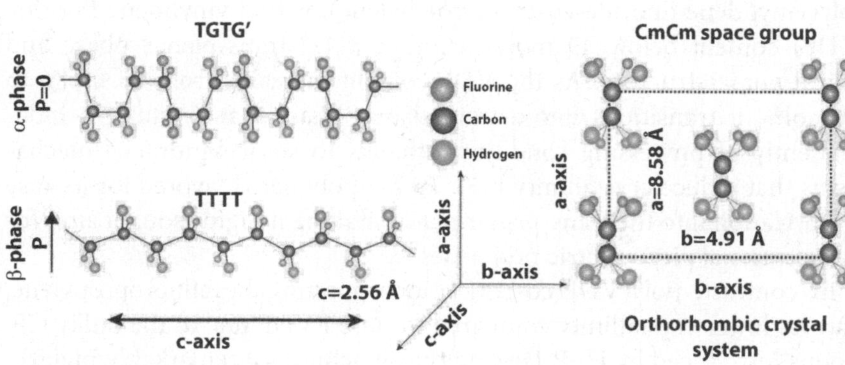

Figure 5.1 A schematic representation of a and b-phase of PVDF polymer, with orthorhombic crystal system.

in various advanced electronic and electromechanical applications, highlighting its versatility and importance in modern materials science [19]. The polar b-phase (crystal form I) exhibits the Cm2m space group with the highest polarization per unit cell 8×10^{-30} Cm [20]. In the beta phase of PVDF, the dipole moment aligns parallel to the polymer chains, driven by the orientation of electronegative fluorine atoms towards electropositive hydrogen atoms along the polymer backbone. Contrastingly, in phases like the α-phase, the dipole moment does not align perpendicular to the polymer chain. This change in alignment stems from the intrinsic dipole nature of PVDF, originating from fluorine and hydrogen interactions. As PVDF shifts between phases, the varying alignment of these dipoles significantly influences the material's properties and its performance across diverse applications [21, 22]. Hence, PVDF displays dipole density piezoelectricity, due to which a negative Longitudinal piezoelectric coefficient is observed. The electroactive phase is widely utilized for integrated capacitors, actuators, sensors, nano- and microgenerators, owing to its outstanding electrical properties [23]. The fabrication and processing conditions affect b-phase formation. The non-polar a-phase that forms the fastest has little practical uses. Through sophisticated processes, it is likely to convert the non-polar α-phase into the electroactive β-phase, despite the associated expenses.

5.2.1.2 *Polyvinylidene Fluoride Copolymer*

Tetrafluoroethylene (VDF-co-TrFE), hexafluoropropylene (VDF-co-HFP), and chlorotrifluoroethylene (VDF-co-CTFE), are the copolymer of PVDF and these has ushered in a new era of electroactive materials characterized by enhanced crystallinity and robust electroactive properties [24]. Poly(vinylidene fluoride-co-trifluoroethylene), with a vinylidene fluoride (VDF) content below 49 mol%, exhibits a 1/3-trans-planar phase and helical phase structure. As the VDF content increases from 55 mol% to 80 mol%, it transitions into a trans-planar phase that crystallizes independently of processing conditions, thanks to steric hindrance mechanisms that reduce crystallinity [25]. This copolymer is favored for its ease of fabrication into the trans-planar phase, making it highly sought after for its exceptional piezoelectric properties.

In contrast, poly(VDF-co-HFP), incorporating hexafluoropropylene, features lower crystallinity compared to pure PVDF due to the bulky CF_3 groups introduced by HFP. Despite this, it achieves a remarkably high d_{31} piezoelectric constant (21 $pC{\cdot}N^{-1}$), the highest within the PVDF family, highlighting its suitability for applications requiring strong electromechanical response.

Poly(VDF-co-CTFE), containing bulky chlorotrifluoroethylene (CTFE) groups, facilitates easier dipole orientation during poling processes. This copolymer exhibits broader ferroelectric hysteresis loops, making it particularly well-suited for applications in energy storage devices where robust ferroelectric properties are essential.

These advancements underscore the versatility and tailored functionalities of PVDF copolymers, each offering distinct advantages for specific applications ranging from piezoelectric sensors to high-performance energy storage systems [26].

5.2.1.3 Polyureas

The aromatic polyureas exhibit piezoelectricity that is stable upto 200°C. The aliphatic polyureas containing odd-numbered methyl groups exhibit piezoelectricity and smaller number of carbons exhibit higher dipoles density leading to higher piezoelectric coefficients. But aliphatic polyureas have piezoelectric coefficients lower than aromatic polyureas.

5.2.1.4 Polyamides

Polyamides containing odd-numbered monomeric units align dipoles to produce a net dipole moment, with dipole density influenced by the quantity of methylene groups. Increasing the number of methylene groups decreases the piezoelectric coefficient; for example, nylon-7 displays 17 $pC{\cdot}N^{-1}$ (d_{33}) whereas nylon-11 shows 14 $pC{\cdot}N^{-1}$ which is slightly less than nylon-7 indicating a decrease in piezoelectric performance. However, the moisture absorption of polyamides at lower temperatures limits their application in piezoelectric devices [27].

5.2.1.5 Liquid-Crystalline Polymers

Liquid crystals exhibit spontaneous polarization based on their symmetry characteristics [28]. In particular, liquid crystals that integrate a carbon atom connecting a mesogenic group and an end alkyl chain exhibit ferroelectric characteristic. Consequently, these liquid crystal phases possess molecular dipoles that contribute to net polarization and enable piezoelectric behavior [29].

5.2.1.6 *Biopolymers*

Biopolymer materials such as collagen, oriented DNA films, chitin, poly-lactic acid, and polypeptides demonstrate polarization, giving rise to piezoelectric coefficients. Collagen, for instance, exhibits a piezoelectric coefficient reported at 2.5 pC/N, relatively lower compared to synthetic polymers. Biopolymers exhibit electroactive properties attributed to polar atomic groups, particularly around asymmetric carbon atoms. Keratin and polypeptides are notable examples, where amide (CONH) dipoles align axially to form a polar β-crystalline structure. This specific molecular arrangement enhances their capability for controlled electrical behavior, making them suitable for applications requiring responsiveness to external stimuli. These biopolymers are integral in diverse fields such as biomedical engineering and materials science, offering opportunities for innovative technologies including sensors, actuators, and biomimetic materials. Their inherent electroactive nature and structural adaptability underline their importance in advancing sustainable and responsive materials for biomaterial.

5.2.2 Amorphous Polymer

Piezoelectricity in amorphous polymers is typically weak, arising from the preservation of molecular dipoles in a quasi-stable state upon freezing-in. Some of the amorphous piezoelectric polymers are poly(vinylidenecyanide vinylacetate) (PVDCN/VAc), polyacrylonitrile (PAN), poly(1-bicyclobutanecarbonitrile), polyphenylethernitrile (PPEN), polyvinyl chloride (PVC) and polyvinyl acetate (PVAc). PVC exhibits piezoelectric coefficients d_{31} in the range of 0.5 to 1.3 pC/N. It arises due to carbon-chlorine dipole [5]. Vinylidene cyanide copolymer exhibit piezoelectric activities that is similar to PVDF. Major properties are dependent on dipole orientation and chain mobility in the glassy state [5].

5.2.2.1 *Blends*

To improve processability and encourage specific crystalline phases beneficial for piezoelectric applications, significant polymer blends are formulated. PVDF and its copolymers are prominently utilized in these blends. Incorporating PMMA into PVDF promotes β-phase formation, enhancing piezoelectric properties. Likewise, blending PVDF with other piezoelectric polymers such as PLLA can also stimulate the formation of the β-phase, thereby augmenting its piezoelectric potential. These tailored blends not only streamline processing but also strategically optimize material characteristics for advanced piezoelectric applications [30].

5.2.3 Composites

Major piezoelectric materials are ceramic in nature with limited mechanical compliance. Thus, composites made of ceramic and polymer are reported to overcome the mechanical compliance limitation [31]. These composites might have less piezoelectric sensitivities than individual ceramic. However, the mechanical stability, ease of design and fabrication and better performance in applications outweigh this reduction in piezoelectric sensitivities. 'piezocomposites' are typically composed of more than one piezoelectric material component but most restricted to no more than two. Piezocomposites are two phase containing typically piezoelectric ceramic material and a polymer material. Piezocomposites are classified based on manner on the which two phases are combined. The notation employed to represent this type of combination is n_1-n_2, where both n_1 and n_2 are integers limited to the values 1, 2, or 3. Typically numbers specify the number of directions in which phase 1 is ceramic material is dispersed in continues phase 2(polymer) in three dimensions. The practical combinations being investigated and reported on are 0-3, 1-3, and 3-3, with particular emphasis on the study of 0-3 and 3-3 combinations. Further, the properties of the composites are Influenced by the following factors: -

- Characteristics of the individual phases
- Loading of phases (Volume fraction)
- Connectivity type and pattern
- Composite fabrication technique

5.2.3.1 0-3 Piezocomposites

In this variation, finely ground ceramic powders are evenly distributed within a continuous polymer matrix. The ceramic form discontinuous phase and polymer phase forms continuous and interconnected in all three directions. Solvent-cast technique is most widely employed for preparation of 0-3 Piezocomposites. Typically, solvent cast involves dissolution of polymer in suitable solvent and mixing ceramic powder in solution and casting on the glass plate. Then the casted solution is allowed to evaporate the solvent lead to formation of films. The limitation of this approach is the inability to precisely control the sheet's thickness, resulting in a sample density that is lower than the anticipated theoretical density. Other techniques for 0-3 Piezocomposites are hot-press and injection molding techniques.

5.2.3.2 3-3 Piezocomposites

In this type of composite, both phases are continually interconnected in all three directions. Composites featuring 3-3 connectivity offer distinct advantages over 0-3 piezocomposite in in applications such as ultrasonic imaging devices. Further, 3-3 composites exhibit improved acoustic impedance matching with water and human tissues.

Among the piezocomposite reported so far, the PVDF based composites are widely used. Researchers have reported the existence of PVDF (Polyvinylidene Fluoride) composites that incorporate a wide variety fillers [32]. Recently, a notable trend in the field of PVDF composites, focusing on the fabrication of composites with the incorporation of multiple fillers, each possessing complementary properties. This approach is aimed at achieving a reduction in the overall filler content while simultaneously enhancing the functional characteristics of the composite materials. A major category of fillers widely used in PVDF and its copolymer-based composites includes magnetic nanoparticles like $CoFe_2O_4$ or Fe_3O_4, barium titanate ($BaTiO_3$), zinc oxide (ZnO), titanium dioxide (TiO_2), and carbon nanotubes (CNTs), etc. [24] PVDF composites present an intriguing aspect, as their final properties can be precisely adjusted by selecting appropriate filler dimensions, morphology, concentration, uniform distribution, interfacial properties, and optimizing the dynamic interaction between filler materials and the polymer matrix, all in coordination with the manufacturing process. Moreover, PVDF-based composites enable the incorporation of fillers with ease of processability. These composites find applications in sensors and actuators, energy generation and storage systems, environmental monitoring and mitigation, and medical applications.

PVDF nanocomposites represent a significant area of interest in current nanotechnology research and development. These composites feature a PVDF polymer matrix infused with nanoparticles (NPs) dispersed throughout. These nanoparticles, which can be made from inorganic or carbon-based materials, are characterized by their nanometer-scale dimensions. Unlike traditional composites, nanocomposites leverage the significant impact of NPs on the interaction between filler and polymer.

PVDF nanocomposites find diverse applications across various fields including membrane technology, piezoelectric energy harvesting, tribology, pyroelectric processes, energy storage in capacitors and supercapacitors, and a range of sensor types such as tactile, pressure, flow, and force sensors. Additionally, they are used in bio-scaffolding, electromagnetic shielding, and Micro-Electro-Mechanical Systems (MEMS) [33–35]. Beyond these applications, PVDF nanocomposites are utilized in coatings, packaging materials, automobiles, aerospace components, injection-molded products, optical integrated circuits, adhesives, fire-retardants, microelectronic packaging, drug delivery systems, and medical devices [36, 37]. Different techniques might be employed in accordance with the application in order to fabricate PVDF nanocomposites. Commonly used methods for producing PVDF-based membranes include electrospinning and phase inversion. PVDF nanocomposites are produced utilizing techniques such solution casting, electrospinning, 3D printing, and compression moulding in the context of energy harvesting applications. The polar -phase of PVDF can be improved by including nanoparticles (NPs) as well as by using fabrication and post-fabrication procedures. The method chosen will depend on how the nanocomposite will be used.

5.3 Energy Harvesting Using Piezoelectricity

Energy harvesting through the piezoelectric effect utilize piezoelectric materials as potential sources of electrical energy to power low-power devices. This method capitalizes on the direct piezoelectric effect, wherein mechanical vibrations are converted into electrical signals. Various mechanical vibrations and movements from ambient conditions can be harnessed for this purpose [31]. Since 2006, there has been a growing trend in the development of piezoelectric energy harvesting systems based on polymers [38–40]. The electrical output from piezoelectric materials is contingent on the dimensions and mass of the device. For higher electrical power output need larger and heavier devices. Thus, restricting to small-scale applications, where miniaturization is essential. Also, frequency range of typical environmental vibrations fall in the range of 50 to 250 Hz [31]. Tuning to these typical environmental vibrations imposing limitations on miniaturization. Nanogenerator (NG) based energy harvesters are sustainable energy source to low powered portable devices.

Flexible nanogenerators based on piezoelectricity transform mechanical energy like vibrations, bending, and stretching into electrical energy. They can be integrated into wearable electronics, sensors, and clothes due to their flexible and lightweight construction. The fundamental component of this nanogenerator comprises flexible piezoelectric polymers, enabling the device to undergo bending and flexing motions while retaining its functionality. The piezoelectric characteristics of PVDF polymer enable a cutting-edge flexible nanogenerator for energy harvesting. PVDF is flexible, lightweight, and can transform mechanical energy into electricity. When stressed or deformed, a piezoelectric PVDF polymer in a flexible nanogenerator generates an electric charge. The flexible piezoelectric nanogenerator was first reported in the year 2009 by Professor Liwei Lin and his team. They fabricated a based flexible nanogenerator using the near-field electrospinning (NFES) method and synthesized the single PVDF nanofiber with piezoelectric properties [41]. In 2010, Prof. Zhong Lin Wang and his team developed the first hybrid energy harvester utilizing PVDF nanofibers for mechanical energy harvesting alongside biochemical energy harvesting from glucose [42]. Further, his team reported the first wearable nanogenerator based on PVDF. The prepared nanogenerator was a hybrid nanogenerator where a zinc oxide (ZnO) nanowire array was used along with PVDF polymer. The prepared device could generate electrical outputs of ~10 nA/cm^2; this work gave a breakthrough for flexible and wearable nanogenerators [43, 44].

5.4 Triboelectric Energy Harvesting Using Electroactive Polymer-Based Composites

Harnessing energy from everyday activities to generate useful electrical power presents a compelling solution for addressing the current energy crisis. This innovative approach offers the potential to significantly reduce our reliance on conventional batteries, which present intricate challenges related to recycling and disposal. In the realm of electrical energy harvesting from vibrational sources, there exist a limited number of energy harvesting devices. However, these devices often entail intricate processes involving fabrication, implementation, and maintenance. Triboelectric nanogenerators (TENGs) stand out due to their cost-effective fabrication, accessible design, and high energy efficiency. TENGs offer a viable alternative to batteries, especially in low-power microelectronic devices [45].

For the last ten years nanogenerators has swiftly advanced owing to the creation of new types of nanogenerators, such as the TENGs. Nanogenerators have a wide range of possible uses, from powering tiny electronic devices and sensors to capturing energy from environmental vibrations and human movements. Triboelectrification (TE) is a frequent occurrence in our daily lives. Since the time of ancient Greek civilization,

TE has been well recognized for more than 2600 years. It is present everywhere and at all times in our lives. Contact electrification (CE), the technical word for TE, is the process by which two distinct materials—even those that have a chemical composition—can be charged electrically after being physically separated from one another. Many readers treat CE and TE similarly, despite the fact that there are significant variations between the two. TE is a "convolution" of friction and CE. Friction is not required for charge transfer, but CE does occur in TE. Debris generation and friction coefficient are both aspects of friction [46, 47]. As a result, TE is an engineering activity while Contact electrification is a scientific phenomenon in which friction is not required [48].

TENGs demonstrate versatility by operating in various modes such as contact-separation, sliding, single-electrode, and free-standing modes. These diverse operational capabilities expand the potential applications of TENGs, making them suitable for use as wearable well as portable power source. To enhance the tribo-charges, micro-patterns (roughness) is incorporated into the design of TENGs within the triboelectric layers. TENGs operate on the foundational concept of the triboelectric effect, where materials of varying electron affinities generate an electrical charge upon contact and subsequent separation. Typically, TENGs consist of two materials that touch and then mechanically separate, creating this charge. Metals, polymers, and inorganic materials are commonly used in constructing TENGs. TENGs offer several advantages, including easy manufacturing and high efficiency. They also have diverse applications, including capturing energy from environmental vibrations and human motion [49].

The development of triboelectric nanogenerators (TENGs) has sparked renewed interest in converting small, irregular mechanical energy into electric power by combining triboelectric (TE) and electrostatic induction mechanisms. In 2012, researchers led by Fan *et al.* pioneered the first TENG, demonstrating its ability to generate electrical power from mechanical actions such as tapping or sliding [50]. Their device utilized a micropatterned plastic film comprising a transparent polyethylene terephthalate (PET) substrate coated with an indium tin oxide (ITO) layer and a micro-patterned polydimethylsiloxane (PDMS) layer, leveraging the triboelectric effect. This groundbreaking work illustrated how TENGs could function as self-powered pressure sensors, capable of monitoring and quantifying applied pressure. It showcased the potential of TENGs as advanced technology for self-powered sensors and harnessing mechanical energy, promising applications in healthcare and environmental monitoring. In 2018, the first 3D printed TENG was developed, featuring an ultra-flexible design suitable for biomechanical energy harvesting.

Constructed from ionic hydrogel and composite resin printed parts serving as electrodes and the electrification layer, this TENG exhibited stable performance with respectable power density and charge density, even at low triggering frequencies enabled by Maxwell's displacement current. The researchers demonstrated several self-powered applications of 3D-printed TENGs, including smart watches, smart lighting shoes, and a temperature sensor [51]. Today, TENGs boast impressive metrics such as a power density of 500 W/m^2 and proven conversion efficiencies exceeding 50% [52]. Their ability to capture various forms of ambient mechanical energy—such as human motion, wind, flowing water, vibrations, and more—positions TENGs as pivotal technologies for distributed energy systems and the Internet of Things (IoT).

5.4.1 Working Modes of TENG

TENGs harness the principles of triboelectrification and electrostatic induction to convert mechanical energy into electrical power. As a burgeoning field in energy harvesting, TENGs have evolved with advancements in nanotechnology, focusing on optimizing their operational modes and materials. Fundamentally, TENGs operate by exploiting the triboelectric effect, where materials with different electron affinities generate static electricity when brought into contact and then separated. This phenomenon creates a potential difference between electrodes, which can be harnessed as an electrical output. In addition to triboelectrification, electrostatic induction plays a crucial role in TENG operation, where charges induced by the movement of materials within an electric field contribute to the generation of electricity.

TENGs operate in several primary configurations, each tailored to specific applications and mechanical environments. These configurations include Vertical Contact Separation, where two materials with different triboelectric properties contact and separate vertically, creating a potential difference between electrodes. In the Linear Sliding mode, materials slide linearly against each other, generating electricity through the triboelectric effect induced by movement. The Single-Electrode configuration involves a single triboelectric layer and an electrode, where movement of the triboelectric layer relative to the electrode induces charges and generates electrical output. In the Free-Standing configuration, triboelectric layers are suspended or flexible, enabling mechanical movement that generates electrical charges. These diverse operational modes highlight the versatility of TENGs in harvesting energy from various mechanical stimuli, paving the way for applications ranging from self-powered sensors to wearable electronics and beyond.

These operational modes necessitate careful selection and integration of triboelectric materials, electrode configurations, and insulating layers to optimize efficiency and reliability. Commonly used materials include polymers like polydimethylsiloxane (PDMS), polytetrafluoroethylene (PTFE), and metals like aluminum or copper. During operation, cyclic mechanical movement induces the displacement of charges, disrupting the electrostatic equilibrium and generating an alternating current (AC) output. This AC signal exhibits positive and negative peaks corresponding to the directional movement of the mechanical stimulus. Ongoing research focuses on enhancing TENG performance by exploring new materials, refining fabrication techniques, and expanding application domains. These efforts aim to leverage TENGs for self-powered sensors, wearable electronics, and energy harvesting in environments where traditional power sources are impractical or inaccessible. As TENG technology continues to evolve, its potential to contribute to sustainable energy solutions grows, promising advancements in renewable energy generation from everyday mechanical actions.

5.4.1.1 *Contact Separation Mode*

In Figure 5.2(a), the contact separation mode of TENG illustrates triboelectrification occurring when two triboelectric materials or layers make contact and then separate. This interaction can involve either dissimilar

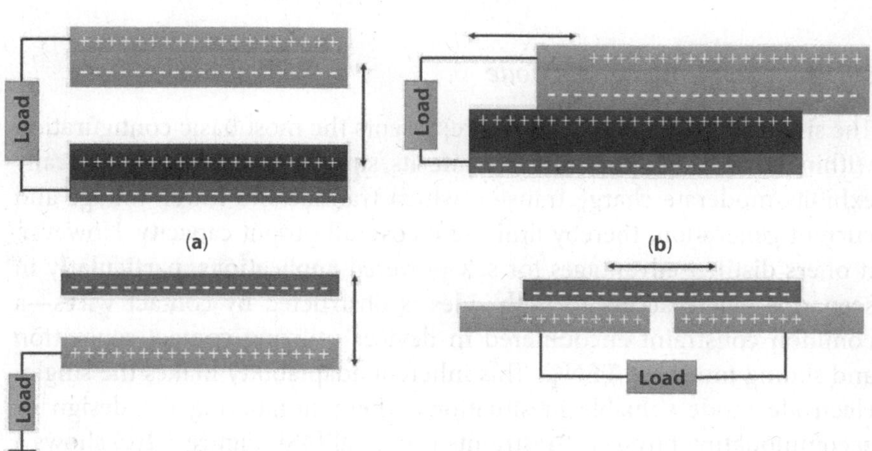

Figure 5.2 Working modes of TENGs (Conceptualized from Wang *et al.* [48]).

dielectric materials or a combination of dielectric and metallic layers. This mode is valued for its straightforward design, cost-effectiveness, and ease of manufacture. Notably, it was among the first developed and effectively utilized to power low-power electronic devices. Additionally, it can be configured in a stacked arrangement of multiple units to boost overall output performance [45].

5.4.1.2 *Linear Sliding Mode*

In the linear sliding mode of TENG, charge generation occurs due to the relative motion between the layers comprising its structure. This mode involves electrodes positioned on the rear surface of the triboelectric layers, similar to the architecture of the contact separation mode. A distinguishing feature of the linear sliding mode, illustrated in Figure 5.2(b), is the lateral displacement of these layers during operation.

Compared to the vertical contact separation method, the sliding mode typically exhibits lower overall performance because it requires greater displacement during sliding. However, its advantage lies in the larger contact area involved in lateral motion, which inherently enables higher charge density generation. Moreover, incorporating additional grating structures holds promise for enhancing the TENG's overall output performance in sliding mode.

It's also noteworthy that TENGs operating in sliding mode can be adapted for rotational applications by employing cylindrical grating structures. This adaptability underscores the versatility of TENG technology across different mechanical energy harvesting scenarios [48].

5.4.1.3 *Single Electrode Mode*

The single-electrode mode TENG represents the most basic configuration within the TENG spectrum. Despite its simplicity, this mode typically exhibits moderate charge transfer, which translates to lower voltage and current generation, thereby limiting its overall output capacity. However, it offers distinct advantages for self-powered applications, particularly in scenarios where access to both sides is obstructed by contact wires—a common constraint encountered in devices utilizing contact separation and sliding modes of TENG. This inherent adaptability makes the single-electrode mode valuable in situations where simplifying the design or accommodating physical constraints is critical [48]. Figure 5.2(c) shows a schematic of single electrode mode TENG.

5.4.1.4 Free Standing Mode

In the free-standing configuration of a TENG, a single electrode has the ability to move independently between two additional electrodes or triboelectric layers. Unlike other modes where electrodes may move together, here the electrodes remain stationary while the triboelectric layers or non-electrode components can pass over them. This unique TENG operating mode demonstrates remarkable performance characteristics, consistently delivering high output efficiency and boasting a notably elevated figure-of-merit. These attributes underscore its exceptional capabilities in energy conversion and highlight its potential for various practical applications [48]. Additionally, the design of this kind of TENG gadget makes it simple to fabricate and integrate into a variety of real-time applications. Figure 5.2(d) shows the setup needed for a TENG operating in free-standing mode.

5.4.2 Fabrication of TENG

The effective design of TENGs hinges significantly on the meticulous arrangement of triboelectric layers and electrodes, a field constantly evolving through refined engineering. A pivotal element in optimizing electrical output involves integrating triboelectric layers characterized by high surface roughness. This attribute triggers a remarkable effect within TENG devices by augmenting the number of contact points. Each micro or nano-scale irregularity engages with the corresponding triboelectric layer during mechanical activation, fostering myriad contact sites. This contrasts sharply with flat surfaces, where contact typically occurs at singular points, thereby substantially enhancing the generated output. Figure 5.3 delineates diverse methodologies for imparting surface roughness to triboelectric layers, encompassing techniques such as lithography, inductively coupled plasma etching, wet etching, 3D printing, and casting. Each approach offers distinct advantages and drawbacks. For instance, advanced methods like lithography and plasma etching necessitate costly equipment and skilled operators, which may pose constraints in certain applications.

The introduction of surface roughness *via* these techniques is pivotal in enhancing the overall performance of TENGs. Lithography allows for precise control over surface topography but requires substantial initial investment and technical expertise. Similarly, inductively coupled plasma etching offers high precision but at the expense of operational complexity and cost. Wet etching provides a simpler alternative but may be limited in achieving precise surface features compared to lithography. On the other

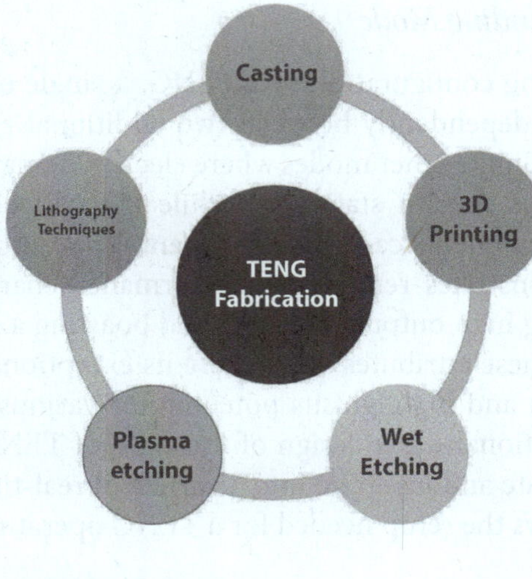

Figure 5.3 Fabrication techniques of TENGs.

hand, 3D printing and casting offer versatility in manufacturing complex geometries and are relatively cost-effective but may lack the precision of lithographic techniques. The strategic incorporation of triboelectric layers with high surface roughness is pivotal for maximizing the efficiency of TENG devices. The choice of surface texturing method should consider the specific requirements of the application, balancing factors such as cost, complexity, and the desired level of precision in surface features. As research and development continue, advancements in these fabrication techniques promise to further optimize the performance of TENGs across various fields and applications [53].

On the other hand, cost-effective methods exist for inducing surface roughness. One such technique utilizes sandpaper as a molding tool. Here, sandpaper with different levels of roughness acts as a template for forming a textured triboelectric layer. Polymers such as PDMS and other silicon elastomer-based materials undergo a curing process when applied onto the sandpaper, which is initiated by applying heat. The resulting polymeric layer from this curing process exhibits microscale surface irregularities. A less expensive option is to modify a petri dish made of the acrylonitrile butadiene styrene (ABS) polymer that has already been used. In order to produce the necessary surface alteration in this instance, acetone is used as the etchant during a wet etching process. This approach has the benefit of being both quick and simple in terms of preparation [53].

In recent years, 3D printing has emerged as a potential production technique for nanogenerators. The combination of three-dimensional printing and nanogenerators may make it possible to create highly effective, individualized, and adaptable power sources for a range of uses. The benefits of 3D printing over traditional production techniques can be used to understand its role in the context of nanogenerators. Well first of all, 3D printing enables the highly accurate and precise fabrication of complicated shapes [54]. This is especially helpful for nanogenerators because their complicated designs are frequently needed to maximize their energy conversion efficiency. A triboelectric nanogenerator (TENG) with a porous structure, can be created *via* 3D printing to improve its output power compared to a TENG with a solid structure. Moreover, piezoelectric nanogenerators (PENGs) with intricate interdigitated electrode designs have been made *via* 3D printing and have shown an increase in output power compared to conventional flat electrode designs. Table 5.1 shows the recent developments in the field of TENGs, highlighting advancements in materials, structures, and fabrication techniques. In addition, with 3D printing it is possible to fabricate nanogenerators with various materials, such as polymers, metals, ceramics, and composites. As a result, the nanogenerator's characteristics, such as its mechanical flexibility, electrical conductivity, and dielectric constant, can be designed to meet particular application needs. For instance, scientists have created a PENG utilizing a piezoelectric polymer and 3D printing. This device demonstrated high output voltage and current density as well as great flexibility and durability [55]. Similar to this, a TENG was created utilizing 3D printing using a composite material made of polydimethylsiloxane (PDMS) and silver nanoparticles, which demonstrated outstanding mechanical properties as well as high output voltage and current density [56, 57].

Finally, 3D printing makes it possible to quickly prototype and iterate nanogenerators, which can cut down on both the cost and time of development. This is especially helpful in research and development, where the design and functionality of the nanogenerator may need to go through several iterations. For instance, scientists have created a TENG with a elastic and flexible structure using 3D printing, which was then iteratively adjusted to produce a high output power and efficiency. Last but not least, 3D printing makes it possible to combine several parts and functions into a single object, which can increase the adaptability and functionality of the nanogenerator. For instance, researchers have created a PENG with an integrated temperature sensor using 3D printing, demonstrating the capability to concurrently measure the temperature and harvest energy from changes in the ambient temperature [58]. Similar to this, a TENG with an integrated wireless communication module was created *via* 3D printing, demonstrating the capability to harness energy from human motion and

Table 5.1 Recent developments in the field of TENGs.

Materials used	3D printing technology	Structure	Working mode	Observation	Applicability demonstrated	Ref.
TENGs						
CNT/ZnO (0.2% by weight) reinforced PLA	FDM	Donut pattern	Vertical contact separation mode	With the reinforcement of CNT-ZnO in PLA, output voltage (V_{oc}) rose from 1 to 8.9 V	LEDs	[60]
Cellulose nanofiber (CNF)/MXene nanosheets (5% by weight)	Freeze-drying assisted 3D printing	Deep Trap hierarchical structure	Single electrode mode	Introduction of Mxene into CNF/TENGs increased charge density by 1.8% (900 mC/cm^3)	Multi directional non-Contact 3D sensor	[61]
MXene/Ecoflex nanocomposites		Pyramidal structure	Single electrode mode	Inorder to capture figure motions for human-machine interaction STTS provided 19.9 V (peak to peak voltage) and high sensitivity (0.088 $Vkpa^{-1}$)	Self-powered Toroidal Triboelectric Sensor (STTS) with smart glove	[62]

(*Continued*)

Table 5.1 Recent developments in the field of TENGs. (*Continued*)

Materials used	3D printing technology	Structure	Working mode	Observation	Applicability demonstrated	Ref.
Zirconium (MOF-525)/Ecoflex (Co-NPC Mxene) nanocomposites	3D printed mask template (process not mentioned)	Multilayered microstructure (ring, band etc. based on applicability)	Single electrode mode	Fabricated TENG delivered power density of 25.7 W/m^2 and sensitivity 149 $VKpa^{-1}$ as compared to previously reported TENGs	Sensing characteristics of human body movements	[63]
Vitrimeric elastomers and MWCNT (12.5 % by weight)	Salt assisted DIW printing	Tube shaped (hierarchical porous structure)	Single electrode mode	Fabricated TENG showed excellent electrical performance and was recyclable in different forms by light assisted fabrication	Electronic watch	[64]

(*Continued*)

Table 5.1 Recent developments in the field of TENGs. (*Continued*)

Materials used	3D printing technology	Structure	Working mode	Observation	Applicability demonstrated	Ref.
PTFE	FDM	Vortex like flexible self-recovery blades	Contact separation mode	Within 150 minutes, the synthetic TENG produced a Voc of 650 V and a power density of 4.4 W/m^2 with degradation efficiencies of 89, 92, and 94% for methylene blue, methyl orange, and malachite green, respectively	Electrochemical degradation systems	[65]
PTFE rollers	Resin based 3D printing	Bearing Structure	Free standing mode	A fabricated TENG was able to charge a capacitor to a voltage of 4V in 80s at 600 rpm and generate a peak output of 0.96 mW under an external load of 8 m	Self-powered and high-speed sensing for vehicle monitoring	[66]

(*Continued*)

Table 5.1 Recent developments in the field of TENGs. (*Continued*)

Materials used	3D printing technology	Structure	Working mode	Observation	Applicability demonstrated	Ref.
PLA and PTFE	FDM	Acrylic tube	Sliding mode	Developed wave driven TENG with ability to deliver power density of $1.05 W/m_2$ at 5hz frequency. It was strong and durable as well	Capacitor charging	[67]
PDMS and Elastomer EcoFlex 0030	DIW	Silicone elastomer shell	Contact separation mode	Power density of $15.59 mW/m^2$ at the resistance of 80MΩ was delivered by fabricated TENG	LED control device	[68]
PDMS, Al and PCL	FDM	Circular coil type	Contact separation mode	The fabricated TENG produced 0.45mW of power at the resistance of 100MΩ	Voice print recognition sensor	[69]

transfer data wirelessly [59]. The ability to create diverse geometries, flexible materials, quick prototyping, and integrated functionality make 3D printing an attractive fabrication technique for nanogenerators.

5.4.3 Applications of TENGs

TENG is applied across a wide spectrum of uses, encompassing four primary domains. Firstly, it serves as a dependable source of sustainable micropower, empowering self-powered systems. Secondly, TENG acts as active sensors in numerous fields such as medical applications, infrastructure monitoring, environmental assessment, and human-machine interfacing (HMI). Additionally, it plays a pivotal role as a fundamental network unit, facilitating the capture of low-frequency water energy to support the generation of large-scale blue energy. Lastly, TENG is employed as a direct power source, particularly valuable for high-voltage (HV) instruments [70]. These diverse applications highlight the versatility and significance of TENG technology in modern innovation and sustainability efforts.

5.4.3.1 TENGs as a Source of Sustainable Micropower

Applications for TENG technology as micro-power generators in self-sustaining systems are numerous. These applications involve the collection of biomechanical or ambient energy sources, such as heartbeats and walking motions in people, wind power and machine vibrations. The efficient performance of TENGs at low frequencies has led to significant research and development efforts in biomechanical energy harvesting from the early stages of TENG technology. Notable applications include smart backpacks [71], self-powered watches [72], and acoustic energy harvesters [73]. Shape-adaptive TENGs for flexible, conformal power sources and fiber-based TENGs for wearable energy sources are two separate design methods that represent recent developments in this sector. It is crucial to emphasize how tightly related these two approaches are, how they work best together rather than against one another. The use of elastic and flexible materials allows for the smooth conformation of devices to the curves of the human body, improving both comfort and efficiency. A team of researchers created a TENG that is shaped like a tube and is made of silicone-based materials. The implantation of this novel device under shoes or garments enables the collection of electrical energy from human walking. Wearable electronics such as fitness trackers and watches can be powered by this gathered energy [74].

5.4.3.2 *TENG as an Active and Self-Powered Sensor*

Without the use of extra transducers, TENG has the unique capacity to convert mechanical impulses directly into electrical signals. With regard to active sensing and self-powered sensors, which require little to no standby power consumption and less complex control circuitry than conventional passive transistors, this characteristic places TENG in a position to have tremendous potential. Research in this domain encompasses various applications such as tactile touch sensors [75], acoustic sensors [73], motion and acceleration sensors [76], as well as chemical sensors [76]. Simultaneously, in the context of the burgeoning IoTs, the development of advanced Human-Machine Interfaces (HMI) has assumed increasing importance to facilitate convenient, secure, and innovative interactions between humans and external devices. With the growing emphasis on cybersecurity, there is a heightened need for reliable and ongoing authentication methods. The advantage of using keystroke dynamics, a behavioural biometric approach based on people's typing patterns, is that it can increase security without interfering with users' usual activity. Chen *et al.* first proposed the idea of keystroke dynamics implementation by TENG-based sensing arrays in 2015. Their device's single-electrode contact separation mode, however, presented difficulties because of outside interferences [77]. However, their device operated based on a single-electrode contact separation mode, which posed challenges due to environmental interferences. Building upon this and drawing inspiration from earlier advancements in triboelectric keyboards, researchers [78] successfully created a two-factor security system that leverages pressure-enhanced keystroke dynamics. This innovative system not only authenticates users but also has the capability to identify them based on their distinct typing patterns. Remarkably, it achieves an impressive accuracy rate of up to 98.7%.

5.4.3.3 *TENG as a Blue Energy Harvester*

The utilization of TENGs for capturing naturally occurring mechanical energy sources, including wind energy, raindrop energy, water wave and ultrasonic energy, has garnered considerable attention [79]. Within this realm, a noteworthy focus has been on harnessing the kinetic energy of water waves in expansive ocean environments, often referred to as "blue energy." This emphasis on blue energy is especially significant due to recent findings indicating that TENGs outperform EMGs (Electromagnetic Generators) in efficiently capturing low frequency vibration energy [80]. A variety of TENG designs, including arc-shaped TENG walls enclosing a rolling ball, enclosed rolling spherical structures, and vertical plates with parallel electrodes, have been proposed for harvesting blue energy. Enclosed rolling spherical structures are more advantageous since they can capture energy from all angles [81]. Sometimes spring-based structures are integrated into TENGs to increase their efficiency in low frequency energy gathering. Impact energy can be quickly converted into elastic potential energy for additional operating cycles [82].

5.4.3.4 TENG as a Direct High Voltage Power Source

Due to their unique characteristics of high voltage and low current, coupled with exceptional portability and safety features, triboelectric nanogenerators (TENGs) present a promising alternative to traditional high voltage power sources. TENGs typically generate voltages ranging from 1 to 10 kV, significantly reducing the necessity for expensive and complex power converters. This simplification not only cuts down on system complexity but also lowers overall costs. In high voltage applications, it is crucial to minimize current demands due to TENGs' limited charge transfer capability per operational cycle. By optimizing current requirements, TENGs can achieve performance levels comparable to or even surpassing those of conventional power sources. Moreover, the reduced current needs enhance safety for both personnel and equipment. Once the limited charge transfer process concludes, the associated risks of high voltage are mitigated. An intriguing application of TENG technology lies in nanoelectrospray ionization for highly sensitive nano-coulomb molecular mass spectrometry. This innovative approach harnesses the high-voltage output of TENGs while skillfully managing the constraint of limited charge transfer, enabling precise control over ion formation. This capability opens new avenues in analytical chemistry and molecular research, where sensitivity and accuracy are paramount. TENGs represent a versatile and efficient solution for high voltage requirements, offering benefits in terms of cost-effectiveness, simplicity, and safety. As research progresses and technology advances, the potential applications of TENGs in diverse fields, including analytical instrumentation and portable electronics, continue to expand, promising further advancements in both performance and functionality [83]. TENGs' high voltage output can be used as an electrostatic actuator's power source. A TENG-powered actuation system using dielectric elastomer actuators was shown by Chen *et al.* This technology could change the spacing between tuneable optical gratings or regulate the on/off status of an intelligent switch [84].

In summary, Triboelectric nanogenerators (TENGs) are poised to exert significant influence across a spectrum of diverse disciplines. TENG technology stands out as a micro/nanopower source ideally suited for small and mobile electronic devices. Its exceptional performance, coupled with the perpetual availability of mechanical vibrations, enables seamless integration with existing sensor systems. This integration facilitates the development of self-powered networks capable of real-time, *in-situ* sensing, offering extensive applications, including environmental monitoring, forest fire prevention, demographic mapping, and wildlife tracking. TENGs demonstrate considerable promise as active sensors, proficiently detecting mechanical motion and disturbances. This functionality extends their utility to fields such as artificial intelligence, Human-Machine Interfaces (HMI), robotics, and security. The advent of self-powered sensors signifies a transformative leap akin to the evolution from wired to wireless communication, ushering in enhanced capabilities. The potential exists to construct large-scale energy harvesting systems by integrating numerous TENG units into a cohesive network and optimizing their performance through various materials and designs. These systems have the capacity to harness the energy inherent in water waves and wind, commonly denoted as "blue energy." TENGs may serve as high-voltage sources, catering specifically to high-voltage applications such as plasma excitation and field emission. The multifaceted nature of TENG technology underlines its potential to revolutionize diverse domains, catalyzing innovative advancements across the board.

5.5 Summary

The chapter summarizes about the significance of electroactive polymers in energy harvesting application. It describes the undermining piezoelectric and triboelectric polymers mechanism and various methods to tailor the properties by compositing. This chapter explores the EAP's energy harvesting using piezoelectric and triboelectric polymers. It underlines their potential in advancing the development of sustainable energy solutions and the integration of energy harvesting technologies into various sectors.

Acknowledgments

The authors want to acknowledge the financial support from Faculty Research Seed Grant (2022), Indian Institute of Technology Bombay (IITB) Powai, Mumbai, Maharashtra, India.

AI disclosure statement: The author acknowledges the use of AI-based tools for minor language editing or grammatical correction. However, all intellectual contributions, analyses, and conclusions presented in this chapter are entirely the author's own.

References

1. Assembly, G., Resolution adopted by the General Assembly on 6 July 2017, Technical Report A/RES/71/313, 2017.
2. Sripadmanabhan Indira, S., Aravind Vaithilingam, C., Oruganti, K.S.P., Mohd, F., Rahman, S., Nanogenerators as a sustainable power source: state of art, applications, and challenges. *Nanomaterials*, 9, 5, 773, 2019.
3. Chen, B. and Wang, Z.L., Toward a New Era of Sustainable Energy: Advanced Triboelectric Nanogenerator for Harvesting High Entropy Energy. *Small*, 18, 43, 2107034, 2022.
4. Kim, W.-G., Kim, D.-W., Tcho, I.-W., Kim, J.-K., Kim, M.-S., Choi, Y.-K., Triboelectric nanogenerator: Structure, mechanism, and applications. *ACS Nano*, 15, 1, 258–287, 2021.
5. Harrison, J., Ounaies, Z., *Piezoelectric Polymers*. NASA Langley Research Center, Institute for Computer Applications in Science, 2001.
6. Lohmann, R., *et al.*, Are Fluoropolymers Really of Low Concern for Human and Environmental Health and Separate from Other PFAS? *Environ. Sci. Technol.*, 54, 20, 12820–12828, Oct. 2020, doi: 10.1021/acs.est.0c03244.
7. Ford, T.A. and Edward, H.W., Polyvinylidene fluoride and process for obtaining the same, US2435537A, Feb. 03, 1948, Accessed: Oct. 02, 2023. [Online]. Available: https://patents.google.com/patent/US2435537A/en.
8. Ebnesajjad, S. (Ed.), 6 - Introduction to Fluoropolymers*, in: *Introduction to Fluoropolymers*, pp. 63–89, William Andrew Publishing, Oxford, 2013, doi: 10.1016/B978-1-4557-7442-5.00006-1.
9. Ebnesajjad, S., 3 - Introduction to Fluoropolymers, in: *Applied Plastics Engineering Handbook (Second Edition)*, M. Kutz (Ed.), pp. 55–71, Plastics Design Library, William Andrew Publishing, 2017, doi: 10.1016/B978-0-323-39040-8.00003-1.
10. Liu, F. and Grainger, D.W., C - Fluorinated Biomaterials, in: *Biomaterials Science (Third Edition)*, B.D. Ratner, A.S. Hoffman, F.J. Schoen, J.E. Lemons (Eds.), pp. 92–103, Academic Press, 2013, doi: 10.1016/B978-0-08-087780-8.00011-5.

11. Kawai, H., The Piezoelectricity of Poly (vinylidene Fluoride). *Jpn. J. Appl. Phys.*, 8, 7, 975, Jul. 1969, doi: 10.1143/JJAP.8.975.
12. Marutake, M., The days when piezoelectric PVDF was discovered. *Ferroelectrics*, 171, 1, 5–6, Sep. 1995, doi: 10.1080/00150199508018416.
13. Oshiki, M. and Fukada, E., Piezoelectric Effect in Stretched and Polarized Polyvinylidene Fluoride Film. *Jpn. J. Appl. Phys.*, 15, 1, 43, Jan. 1976, doi: 10.1143/JJAP.15.43.
14. Tasaka, S. and Miyata, S., The origin of piezoelectricity in poly(vinylidene fluoride). *Ferroelectrics*, 32, 1, 17–23, Jan. 1981, doi: 10.1080/00150198108238668.
15. Nishiyama, T., Sumihara, T., Sasaki, Y., Sato, E., Yamato, M., Horibe, H., Crystalline structure control of poly(vinylidene fluoride) films with the antisolvent addition method. *Polym. J.*, 48, 10, 1035–1038, Oct. 2016, doi: 10.1038/pj.2016.62.
16. Jayakumar, O.D., *et al.*, Fabrication of flexible and self-standing inorganic-organic three phase magneto-dielectric PVDF based multiferroic nanocomposite films through a small loading of graphene oxide (GO) and Fe_3O_4 nanoparticles. *Dalton Trans.*, 44, 36, 15872–15881, 2015.
17. Cai, X., Lei, T., Sun, D., Lin, L., A critical analysis of the α, β and γ phases in poly(vinylidene fluoride) using FTIR. *RSC Adv.*, 7, 25, 15382–15389, 2017, doi: 10.1039/C7RA01267E.
18. El Mohajir, B.-E. and Heymans, N., Changes in structural and mechanical behaviour of PVDF with processing and thermomechanical treatments. 1. Change in structure. *Polymer*, 42, 13, 5661–5667, Jun. 2001, doi: 10.1016/S0032-3861(01)00064-7.
19. Thakur, P., *et al.*, Improving the thermal stability, electroactive β phase crystallization and dielectric constant of NiO nanoparticle/C–NiO nanocomposite embedded flexible poly (vinylidene fluoride) thin films. *RSC Adv.*, 6, 31, 26288–26299, 2016.
20. Badatya, S., Bharti, D.K., Srivastava, A.K., Gupta, M.K., Solution processed high performance piezoelectric eggshell membrane – PVDF layer composite nanogenerator *via* tuning the interfacial polarization. *J. Alloys Compd.*, 863, 158406, May 2021, doi: 10.1016/j.jallcom.2020.158406.
21. Lovinger, A.J., Ferroelectric Polymers. *Science*, 220, 4602, 1115–1121, Jun. 1983, doi: 10.1126/science.220.4602.1115.
22. Zhu, L. and Wang, Q., Novel Ferroelectric Polymers for High Energy Density and Low Loss Dielectrics. *Macromolecules*, 45, 7, 2937–2954, Apr. 2012, doi: 10.1021/ma2024057.
23. Karan, S.K., *et al.*, Effect of γ-PVDF on enhanced thermal conductivity and dielectric property of Fe-rGO incorporated PVDF based flexible nanocomposite film for efficient thermal management and energy storage applications. *RSC Adv.*, 6, 44, 37773–37783, 2016.
24. Costa, C.M., *et al.*, Smart and Multifunctional Materials Based on Electroactive Poly(vinylidene fluoride): Recent Advances and Opportunities

in Sensors, Actuators, Energy, Environmental, and Biomedical Applications. *Chem. Rev.*, 123, 19, 11392–11487, 2023.
25. Yang, L., Li, X., Allahyarov, E., Taylor, P.L., Zhang, Q.M., Zhu, L., Novel polymer ferroelectric behavior *via* crystal isomorphism and the nanoconfinement effect. *Polymer*, 54, 7, 1709–1728, 2013.
26. Li, Z., Wang, Y., Cheng, Z.-Y., Electromechanical properties of poly (vinylidene-fluoride-chlorotrifluoroethylene) copolymer. *Appl. Phys. Lett.*, 88, 6, 062904, 2006.
27. Hattori, T., Takahashi, Y., Iijima, M., Fukada, E., Piezoelectric and ferroelectric properties of polyurea-5 thin films prepared by vapor deposition polymerization. *J. Appl. Phys.*, 79, 3, 1713–1721, 1996.
28. Meyer, R.B., Liebert, L., Strzelecki, L., Keller, P., Ferroelectric liquid crystals. *J. Phys. Lett.*, 36, 3, 69–71, 1975.
29. Nalwa, H.S., *Ferroelectric polymers: chemistry: physics, and applications*, CRC Press, 1995.
30. Dong, W., *et al.*, Dramatic improvement in toughness of PLLA/PVDF blends: the effect of compatibilizer architectures. *ACS Sustain. Chem. Eng.*, 4, 8, 4480–4489, 2016.
31. Vijaya, M.S., *Piezoelectric Materials and Devices*, CRC Press, 2016, doi: 10.1201/b12709.
32. Mooti, A., *et al.*, Magnetic and high-dielectric-constant nanoparticle polymer tri-composites for sensor applications. *J. Mater. Sci.*, 55, 16234–16246, 2020.
33. Stadlober, B., Zirkl, M., Irimia-Vladu, M., Route towards sustainable smart sensors: ferroelectric polyvinylidene fluoride-based materials and their integration in flexible electronics. *Chem. Soc. Rev.*, 48, 6, 1787–1825, Mar. 2019, doi: 10.1039/C8CS00928G.
34. Mokhtari, F., Cheng, Z., Raad, R., Xi, J., Foroughi, J., Piezofibers to smart textiles: a review on recent advances and future outlook for wearable technology. *J. Mater. Chem. A*, 8, 19, 9496–9522, May 2020, doi: 10.1039/D0TA00227E.
35. Wan, C. and Bowen, C.R., Multiscale-structuring of polyvinylidene fluoride for energy harvesting: the impact of molecular-, micro- and macro-structure. *J. Mater. Chem. A*, 5, 7, 3091–3128, Feb. 2017, doi: 10.1039/C6TA09590A.
36. Global Polyvinylidene Fluoride Market Size, PVDF Industry Report 2025, Accessed: Jun. 01, 2022. [Online]. Available: https://www.grandviewresearch.com/industry-analysis/polyvinylidene-fluoride-pvdf-market.
37. Holding, S., Technology of Fluoropolymers. *Chromatographia*, 71, 1, 173–173, Jan. 2010, doi: 10.1365/s10337-009-1422-3.
38. Fukada, E., Recent developments of polar piezoelectric polymers. *IEEE Trans. Dielectr. Electr. Insul.*, 13, 5, 1110–1119, 2006.
39. Chandran, A.M., Varun, S., Mural, P.K.S., Fabrication of novel nanogenerator from PVDF nanocomposites encompassing hybrid silanized MWCNTs. *AIP Conf. Proc.*, 2289, 1, 020016, AIP Publishing LLC, 2020.

40. Chandran, A.M., Varun, S., Mural, P.K.S., Flexible electroactive PVDF/ZnO nanocomposite with high output power and current density. *Polym. Eng. Sci.*, 61, 6, 1829–1841, 2021, doi: 10.1002/pen.25704.
41. A direct-write piezoelectric PVDF nanogenerator, Accessed: Oct. 04, 2023. [Online]. Available: https://ieeexplore.ieee.org/abstract/document/5285796/.
42. Hansen, B.J., Liu, Y., Yang, R., Wang, Z.L., Hybrid Nanogenerator for Concurrently Harvesting Biomechanical and Biochemical Energy. *ACS Nano*, 4, 7, 3647–3652, Jul. 2010, doi: 10.1021/nn100845b.
43. Lee, M., *et al.*, A Hybrid Piezoelectric Structure for Wearable Nanogenerators. *Adv. Mater.*, 24, 13, 1759, 2012, doi: https://doi.org/10.1002/adma.201200150.
44. Cha, S., *et al.*, Porous PVDF As Effective Sonic Wave Driven Nanogenerators. *Nano Lett.*, 11, 12, 5142–5147, Dec. 2011, doi: 10.1021/nl202208n.
45. Vivekananthan, V., Chandrasekhar, A., Rao Alluri, N., Purusothaman, Y., Khandelwal, G., Kim, S.-J., Triboelectric Nanogenerators: Design, Fabrication, Energy Harvesting, and Portable-Wearable Applications, in: *Nanogenerators*, S. Jae Kim, A. Chandrasekhar, N. Rao Alluri (Eds.), IntechOpen, 2020, doi: 10.5772/intechopen.90951.
46. Bhat, A.R., Kumar, R., Mural, P.K.S., Natural Fiber Reinforced Polymer Composites: A Comprehensive Review of Tribo-Mechanical Properties. *Tribol. Int.*, 189, 108978, 2023.
47. Bhat, A.R. and Saxena, P., Natural fiber–reinforced polymer nanocomposites, in: *Nanomaterials for Sustainable Tribology*, pp. 115–141, CRC Press, 2023.
48. Wang, Z.L., From contact electrification to triboelectric nanogenerators. *Rep. Prog. Phys.*, 84, 9, 96502, 2021.
49. Chen, B., Tang, W., Wang, Z.L., Advanced 3D printing-based triboelectric nanogenerator for mechanical energy harvesting and self-powered sensing. *Mater. Today*, 50, 224–238, 2021.
50. Fan, F.-R., Lin, L., Zhu, G., Wu, W., Zhang, R., Wang, Z.L., Transparent triboelectric nanogenerators and self-powered pressure sensors based on micropatterned plastic films. *Nano Lett.*, 12, 6, 3109–3114, 2012.
51. Chen, B., *et al.*, Three-dimensional ultraflexible triboelectric nanogenerator made by 3D printing. *Nano Energy*, 45, 380–389, 2018.
52. Wang, Z.L., Triboelectric nanogenerators as new energy technology for self-powered systems and as active mechanical and chemical sensors. *ACS Nano*, 7, 11, 9533–9557, 2013.
53. Triboelectric Nanogenerators: Design, Fabrication, Energy Harvesting, and PortableWearable Applications.
54. Bhat, A.R., Gupta, V., Bankapalli, N.K., Saxena, P., Raina, A., Haq, M.I.U., 3D Printing and New Product Development: Opportunities and Challenges, in: *3D Print. Sustain. Prod. Dev*, pp. 1–20, 2024.
55. Chang, S.-M., *et al.*, Optimization of piezoelectric polymer composites and 3D printing parameters for flexible tactile sensors. *Addit. Manuf.*, 67, 103470, 2023.

56. Zheng, R., Chen, Y., Chi, H., Qiu, H., Xue, H., Bai, H., 3D printing of a polydimethylsiloxane/polytetrafluoroethylene composite elastomer and its application in a triboelectric nanogenerator. *ACS Appl. Mater. Interfaces*, 12, 51, 57441–57449, 2020.
57. Malik, A., Jammoria, N.S., Bhat, A.R., Saxena, P., Haq, M.I.U., Raina, A., Nanocomposites and tribology, in: *Nanomater. Sustain. Tribol*, p. 25, 2023.
58. Yan, A., Yuan, X., Li, Z., Yang, J., Ren, K., Dong, S., 3D-printed flexible, multilayered ceramic-polymer composite grid with integrated structural-self-sensing function. *Sens. Actuators Phys.*, 332, 113187, 2021.
59. Yang, P., *et al.*, Monitoring the degree of comfort of shoes in-motion using triboelectric pressure sensors with an ultrawide detection range. *ACS Nano*, 16, 3, 4654–4665, 2022.
60. Mohapatra, A., Divakaran, N., Alex, Y., Ajay Kumar, P.V., Mohanty, S., The significant role of CNT-ZnO core-shell nanostructures in the development of FDM-based 3D-printed triboelectric nanogenerators. *Mater. Today Nano*, 22, 100313, 2023, doi: 10.1016/j.mtnano.2023.100313.
61. Wang, B., *et al.*, 3D printing deep-trap hierarchical architecture-based non-contact sensor for multi-direction motion monitoring. *Nano Energy*, 107, October 2022, 108135, 2023, doi: 10.1016/j.nanoen.2022.108135.
62. Zhang, S., *et al.*, 3D printed smart glove with pyramidal MXene/Ecoflex composite-based toroidal triboelectric nanogenerators for wearable human-machine interaction applications. *Nano Energy*, 106, October 2022, 108110, 2023, doi: 10.1016/j.nanoen.2022.108110.
63. Rana, S.M.S., *et al.*, Zirconium metal-organic framework and hybridized Co-NPC@MXene nanocomposite-coated fabric for stretchable, humidity-resistant triboelectric nanogenerators and self-powered tactile sensors. *Nano Energy*, 104, PA, 107931, 2022, doi: 10.1016/j.nanoen.2022.107931.
64. Chen, S., *et al.*, Bio-based vitrimeric elastomers enable completely sustainable triboelectric nanogenerator. *Sci. China Mater.*, 66, 3, 1089–1096, 2022, doi: 10.1007/s40843-022-2204-8.
65. Liu, S., Liu, Y., Chen, Y., Wang, S., Men, C., Gao, S., Novel 3D Printed Vortex-like Flexible Roller-Compacted Triboelectric Nanogenerator for Self-Powered Electrochemical Degradation of Organic Contaminants. *ACS Appl. Mater. Interfaces*, 14, 15, 17426–17433, 2022, doi: 10.1021/acsami.2c01750.
66. Yang, J., Sun, Y., Zhang, J., Chen, B., Wang, Z.L., 3D-printed bearing structural triboelectric nanogenerator for intelligent vehicle monitoring. *Cell Rep. Phys. Sci.*, 2, 12, 0–14, 2021, doi: 10.1016/j.xcrp.2021.100666.
67. Wang, Y., Pham, A.T.T., Han, X., Du, D., Tang, Y., Design and evaluate the wave driven- triboelectric nanogenerator under external wave parameters: Experiment and simulation. *Nano Energy*, 93, December 2021, 106844, 2022, doi: 10.1016/j.nanoen.2021.106844.
68. Wang, Z., *et al.*, Integrated and shape-adaptable multifunctional flexible triboelectric nanogenerators using coaxial direct ink writing 3D printing. *Nano Energy*, 90, PA, 106534, 2021, doi: 10.1016/j.nanoen.2021.106534.

69. Guo, R., *et al.*, A Voiceprint Recognition Sensor Based on a Fully 3D-Printed Triboelectric Nanogenerator *via* a One-Step Molding Route. *Adv. Eng. Mater.*, 22, 5, 1–5, 2020, doi: 10.1002/adem.201901560.
70. Wu, C., Wang, A.C., Ding, W., Guo, H., Wang, Z.L., Triboelectric Nanogenerator: A Foundation of the Energy for the New Era. *Adv. Energy Mater.*, 9, 1, 1802906, Jan. 2019, doi: 10.1002/aenm.201802906.
71. Yang, W., *et al.*, Harvesting energy from the natural vibration of human walking. *ACS Nano*, 7, 12, 11317–11324, 2013.
72. Quan, T., Wang, X., Wang, Z.L., Yang, Y., Hybridized electromagnetic-triboelectric nanogenerator for a self-powered electronic watch. *ACS Nano*, 9, 12, 12301–12310, 2015.
73. Yang, J., Chen, J., Liu, Y., Yang, W., Su, Y., Wang, Z.L., Triboelectrification-based organic film nanogenerator for acoustic energy harvesting and self-powered active acoustic sensing. *ACS Nano*, 8, 3, 2649–2657, 2014.
74. Wang, J., *et al.*, Sustainably powering wearable electronics solely by biomechanical energy. *Nat. Commun.*, 7, 1, 12744, 2016.
75. Lin, L., *et al.*, Triboelectric active sensor array for self-powered static and dynamic pressure detection and tactile imaging. *ACS Nano*, 7, 9, 8266–8274, 2013.
76. Zhou, Y.S., *et al.*, Nanometer resolution self-powered static and dynamic motion sensor based on micro-grated triboelectrification. *Adv. Mater.*, 26, 11, 1719–1724, 2014.
77. Chen, J., *et al.*, Personalized keystroke dynamics for self-powered human-machine interfacing. *ACS Nano*, 9, 1, 105–116, 2015.
78. Wu, C., *et al.*, Keystroke dynamics enabled authentication and identification using triboelectric nanogenerator array. *Mater. Today*, 21, 3, 216–222, 2018.
79. Yang, Y., Zhang, H., Liu, R., Wen, X., Hou, T.-C., Wang, Z.L., Fully enclosed triboelectric nanogenerators for applications in water and harsh environments. *Adv. Energy Mater.*, 3, 12, 1563–1568, 2013.
80. Zi, Y., Guo, H., Wen, Z., Yeh, M.H., Hu, C., Wang, Z.L., Harvesting low-frequency (< 5 Hz) irregular mechanical energy: a possible killer application of triboelectric nanogenerator. *ACS Nano*, 10, 4797–4805, 2016.
81. Wang, X., Niu, S., Yin, Y., Yi, F., You, Z., Wang, Z.L., Triboelectric nanogenerator based on fully enclosed rolling spherical structure for harvesting low-frequency water wave energy. *Adv. Energy Mater.*, 5, 24, 1501467, 2015.
82. Jiang, T., Yao, Y., Xu, L., Zhang, L., Xiao, T., Wang, Z.L., Spring-assisted triboelectric nanogenerator for efficiently harvesting water wave energy. *Nano Energy*, 31, 560–567, 2017.
83. Li, A., Zi, Y., Guo, H., Wang, Z.L., Fernández, F.M., Triboelectric nanogenerators for sensitive nano-coulomb molecular mass spectrometry. *Nat. Nanotechnol.*, 12, 5, 481–487, 2017.
84. Chen, X., Jiang, T., Yao, Y., Xu, L., Zhao, Z., Wang, Z.L., Stimulating acrylic elastomers by a triboelectric nanogenerator–toward self-powered electronic skin and artificial muscle. *Adv. Funct. Mater.*, 26, 27, 4906–4913, 2016.

6

Advancing Piezoelectric and Triboelectric Energy Harvesting Using Piezoelectric Polymer Composites

Sayyid Abdul Basith[1,2], Arunkumar Chandrasekhar[2]* and A. Arockiarajan[1,3]

[1]*Department of Applied Mechanics and Biomedical Engineering, Indian Institute of Technology Madras (IIT Madras), Chennai, Tamil Nadu, India*
[2]*Nanosensors and Nanoenergy Lab, Department of Sensors and Biomedical Technology, School of Electronics Engineering, Vellore Institute of Technology, Vellore, Tamilnadu, India*
[3]*Centre of Excellence in Ceramics Technologies for Futuristic Mobility, Indian Institute of Technology Madras (IIT Madras), Chennai, Tamil Nadu, India*

Abstract

Piezoelectric polymer composites are pivotal due to their ability to transform mechanical energy into electrical energy, making them essential for triboelectric and piezoelectric energy harvesting. Their cost-effectiveness and scalability support widespread manufacturing, while their flexible, lightweight, and versatile characteristics allow for easy incorporation into diverse devices and systems, broadening the scope of possible uses. This chapter explores the fundamental properties, advantages, and limitations of piezoelectric polymers, their mechanisms of energy harvesting, and their potential applications in piezoelectric, triboelectric, and hybrid energy-harvesting technologies. Combining the piezoelectric and triboelectric mechanisms into hybrid systems allows these composites to harvest energy from multiple sources simultaneously. It maximizes overall energy conversion capabilities and opens up new possibilities for self-powered systems in various fields.

Keywords: Polymer composites, piezoelectric, triboelectric, hybrid nanogenerators, nanosensors, energy harvesting

**Corresponding author*: arunkumar.c@vit.ac.in

Raneesh Balakrishnan, Sobi K. Chacko and Visakh P.M. (eds.) Electroactive Polymer-Based Smart Materials and Applications, (159–184) © 2026 Scrivener Publishing LLC

6.1 Introduction

Researchers and engineers have turned their attention to energy harvesting in the quest for sustainable and efficient energy sources [1]. Energy harvesting technologies present a viable option for capturing ambient energy from a variety of sources, including mechanical vibrations, human activity, and environmental forces, and transforming it into practical electrical power. Within the wide range of materials and systems utilized for energy harvesting, piezoelectric polymer composites have proven to be a valuable choice, distinguished by their distinctive properties and wide-ranging potential uses [2]. This section delves into the uses of piezoelectric polymer composites for both triboelectric and piezoelectric energy harvesting. These polymers possess the impressive capability to transform mechanical energy into electrical energy and back, rendering them highly effective for energy harvesting purposes. By integrating these piezoelectric materials into composites, their functionality and performance can be further enhanced, opening up exciting possibilities for capturing energy from diverse sources.

This opening section provides an outline of the chapter and examines the importance of piezoelectric polymer composites in energy harvesting applications. The first section of the chapter explores the properties of piezoelectric polymers, delves into the principles of piezoelectricity, and the phenomenon by which these materials generate an electric charge under mechanical stress. Furthermore, the discussion includes different kinds of piezoelectric polymers, such as polyvinylidene fluoride (PVDF), polyvinylidene fluoride-trifluoroethylene (PVDF-TrFE), and polyvinylidene fluoride-tetrafluoroethylene (PVDF-TFE). Grasping the distinctive properties and behaviors of these materials is essential for enhancing their utilization in energy harvesting applications. The second section is devoted to the use of piezoelectricity in energy harvesting and discusses the fundamental concepts of how mechanical vibrations or strains are transformed into electrical energy. It examines the operational mechanisms of piezoelectric energy harvesters, including the electromechanical coupling process and the efficiency of energy conversion. This section also covers the potential uses of piezoelectric energy harvesting in wearable technology, wireless sensors, and systems for monitoring structural health and reviews the benefits and drawbacks of these technologies.

The following section delves into triboelectric energy harvesting utilizing electroactive polymer-based composites. This process generates electricity through friction between materials possessing different triboelectric properties. The discussion includes the enhancement of energy-harvesting performance through the incorporation of piezoelectric polymers into triboelectric composites. It outlines the operational principles of triboelectric energy harvesters made from electroactive polymer composites and highlights potential applications such as self-powered sensors, energy-autonomous devices, and interfaces for human-machine interaction. By exploiting ambient energy with these advanced materials, pathways to more sustainable and autonomous energy solutions are opened.

6.2 Piezoelectric Polymers and Their Properties

Piezoelectric polymers have attracted considerable interest in the field of energy harvesting due to their impressive ability to transform mechanical energy into electrical energy. These materials are crucial in energy conversion systems, which capture ambient mechanical energy and convert it to electrical power. This distinctive feature renders piezoelectric polymers extremely useful for a range of applications, especially in energy harvesting. The conversion of mechanical energy to electrical energy in these polymers is based on the piezoelectric effect. When these polymers are subjected to mechanical stress or strain, they produce electric charges, which can then be harvested and used as a source of electrical power. Conversely, when an electric field is applied, the material can experience mechanical deformation, serving as an actuator or sensor. This two-way capability of energy conversion enhances the versatility of piezoelectric polymers in various technological applications [3].

A major benefit of piezoelectric polymers is their flexibility, allowing them to adapt to various shapes and be incorporated into a wide range of devices and structures [4]. Contrary to conventional stiff piezoelectric materials like ceramics, piezoelectric polymers can be produced as thin films or flexible sheets. This versatility broadens their applicability to include wearable technology, smart textiles, and devices with curved surfaces. This flexibility extends their usability in energy harvesting applications, as they can be seamlessly integrated into different systems without compromising their performance [5]. Another significant attribute of piezoelectric polymers is their light weight. This quality is especially valuable in contexts where reducing weight is important, like in portable electronics or devices for wearable energy harvesting [6]. The low mass of these polymers allows for efficient energy conversion without adding excessive burden to the overall system. Additionally, their lightweight nature contributes to improved mechanical compliance, making them more responsive to even subtle mechanical vibrations or strains [7].

Piezoelectric polymers also exhibit excellent compatibility with various substrates, enabling their integration into various materials and systems [8]. They can be easily deposited onto different surfaces, including metals, ceramics, or flexible substrates, without significantly affecting their piezoelectric properties. This compatibility expands the potential applications of piezoelectric polymers, enabling their use in diverse fields, such as structural health monitoring [9], biomedical devices [10], and energy harvesting [11] from unconventional sources.

6.2.1 Principles of Piezoelectricity

Piezoelectricity is a fundamental phenomenon that underlies the unique behavior of piezoelectric materials. Understanding the principles of piezoelectricity is essential for comprehending how these materials convert mechanical energy into electrical energy and vice versa. Piezoelectricity is based on the principle that certain materials exhibit spontaneous electric polarization when subjected to mechanical stress or strain [12]. This phenomenon arises from the asymmetrical arrangement of atoms or molecules within the crystal lattice of the materials. When mechanical stress is applied to a piezoelectric material, it causes a deformation or displacement of the crystal structure, redistributing charges and generating an electric field [13]. This direct piezoelectric effect generates electric charges in response to applied mechanical stress or strain. When a piezoelectric material is subjected to a mechanical force, such as compression or tension, the lattice structure of the material undergoes a deformation [14]. This deformation separates and redistributes positive and negative charges within the material, creating an electric potential difference or voltage across its surfaces. This generated electric field is proportional to the applied mechanical stress or strain and can be measured as an output voltage [15].

Conversely, the converse piezoelectric effect describes the phenomenon in which a piezoelectric material deforms or experiences mechanical strain in response to an applied electric field. When an electric field is applied to a piezoelectric material, the charges within the material experience a force that causes a rearrangement of the crystal structure. This rearrangement results in a mechanical deformation or strain, such as elongation or contraction, proportional to the applied electric field [20]. This property allows piezoelectric materials to act as actuators, converting electrical energy into mechanical motion. Figure 6.1 illustrates the concept of piezoelectricity and the relationship between strain, stress, and electric field in piezoelectric materials. The direct and converse piezoelectric effects are

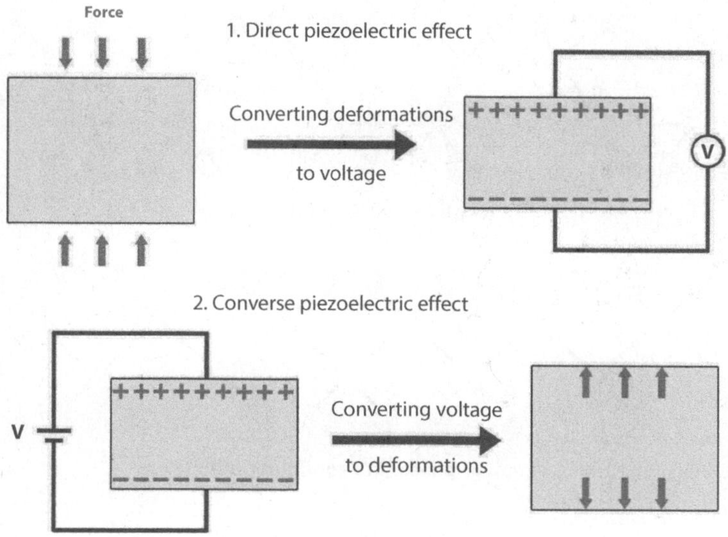

Figure 6.1 Mechanism of direct and converse piezoelectric effects.

interconnected and governed by the piezoelectric coefficients of the material [16]. Piezoelectric coefficients are a measure of how much a material responds to an electric field or applied stress. These coefficients are denoted as d for the direct effect and g for the converse effect. They quantify the relationship between strain or displacement (direct effect) and electric field (converse effect). One common representation is the d33 coefficient, corresponding to the induced charge per unit stress along a specific direction [17, 18]. Another representation is the strain versus electric field graph, demonstrating the linear relationship between these parameters [19].

6.2.2 Commonly Used Piezoelectric Polymers

PVDF is one of the most commonly used piezoelectric polymers, renowned for its excellent piezoelectric properties and suitability for various applications [21]. This section explores PVDF in detail, including its molecular structure, crystalline phases, piezoelectric properties, and the impact of processing techniques on its performance. PVDF has a molecular structure composed of repeating vinylidene fluoride units (CH_2CF_2). This polymer chain is characterized by alternating carbon and fluorine atoms. The arrangement of these atoms gives them a polar nature, contributing to their piezoelectric properties. PVDF exhibits different crystalline phases,

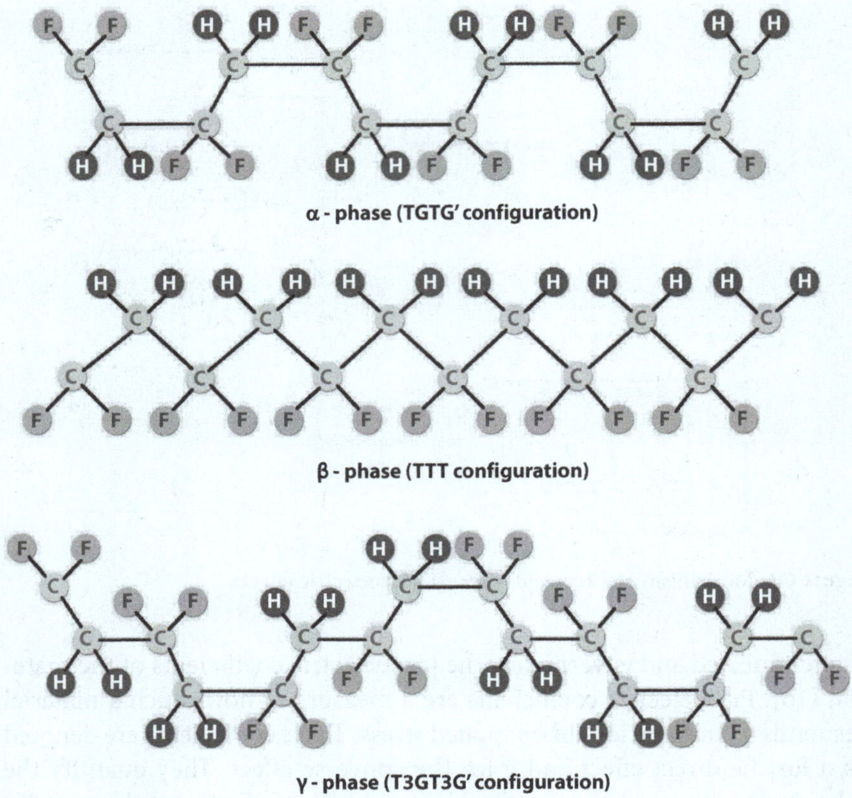

Figure 6.2 α, β, and γ phases of PVDF.

including the α, β, and γ phases, as shown in Figure 6.2, each with distinct properties and piezoelectric behavior [22]. The α phase is non-polar, exhibiting a disordered arrangement of the polymer chains. In this phase, it does not possess any significant piezoelectric response.

On the other hand, the β phase is the most desirable for piezoelectric applications. It is a polar crystalline phase with a highly ordered arrangement of polymer chains. The β phase exhibits excellent piezoelectric properties and is commonly achieved through mechanical stretching or electrical poling techniques [23]. The piezoelectric properties of PVDF are determined by its molecular structure and crystalline phases [24]. PVDF has notable piezoelectric coefficients, denoted as d and g. The d coefficient represents the direct piezoelectric effect, relating the induced charge per unit stress, while the g coefficient represents the converse piezoelectric effect, relating the strain per unit electric field. PVDF possesses relatively high d and g coefficients, making it an efficient piezoelectric material. The electromechanical coupling factor, denoted as k, is another important parameter characterizing its piezoelectric performance. It represents the efficiency of energy transfer between mechanical and electrical domains. PVDF has a relatively high k value, indicating good energy conversion efficiency [25]. PVDF also exhibits dielectric properties that influence its overall piezoelectric behavior. Dielectric properties include permittivity of the material, loss tangent, and breakdown strength [26]. These properties affect the electrical energy storage capacity, power dissipation, and voltage withstand capability of PVDF-based devices [27].

In addition to PVDF, there are other piezoelectric polymers that exhibit promising properties for energy harvesting applications. Two notable examples are PVDF-TrFE and PVDF-TFE. PVDF-TrFE is a copolymer composed of vinylidene fluoride (VDF) and trifluoroethylene (TrFE) monomers. This copolymer exhibits a unique combination of properties derived from the presence of TrFE, which enhances its piezoelectric performance compared to pure PVDF. PVDF-TrFE has a higher crystallinity and greater piezoelectric coefficients (d and g) than PVDF, making it more efficient in energy conversion applications. Adding TrFE also improves the thermal stability of the polymer, allowing it to withstand higher temperatures without significant degradation [28]. PVDF-TFE is another copolymer that combines the VDF monomer with tetrafluoroethylene (TFE). Similar to PVDF-TrFE, PVDF-TFE exhibits enhanced piezoelectric properties compared to PVDF. The incorporation of TFE in the copolymer structure leads to increased crystallinity and improved piezoelectric coefficients. PVDF-TFE also demonstrates good thermal stability, making it suitable for high-temperature applications [29].

When comparing these polymers to PVDF, PVDF-TrFE, and PVDF-TFE offer advantages in terms of improved piezoelectric performance and enhanced thermal stability. Their higher piezoelectric coefficients result in more efficient energy conversion, generating higher output voltages and power. Their enhanced thermal stability also ensures reliable operation under elevated temperatures, expanding their application range. However, it is essential to consider the limitations and trade-offs associated with these polymers. One consideration is their processability. PVDF-TrFE and PVDF-TFE have more complex processing requirements than PVDF due to the incorporation of additional monomers. Achieving the desired crystalline phase and alignment of polymer chains may require more controlled processing techniques. Cost is another factor to consider. PVDF-TrFE and PVDF-TFE tend to be more expensive compared to PVDF due to the additional monomers involved in their synthesis. The cost consideration may impact the scalability and commercial viability of devices utilizing these polymers for energy-harvesting applications. Environmental stability is also a crucial aspect to assess. While PVDF has good environmental stability, PVDF-TrFE and PVDF-TFE may exhibit varying degrees of vulnerability to degradation under certain conditions. It is important to evaluate their performance and stability in specific operating environments to ensure long-term reliability.

6.2.3 Enhancing Piezoelectric Performance

Several strategies and techniques have been developed to maximize the piezoelectric performance of polymers. These techniques include mechanical stretching, electric poling, and composite fabrication with the inclusion of nanoparticles or fillers. Mechanical stretching is a widely used technique to enhance the piezoelectric performance of polymers. The polymer chains are aligned by subjecting the polymer to controlled mechanical strain or stretching, improving the crystallinity. The alignment of polymer chains increases the piezoelectric coefficients, resulting in higher output voltages and improved energy conversion efficiency. Mechanical stretching can be applied through various methods, such as uniaxial or biaxial stretching [30]. Researchers have employed specialized stretching devices in mechanical stretching to achieve controlled alignment of polymer chains and improved piezoelectric properties. Electric poling is another effective technique to enhance the piezoelectric properties of polymers. In this process, a high electric field is applied to the polymer, inducing the alignment of polar domains and the orientation of dipoles. Electric poling improves the overall piezoelectric response of the polymer, particularly in achieving the desired crystalline phase. The poling process helps to align the polymer chains and enhance the polarization, resulting in improved piezoelectric coefficients [31]. Composite fabrication is a versatile approach to enhance the piezoelectric properties of polymers. Incorporating nanoparticles or fillers, such as carbon nanotubes, graphene, and piezoelectric ceramics, into the polymer matrix can serve as nucleation sites for crystallization, leading to improved crystallinity. They induce strain or stress within the polymer matrix to improve mechanical and piezoelectric properties, enabling the development of more efficient and sensitive sensors and transducers [32].

6.3 Energy Harvesting Using Piezoelectricity

The demand for autonomous and self-powered devices has been steadily increasing in recent years. These devices, ranging from wearable electronics to wireless sensor networks, require a reliable and sustainable power source to operate efficiently. Traditional power sources such as batteries or grid connections are often impractical or unfeasible due to limited lifespan, maintenance requirements, or inaccessible locations. As a result, researchers have been exploring alternative methods of generating electrical energy, leading to the emergence of energy harvesting as a promising solution. Energy harvesting involves capturing and converting ambient energy from the surrounding environment into usable electrical power. This approach enables the creation of autonomous systems that can operate indefinitely without needing external power sources or frequent battery replacements. By harnessing energy sources such as vibrations, mechanical strain, thermal gradients, or solar radiation, energy harvesting technologies offer a pathway to sustainable and environmentally friendly power generation [33].

Piezoelectricity has gained significant attention among various energy harvesting mechanisms as an up-and-coming technology. Piezoelectric materials possess a unique property where they can convert mechanical strain or vibrations directly into electrical charges. This phenomenon, known as the direct piezoelectric effect, efficiently converts ambient mechanical energy into usable electrical power. Piezoelectricity offers several advantages, including high power density, simplicity, and compatibility with small-scale applications [34]. The potential of piezoelectricity in energy harvesting applications is vast. It holds promise for powering a wide range of autonomous devices, from small sensors and wearable electronics to Internet of Things (IoT) devices and even biomedical implants. The ability to harvest energy from ambient mechanical vibrations or deformations opens up new possibilities for powering gadgets in various settings, including industrial environments, infrastructure monitoring, healthcare, and consumer electronics.

6.3.1 Working Principles of Piezoelectric Energy Harvesting Systems

Piezoelectric energy harvesting systems rely on the electromechanical coupling mechanism of piezoelectric materials, which convert mechanical strain or vibrations into electrical energy. These systems incorporate piezoelectric transducers to capture and convert ambient mechanical energy into usable electrical power. The electromechanical coupling mechanism of piezoelectric materials allows them to generate electrical charges when subjected to mechanical strain or vibrations. This property arises from the asymmetry of their crystal structure, leading to the separation of positive and negative charges. The energy conversion process in piezoelectric energy harvesting involves the conversion of ambient mechanical vibrations or deformations into electrical energy. When the piezoelectric material experiences mechanical inputs, it deforms and undergoes compression and tension on the material, and through the direct piezoelectric effect, it generates an alternating flow of current through the external circuit, as shown in Figure 6.3.

Piezoelectric transducers play a crucial role in energy conversion by capturing and converting mechanical energy into usable electrical power. These transducers are designed in specific configurations, such as cantilevers, beams, or bimorphs, to convert mechanical strain or vibrations into electrical charges efficiently. When subjected to ambient mechanical vibrations, the transducer deforms, inducing strain on the piezoelectric material. This strain generates electrical charges collected by electrodes connected to the transducer. The collected electrical power can then be utilized to power various devices or stored for future use [35].

The effectiveness of piezoelectric energy harvesting systems depends on factors such as the properties of the piezoelectric material, the design of the transducer, and the optimization of the energy conversion process. Ongoing advancements in material engineering, transducer design, and system integration contribute to improving the efficiency and practicality of these systems. Piezoelectric energy harvesting finds applications in various fields, including healthcare, environmental monitoring, and IoT

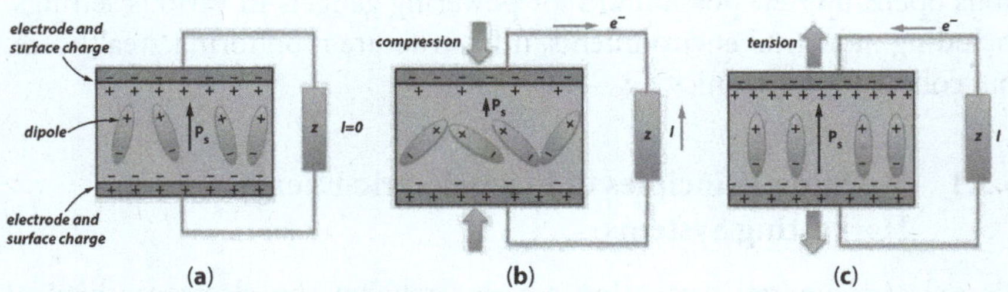

Figure 6.3 (a-c) Piezoelectric energy harvesting mechanism under its normal, compressed, and stretched states. Reproduced from ref. [36] with permission from the Royal Society of Chemistry.

devices. By leveraging the electromechanical coupling mechanism of piezoelectric materials and employing effective transducer designs, these systems offer a promising solution for generating electrical power from ambient mechanical vibrations. They contribute to the development of self-powered and sustainable autonomous devices.

6.3.2 Techniques and Designs for Efficient Energy Harvesting

Efficient energy harvesting plays a crucial role in maximizing the performance and practicality of piezoelectric energy harvesting systems. One key aspect is the importance of optimizing the mechanical resonance of the system. The energy harvesting efficiency can be significantly improved by designing the system to operate at or near its mechanical resonance. Understanding and leveraging the resonant characteristics of the system is essential for maximizing power generation [37]. Impedance matching is another technique used to enhance energy harvesting efficiency. Matching the impedance of the piezoelectric transducer with that of the load ensures maximum power transfer [38]. This can be achieved through the use of matching networks or resonant circuits, improving overall power transfer efficiency.

Energy storage devices, such as capacitors or batteries, are often incorporated into energy harvesting systems to store and regulate the harvested electrical energy. These devices enable the system to accumulate energy over time, even when the harvested power fluctuates. Proper selection and utilization of energy storage devices contribute to overall efficiency improvement [39]. Power management circuits play a vital role in efficiently harvesting and utilizing electrical energy. These circuits regulate the harvested power, convert it to the required voltage levels, and optimize its distribution to the load or storage devices. The integration of power management circuits ensures efficient utilization of the harvested energy, minimizing energy losses and maximizing overall energy harvesting efficiency.

6.3.3 Challenges and Limitations of Piezoelectric Energy Harvesting

Piezoelectric energy harvesting encounters several challenges that hinder its widespread implementation. Low power output, narrow bandwidth, and mechanical fatigue are among the key challenges faced in piezoelectric energy harvesting [40]. Additionally, the environment imposes limitations on energy harvesting systems, with variable vibrations and unpredictable ambient conditions posing significant obstacles. Ongoing research focuses on potential solutions by exploring materials and system design advancements to overcome these challenges. Advanced materials, such as piezoelectric composites or hybrid structures, offer improved performance characteristics, including higher power output and broader bandwidth [41]. System design optimizations, such as integrating multiple piezoelectric elements or developing adaptive harvesting systems, aim to enhance overall efficiency and adaptability.

Furthermore, advancements in energy management and storage technologies play a crucial role in addressing the challenges of low power output and fluctuating ambient conditions. Energy management techniques, including optimization algorithms and efficient power conditioning circuits, optimize the extraction and utilization of harvested energy. Enhanced energy storage devices contribute to overcoming the limitations posed by variable energy availability.

6.3.4 Applications of Piezoelectric Energy Harvesting

Piezoelectric polymers have gained significant attention in the field of energy harvesting. They can efficiently convert mechanical energy into electrical energy, making them ideal for powering small electronic devices and sensors. One notable application is in energy-harvesting shoe insoles. By embedding piezoelectric polymer elements within the insole, the mechanical energy generated during walking or running can be converted into electrical energy, which can then be used to charge batteries or power portable devices [42]. Piezoelectric polymers are also widely used in sensor applications. Their ability to convert mechanical stimuli into electrical signals enables the development of sensitive and responsive sensors. For instance, in wearable motion sensors, piezoelectric polymers can detect and measure body movements, enabling applications in sports performance monitoring, rehabilitation, and virtual reality. These sensors can capture and translate motion into electrical signals, providing valuable data for various applications [43]. Actuators based on piezoelectric polymers have demonstrated excellent performance in applications such as robotics, microfluidics, and precision positioning systems [44, 45]. By applying an electric field to the polymer, the piezoelectric effect causes dimensional changes, enabling precise and controlled movements. This feature is crucial in microactuators for precise fluid control, haptic feedback devices, and adaptive optics systems [46].

Due to the biocompatibility, flexibility, and ability to generate electrical signals in response to mechanical deformation, piezoelectric energy harvesting plays a crucial role in healthcare, particularly in self-powered medical implants and wearable devices. By harvesting energy from the body's mechanical movements or vibrations, piezoelectric energy harvesters can provide a sustainable power source for medical implants such as pacemakers, glucose monitors, and drug delivery systems [47]. Moreover, wearable devices, including health trackers and smartwatches, can benefit from piezoelectric energy harvesting to enhance their battery life or enable self-powered operation [48]. Piezoelectric energy harvesting finds significant use in environmental monitoring applications. For instance, energy can be harvested from vibrations in bridges or machinery for structural health monitoring purposes. Piezoelectric sensors integrated into infrastructure can capture vibrations caused by traffic or other external forces, converting them into electrical energy to power sensors for real-time structural integrity monitoring. This application enhances the efficiency

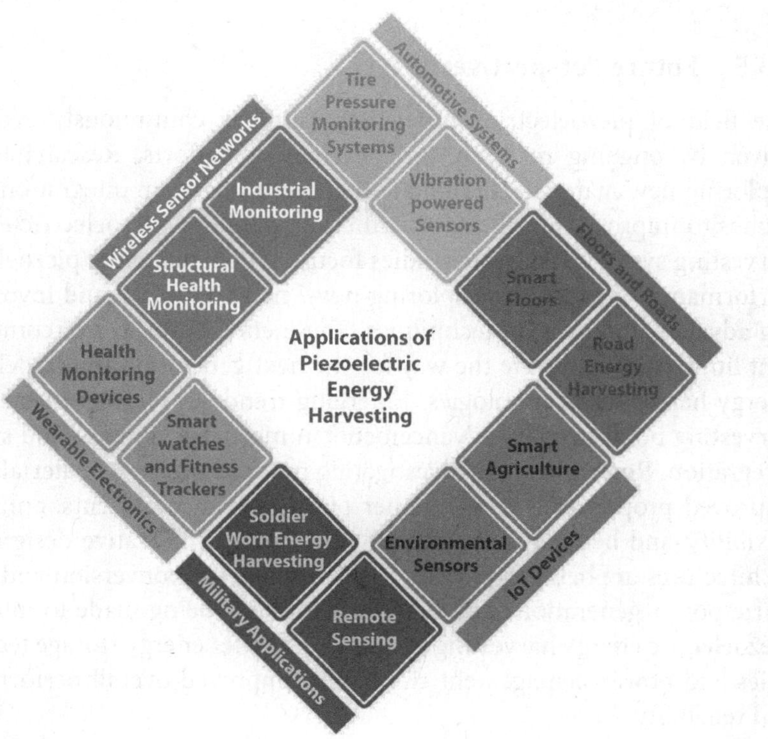

Figure 6.4 Piezoelectric energy harvesting systems in various application domains.

and sustainability of structural health monitoring systems [49]. Integrating piezoelectric energy harvesting into IoT devices opens up new possibilities for self-powered wireless sensor networks and smart systems [50]. IoT devices, including sensors and actuators, require a reliable power source to operate autonomously. Piezoelectric energy harvesters can provide the necessary power by converting ambient mechanical vibrations or deformations into electrical energy. This integration eliminates the need for external power sources or frequent battery replacements, making IoT devices more efficient and sustainable [51]. A graphical representation of various applications of piezoelectric energy harvesting applications is illustrated in Figure 6.4.

Piezoelectric polymers offer several advantages for various applications. Their conformability allows them to be easily integrated into curved or flexible surfaces, expanding design possibilities. They exhibit low power consumption, making them suitable for energy-efficient devices and reducing reliance on external power sources. The potential for integration into various devices, from wearables to biomedical implants, highlights their versatility and adaptability.

6.3.5 Future Perspectives

The field of piezoelectric energy harvesting is continuously evolving, driven by ongoing research and development efforts. Researchers are exploring new materials, design approaches, and system integration techniques to improve the efficiency and effectiveness of piezoelectric energy harvesting systems. Ongoing studies focus on enhancing the piezoelectric performance of polymers, exploring novel nanomaterials, and investigating advanced fabrication techniques. These efforts aim to overcome current limitations and pave the way for the next generation of piezoelectric energy harvesting technologies. Emerging trends in piezoelectric energy harvesting point towards advancements in materials, designs, and system integration. Researchers are investigating new piezoelectric materials with improved properties, such as higher piezoelectric coefficients, enhanced flexibility, and better thermal stability. Moreover, innovative designs and architectures are being explored to optimize energy conversion and maximize power generation. Additionally, efforts are being made to integrate piezoelectric energy harvesting systems with other energy storage technologies and power management circuits for improved overall performance and reliability.

6.4 Triboelectric Energy Harvesting Using Electroactive Polymer-Based Composites

The triboelectric effect is a phenomenon that involves the generation of electrical charges through friction between m aterials having different electron affinities. It is based on the principles of electrostatics and plays a crucial role in energy harvesting applications. The triboelectric effect occurs when two materials with different electron affinities come into contact and then separate or slide one over the other. During contact, electrons from one material transfer to the other material, resulting in a charge imbalance. This charge separation creates an electric potential between the two materials [52]. In the context of energy harvesting, the triboelectric effect provides a means to convert mechanical energy into electrical energy. When mechanical forces, such as friction, compression, or bending, are applied to a triboelectric nanogenerator, the contact and separation or sliding process occurs between the materials, leading to the generation of electric charges. This charge generation and separation mechanism can be attributed to the difference in electron affinities and surface properties of the materials involved. Electrons flow from the material with a lower electron affinity to the material with a higher electron affinity, generating positive and negative charges on the respective surfaces. The contact area and duration of contact play a crucial role in charge transfer [53].

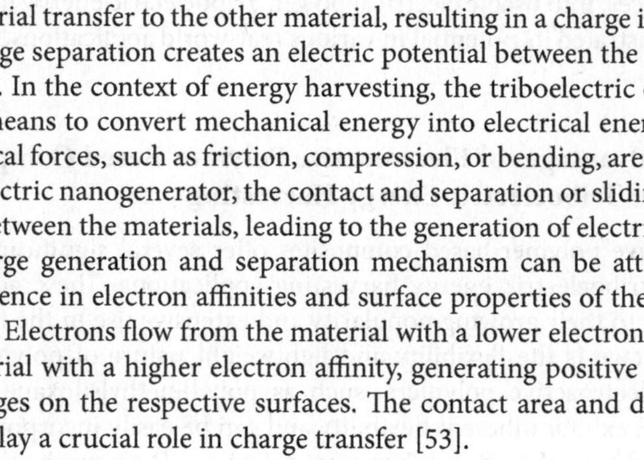

Figure 6.5 Triboelectric energy harvesters in real-world applications. Reproduced with permission from Ref. [54].

Careful selection of materials with contrasting electron affinities and surface properties is necessary to enhance energy harvesting efficiency through the triboelectric effect. The choice of materials impacts the magnitude of the generated charge and the overall performance of the triboelectric energy harvesting system. Understanding the mechanism of charge generation and separation in the triboelectric effect is essential for designing and optimizing energy harvesting devices. By harnessing the triboelectric effect, it becomes possible to convert mechanical energy from various sources, such as human motion, ambient vibrations, or environmental forces, into usable electrical power. Triboelectric energy harvesting has demonstrated its potential in various real-world applications, as shown in Figure 6.5.

6.4.1 Advantages of Electroactive Polymer-Based Composites for Triboelectric Energy Harvesting

Electroactive polymer-based composites offer several significant advantages for triboelectric energy harvesting applications. These advantages contribute to their growing popularity and extensive use in the field. One key advantage is the flexibility and lightweight nature of polymer composites. Electroactive polymers, such as polydimethylsiloxane (PDMS) and PVDF, exhibit inherent flexibility and can be easily incorporated into flexible and wearable energy harvesting devices. Their mechanical properties allow for conformable integration onto various surfaces, enabling seamless integration into different applications. Another advantage is the compatibility of polymer composites with various substrates. These composites can be fabricated and deposited onto a wide range of materials, including textiles, plastics, and metals. This versatility allows for integrating triboelectric energy harvesting technology into diverse environments and applications, such as clothing, electronic devices, and infrastructure. Electroactive polymer composites also possess high dielectric constants, which enhance their charge storage capacity during the triboelectric process. The significant deformation capabilities of these materials enable efficient charge generation and separation, resulting in improved energy harvesting performance. These properties make them particularly suitable for applications that involve repetitive mechanical motion or vibrations. Moreover, the dielectric properties of polymer composites can be tailored by modifying their composition, structure, and processing techniques. Adjusting the filler content, surface roughness, or thickness of the composite layers allows the dielectric constant and triboelectric performance to meet specific energy harvesting requirements.

6.4.2 Challenges of Electroactive Polymer-Based Composites for Triboelectric Energy Harvesting

While electroactive polymer-based composites offer numerous advantages for triboelectric energy harvesting, they also face several challenges that need to be addressed for their widespread application and improved performance [55]. One significant challenge is the relatively low output power density generated by these composites. The triboelectric effect in polymer composites tends to produce lower current levels than other energy-harvesting technologies, such as piezoelectric materials. This limitation hinders the practical application of electroactive polymer composites, particularly in scenarios that require higher power outputs. Another challenge is the limited durability and mechanical fatigue of polymer composites. Continuous cyclic loading during energy harvesting operations can lead to material degradation, resulting in reduced performance and eventual failure. Repeated mechanical stress can cause microcracks, delamination, and loss of adhesion within the composite structure. Addressing these durability issues is crucial to ensure the long-term reliability and sustainability of thermoelectric energy harvesting devices. Electroactive polymer composites are also sensitive to environmental conditions, which poses another challenge. Factors such as temperature variations, humidity, and moisture exposure can affect the electrical and mechanical properties of the composites. These environmental influences can alter the charge generation and separation mechanisms, leading to inconsistent energy harvesting performance. Developing strategies to enhance the environmental stability of polymer composites is essential for maintaining their reliable operation in diverse conditions.

6.4.3 Strategies to Enhance Triboelectric Performance

Various strategies and techniques have been developed and employed to improve the triboelectric performance of electroactive polymer-based composites. These approaches aim to optimize the surface characteristics, interface properties, and overall composition of the composites, enhancing their charge generation and separation capabilities [56]. One strategy involves surface modification techniques, which alter the surface properties of the polymer composites. Chemical treatments, such as the use of acids, bases, or solvents, can modify surface chemistry, leading to enhanced triboelectric performance. Plasma treatment is another effective method that involves exposing the composite surface to plasma-generated species, which can induce surface functionalization and increase surface energy.

Additionally, nanomaterial deposition techniques, such as the incorporation of nanoparticles or nanofibers, can enhance surface roughness and create nanoscale features, promoting improved triboelectric properties. Interface engineering approaches are also employed to enhance the performance of electroactive polymer composites. Interfacial engineering layers, such as thin films or coatings, are introduced at the interfaces between the polymer matrix and the conducting or triboelectric materials. These interfacial layers serve as barriers, preventing charge recombination and promoting efficient charge transfer across the interface. Nanoscale engineering techniques involve designing and manipulating nanoscale features at the interface to encourage charge separation and increase the contact area between the materials, thereby enhancing the triboelectric effect.

Optimizing the composition, morphology, and interface of the polymer composites is another critical strategy to enhance triboelectric performance. The selection of appropriate polymers, fillers, and additives and their proper combination and dispersion within the composite matrix can significantly impact the charge generation and separation mechanisms. Tailoring the morphology of the composite, such as controlling the alignment of polymer chains or the dispersion of fillers, can improve the overall performance. Optimizing the interfaces between the polymer matrix and fillers or electrodes can enhance the charge transfer and reduce energy loss. These strategies to enhance triboelectric performance are continually being explored and refined through ongoing research and development efforts. Understanding the underlying mechanisms and applying advanced characterization techniques enable researchers to optimize surface modification techniques, interface engineering approaches, and composite composition, morphology, and interface. By employing these strategies, researchers aim to achieve higher triboelectric outputs, increased energy conversion efficiency, and improved stability of electroactive polymer-based composites for triboelectric energy harvesting.

6.4.4 Integration of Piezoelectric and Triboelectric Effects in Hybrid Energy Harvesting Systems

In recent years, the integration of piezoelectric and triboelectric effects in hybrid energy harvesting systems has gained significant attention due to its potential to enhance energy generation and improve overall efficiency [57]. This approach combines the energy conversion capabilities of both piezoelectric and triboelectric mechanisms, harnessing mechanical forces and contact-induced motion to maximize energy harvesting capabilities.

Hybrid energy harvesting systems can take advantage of different energy conversion mechanisms by incorporating piezoelectric and triboelectric materials within a single system. The piezoelectric effect allows the conversion of mechanical strain or deformation into electrical charges, while the triboelectric effect enables the generation of electrical energy through the contact and separation of materials with different electronegativities. This combination enables the system to capture energy from various sources, including vibrations, motion, and touch. One of the key advantages of hybrid energy harvesting systems is the improved energy output and efficiency compared to individual piezoelectric or triboelectric systems. By utilizing both effects, the system can generate higher electrical power and achieve a broader bandwidth of energy harvesting. The complementary nature of piezoelectric and triboelectric mechanisms allows for better utilization of available mechanical energy, leading to increased energy output and enhanced performance.

Moreover, hybrid energy harvesting systems offer several advantages in maximizing energy harvesting capabilities. Integrating piezoelectric and triboelectric materials allows for a broader range of material options and design flexibility. Different combinations of materials can be explored to

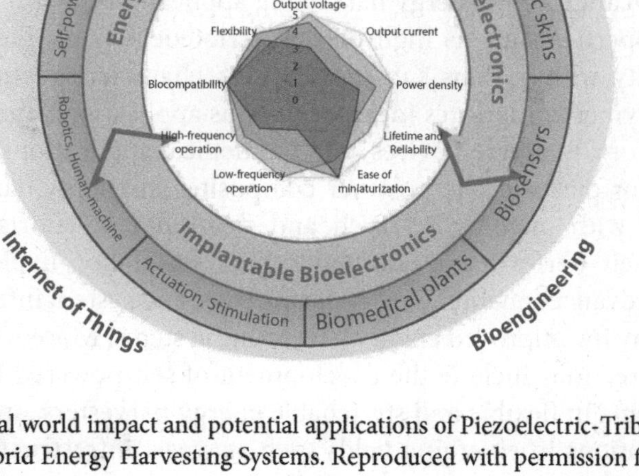

Figure 6.6 Real world impact and potential applications of Piezoelectric-Triboelectric Integrated Hybrid Energy Harvesting Systems. Reproduced with permission from Ref. [57].

optimize the energy conversion efficiency and tailor the system for specific application requirements. Additionally, hybrid systems can overcome the limitations and constraints associated with individual energy harvesting mechanisms, such as the narrow bandwidth of piezoelectric devices or the sensitivity of triboelectric systems to environmental conditions.

The integration of piezoelectric and triboelectric effects in hybrid energy harvesting systems holds great promise for various applications, as shown in Figure 6.6. These systems can be employed in wearable devices, smart structures, Internet of Things (IoT) devices, and other autonomous systems, enabling self-powered operation and reducing reliance on external power sources. Combining piezoelectric and triboelectric effects offers new avenues for energy harvesting research and development, fostering advancements in materials, device designs, and system integration strategies.

6.5 Conclusion

This chapter explores the exciting field of energy harvesting using piezoelectric polymer composites. Throughout the chapter, various aspects of both triboelectric and piezoelectric energy harvesting are discussed, highlighting the advantages, challenges, and potential applications of these technologies. The main points discussed include the working principles of piezoelectric energy harvesting systems, the explanation of the electromechanical coupling mechanism and energy conversion process, and the examination of different techniques and designs employed to enhance efficiency. The focus has been on piezoelectric polymer composites and their significant role in energy harvesting applications, emphasizing their unique properties such as high piezoelectric coefficients, flexibility, and compatibility with various substrates. These characteristics make piezoelectric polymer composites ideal for various applications, including sensors, actuators, wearable devices, and biomedical applications. The future prospects for piezoelectric polymer composites in energy harvesting are promising, with ongoing research and development efforts aiming to enhance their performance, durability, and scalability. Exploring novel materials, advanced manufacturing techniques, and system integration will pave the way for improved energy harvesting systems. Moreover, potential research directions include the development of self-powered IoT devices, advancements in flexible and stretchable energy harvesters, and exploring new applications in emerging fields such as smart infrastructure and the

Internet of Medical Things (IoMT). These directions will contribute to the realization of autonomous and sustainable energy solutions.

Energy harvesting using piezoelectric polymer composites holds immense significance for sustainable energy solutions. These composites offer a promising pathway towards achieving self-sufficient and environmentally friendly energy systems by harnessing ambient mechanical energy and converting it into electrical power. The integration of piezoelectric polymer composites in various applications has the potential to revolutionize industries such as healthcare, robotics, and smart infrastructure. As the pursuit of a sustainable future continues, the importance of energy harvesting cannot be overstated. By tapping into the abundant sources of ambient energy surrounding us, dependence on traditional power sources can be reduced, contributing to a cleaner and more sustainable world. Energy-harvesting technologies, including those based on piezoelectric polymer composites, offer a compelling solution in this endeavor. This chapter provides a comprehensive overview of piezoelectric polymer composites in triboelectric and piezoelectric energy harvesting applications, discussing their working principles and advantages. It is hoped that this chapter will inspire further research, innovation, and collaboration in the field of energy harvesting, ultimately driving towards a more sustainable and energy-efficient future.

AI disclosure statement: The author confirms that no artificial intelligence (AI)-generated content or AI-assisted writing tools were used in the preparation of this chapter (including summaries and the abstract). All text, analysis, and conclusions are the result of the author's original work.

References

1. Su, Y., Li, W., Cheng, X., Zhou, Y., Yang, S., Zhang, X., Chen, C., Yang, T., Pan, H., Xie, G., Chen, G., Zhao, X., Xiao, X., Li, B., Tai, H., Jiang, Y., Chen, L.Q., Li, F., Chen, J., High-performance piezoelectric composites *via* β phase programming. *Nat. Commun.*, 13, 1, 4867, 2022.
2. Sheng, R., Mu, J., Chernozem, R.V., Mukhortova, Y.R., Surmeneva, M.A., Pariy, I.O., Ludwig, T., Mathur, S., Xu, C., Surmenev, R.A., Liu, H.H., Fabrication and Characterization of Piezoelectric Polymer Composites and Cytocompatibility with Mesenchymal Stem Cells. *ACS Appl. Mater. Interfaces*, 15, 3, 3731–3743, 2023.
3. Park, S., Kim, Y., Jung, H., Park, J.Y., Lee, N., Seo, Y., Energy harvesting efficiency of piezoelectric polymer film with graphene and metal electrodes. *Sci. Rep.*, 7, 1, 17290, 2017.

4. Mokhtari, F., Shamshirsaz, M., Latifi, M., Foroughi, J., Nanofibers-based piezoelectric energy harvester for self-powered wearable technologies. *Polymers (Basel)*, 12, 11, 2697, 2020.
5. Zhao, Z., Dai, Y., Dou, S.X., Liang, J., Flexible nanogenerators for wearable electronic applications based on piezoelectric materials. *Mater. Today Energy*, 20, 100690, 2021.
6. Shehzad, M., Wang, S., Wang, Y., Flexible and transparent piezoelectric loudspeaker. *npj Flexible Electron.*, 5, 1, 24, 2021.
7. Maria Joseph Raj, N.P., Alluri, N.R., Khandelwal, G., Kim, S.J., Lead-free piezoelectric nanogenerator using lightweight composite films for harnessing biomechanical energy. *Compos. Part B Eng.*, 161, 608–616, 2019.
8. Smith, M. and Kar-Narayan, S., Piezoelectric polymers: theory, challenges and opportunities. *Int. Mater. Rev.*, 67, 1, 65–88, 2022.
9. Maria Joseph Raj, N.P., Alluri, N.R., Vivekananthan, V., Chandrasekhar, A., Khandelwal, G., Kim, S.J., Sustainable yarn type-piezoelectric energy harvester as an eco-friendly, cost-effective battery-free breath sensor. *Appl. Energy*, 228, 1767–1776, 2018.
10. Alluri, N.R., Selvarajan, S., Chandrasekhar, A., Saravanakumar, B., Jeong, J.H., Kim, S.J., Piezoelectric BaTiO3/alginate spherical composite beads for energy harvesting and self-powered wearable flexion sensor. *Compos. Sci. Technol.*, 142, 65–78, 2017.
11. Vivekananthan, V., Chandrasekhar, A., Alluri, N.R., Purusothaman, Y., Joong Kim, W., Kang, C.N., Kim, S.J., A flexible piezoelectric composite nanogenerator based on doping enhanced lead-free nanoparticles. *Mater. Lett.*, 249, 73–76, 2019.
12. Sezer, N. and Koç, M., A comprehensive review on the state-of-the-art of piezoelectric energy harvesting. *Nano Energy*, 80, 105567, 2021.
13. Surmenev, R.A., Orlova, T., Chernozem, R.V., Ivanova, A.A., Bartasyte, A., Mathur, S., Surmeneva, M.A., Hybrid lead-free polymer-based nanocomposites with improved piezoelectric response for biomedical energy-harvesting applications: A review. *Nano Energy*, 62, 475–506, 2019.
14. Kumar, M., Lim, J., Park, J.Y., Seo, H., Flexoelectric effect driven colossal triboelectricity with multilayer graphene. *Curr. Appl Phys.*, 32, 59–65, 2021.
15. Yan, M., Zhong, J., Liu, S., Xiao, Z., Yuan, X., Zhai, D., Zhou, K., Li, Z., Zhang, D., Bowen, C., Zhang, Y., Flexible pillar-base structured piezocomposite with aligned porosity for piezoelectric energy harvesting. *Nano Energy*, 88, 106278, 2021.
16. Rodrigues-Marinho, T., Perinka, N., Costa, P., Lanceros-Mendez, S., Printable lightweight polymer-based energy harvesting systems: materials, processes, and applications. *Mater. Today Sustain.*, 21, 100292, 2023.
17. Uchino, K., *Piezoelectric Energy Harvesting Systems—Essentials to Successful Developments*, 2018.
18. Sun, H., Zarkadoula, E., Crespillo, M.L., Weber, W.J., Rathod, V., Zinkle, S.J., Ramuhalli, P., Laser Doppler vibrometry for piezoelectric coefficient (d33)

measurements in irradiated aluminum nitride. *Sens. Actuators A. Phys.*, 347, 113886, 2022.

19. Ihlefeld, J.F., Fundamentals of Ferroelectric and Piezoelectric Properties, in: *Ferroelectricity in Doped Hafnium Oxide: Materials, Properties and Devices*, pp. 1–24, Woodhead Publishing, Cambridge, UK, 2019.
20. Manjón-Sanz, A.M. and Dolgos, M.R., Applications of Piezoelectrics: Old and New. *Chem. Mater.*, 30, 24, 8718–8726, 2018.
21. Lu, L., Ding, W., Liu, J., Yang, B., Flexible PVDF based piezoelectric nanogenerators. *Nano Energy*, 78, 105251, 2020.
22. Koseki, Y., Aimi, K., Ando, S., Crystalline structure and molecular mobility of PVDF chains in PVDF/PMMA blend films analyzed by solid-state 19F MAS NMR spectroscopy. *Polym. J.*, 44, 8, 757–763, 2012.
23. Zhou, W., Lin, Y., Zou, K., Zhou, C., Gong, X., Cao, Y., Jiang, S., Enhancement of piezoelectricity in polymer PVDF based on molecular chain structure. *J. Mater. Sci. - Mater. Electron.*, 32, 24, 28708–28717, 2021.
24. Saxena, P. and Shukla, P., A comprehensive review on fundamental properties and applications of poly(vinylidene fluoride) (PVDF). *Adv. Compos. Hybrid Mater.*, 4, 1, 8–26, 2021.
25. Liu, Z.H., Pan, C.T., Lin, L.W., Lai, H.W., Piezoelectric properties of PVDF/MWCNT nanofiber using near-field electrospinning. *Sens. Actuators, A*, 193, 13–24, 2013.
26. Xia, W. and Zhang, Z., PVDF-based dielectric polymers and their applications in electronic materials. *IET Nanodielectr.*, 1, 1, 17–31, 2018.
27. Mishra, S., Unnikrishnan, L., Nayak, S.K., Mohanty, S., Advances in Piezoelectric Polymer Composites for Energy Harvesting Applications : A Systematic Review. *Macromol. Mater. Eng.*, 304, 1, 1800463, 2019.
28. Lindemann, W.R., Philiph, R.L., Chan, D.W.W., Ayers, C.T., Perez, E.M., Beckman, S.P., Strzalka, J., Chaudhary, S., Vaknin, D., Oriented polyvinylidene fluoride-trifluoroethylene (P(VDF-TrFE)) films by Langmuir-Blodgett deposition: A synchrotron X-ray diffraction study. *Phys. Chem. Chem. Phys.*, 17, 43, 29335–29339, 2015.
29. MacKey, M., Flandin, L., Hiltner, A., Baer, E., Confined crystallization of PVDF and a PVDF-TFE copolymer in nanolayered films. *J. Polym. Sci. Part B Polym. Phys.*, 49, 24, 1750–1761, 2011.
30. Yang, J., Yao, X., Meng, Z., Investigation of molecular mechanisms of polyvinylidene fluoride under the effects of temperature, electric poling, and mechanical stretching using molecular dynamics simulations. *Polymer (Guildf).*, 245, 124691, 2022.
31. Kumar, R.S., Sarathi, T., Venkataraman, K.K., Bhattacharyya, A., Enhanced piezoelectric properties of polyvinylidene fluoride nanofibers using carbon nanofiber and electrical poling. *Mater. Lett.*, 255, 126515, 2019.
32. Tripathy, A., Maria Joseph Raj, N.P., Saravanakumar, B., Kim, S.J., Ramadoss, A., Tuning of highly piezoelectric bismuth ferrite/PVDF-copolymer flexible

films for efficient energy harvesting performance. *J. Alloys Compd.*, 932, 167569, 2023.
33. Chandrasekhar, A., Alluri, N.R., Vivekananthan, V., Purusothaman, Y., Kim, S.J., A sustainable freestanding biomechanical energy harvesting smart backpack as a portable-wearable power source. *J. Mater. Chem. C*, 5, 6, 1488–1493, 2017.
34. Selvarajan, S., Alluri, N.R., Chandrasekhar, A., Kim, S.J., Biocompatible electronic platform for monitoring protein-drug interactions with potential in future theranostics. *Sens. Actuators, B*, 305, 127497, 2020.
35. Safaei, M., Sodano, H.A., Anton, S.R., A review of energy harvesting using piezoelectric materials : state-of-the-art a decade later (2008–2018). *Smart Mater. Struct.*, 28, 113001, 2019.
36. Wan, C. and Bowen, C.R., Multiscale-structuring of polyvinylidene fluoride for energy harvesting: the impact of molecular-, micro- and macrostructure. *J. Mater. Chem. A*, 5, 7, 3091–3128, 2017.
37. Karadag, C.V., Ertarla, S., Topaloglu, N., Optimization of beam profiles for improved piezoelectric energy harvesting efficiency. *Struct. Multidiscip. Optim.*, 63, 2, 631–643, 2021.
38. Liang, J. and Liao, W.-H., Impedance matching for improving piezoelectric energy harvesting systems. *Act. Passiv. Smart Struct. Integr. Syst. 2010*, vol. 7643, p. 76430K, 2010.
39. Yang, Y., Chen, Z., Kuai, Q., Liang, J., Liu, J., Zeng, X., Circuit Techniques for High Efficiency Piezoelectric Energy Harvesting. *Micromachines*, 13, 7, 1044, 2022.
40. Toprak, A. and Tigli, O., Piezoelectric energy harvesting: State-of-the-art and challenges. *Appl. Phys. Rev.*, 1, 3, 031104, 2014.
41. Bouhamed, A., Binyu, Q., Böhm, B., Jöhrmann, N., Behme, N., Goedel, W.A., Wunderle, B., Hellwig, O., Kanoun, O., A hybrid piezoelectric composite flexible film based on PVDF-HFP for boosting power generation. *Compos. Sci. Technol.*, 208, 108769, 2021.
42. Chaudhary, P. and Azad, P., Energy Harvesting Using Shoe Embedded with Piezoelectric Material. *J. Electron. Mater.*, 49, 11, 6455–6464, 2020.
43. De Fazio, R., Mastronardi, V.M., De Vittorio, M., Visconti, P., Wearable Sensors and Smart Devices to Monitor Rehabilitation Parameters and Sports Performance : An Overview. *Sensors*, 23, 4, 1856, 2023.
44. Hilber, W., Stimulus-active polymer actuators for next-generation microfluidic devices. *Appl. Phys. A Mater. Sci. Process.*, 122, 8, 751, 2016.
45. Chi, Y., Zhao, Y., Hong, Y., Li, Y., Yin, J., A Perspective on Miniature Soft Robotics: Actuation, Fabrication, Control, and Applications. *Adv. Intell. Syst.*, 6, 2, 2300063, 2024.
46. Chen, J., Teo, E.H.T., Yao, K., Electromechanical Actuators for Haptic Feedback with Fingertip Contact. *Actuators*, 12, 3, 104, 2023.
47. Ali, F., Raza, W., Li, X., Gul, H., Kim, K.H., Piezoelectric energy harvesters for biomedical applications. *Nano Energy*, 57, 879–902, 2019.

48. Liu, Y., Khanbareh, H., Halim, M.A., Feeney, A., Zhang, X., Heidari, H., Ghannam, R., Piezoelectric energy harvesting for self-powered wearable upper limb applications. *Nano Sel.*, 2, 8, 1459, 2021.
49. Guo, L., Wang, H., Braley, J., Venkiteela, G., Field Evaluation of Piezoelectric Energy Harvesters on Bridge Structure. *Machines*, 11, 4, 462, 2023.
50. Asthana, P. and Khanna, G., A broadband piezoelectric energy harvester for IoT based applications. *Microelectron. J.*, 93, 104635, 2019.
51. Kusagur, S.M., Arunkumar, G., Manjunath, T.C., Use of smart intelligent sensor & actuator mechanical materials (PVDF / PZT) in developing MIMO mathematical model of the smart structure and its use to control the active vibrations using discrete sliding mode control theory with output samples. *Mater. Today Proc.*, 37, 1592–1602, 2021.
52. Basith, S.A. and Chandrasekhar, A., COVID-19 clinical waste reuse: A triboelectric touch sensor for IoT-cloud supported smart hand sanitizer dispenser. *Nano Energy*, 108, 108183, 2023.
53. Basith, S.A. and Chandrasekhar, A., Synergistic Effect of Dual Surface Modified Ecoflex Polymer for Dynamic Triboelectric Nanogenerator Towards Sustainable Battery-Free Tally Counter. *Adv. Mater. Technol.*, 8, 18, 2300495, 2023.
54. Cao, X., Xiong, Y., Sun, J., Xie, X., Sun, Q., Wang, Z.L., Multidiscipline Applications of Triboelectric Nanogenerators for the Intelligent Era of Internet of Things. *Nano Micro Lett.*, 15, 1, 14, 2022.
55. Wang, A.C., Wu, C., Pisignano, D., Wang, Z.L., Persano, L., Polymer nanogenerators: Opportunities and challenges for large-scale applications. *J. Appl. Polym. Sci.*, 135, 24, 45674, 2018.
56. Muktar Ahmed, R.F.S., Mohan, S.B., Ankanathappa, S.M., Shivanna, M., Basith, S.A., Chandrashekara Shastry, M.H., Chandrasekhar, A., Sannathammegowda, K., Sucrose assisted chemical-free synthesis of rGO for triboelectric nanogenerator: Green energy source for smart-water dispenser. *Nano Energy*, 106, 108085, 2023.
57. Mariello, M., Recent Advances on Hybrid Piezo-Triboelectric Bio-Nanogenerators: Materials, Architectures and Circuitry. *Nanoenergy Adv.*, 2, 1, 64–109, 2022.

7

Fluorinated Electroactive Polymers as Electrolytes in Cells and Battery Applications

Anna M. Abraham[1,2] and Soney C. George[1]*

[1]*Center for Nanoscience and Technology, Department of Basic Sciences, Amal Jyothi College of Engineering, Koovappally P.O., Kerala, India*
[2]*APJ Abdul Kalam Technological University, CET Campus, Thiruvananthapuram, Kerala, India*

Abstract

Fluorinated electroactive polymers form a class of material that has recently found attention solely because of the unique features they possess. Incorporation of fluorine atoms into the backbone of this class of materials imparts quite a few desirable features, including chemical and thermal strength endurance, excellent toughness, and also low dielectric constant. Moreover, fluorinated electroactive polymers can be engineered to possess a variety of different electrical properties, including high conductivity, a high dielectric constant, and ferroelectricity. The application sectors for these fluorinated polymers are broad and include the majority of areas associated with actuators, sensors, batteries, electrochromic displays, and fuel cells. Fluorinated electroactive polymers are a relatively new topic of research but with immense potential for future development. Future work will continue on fluorinated monomers, as well as development of novel fluorinated electroactive polymers and architecture in electroactive polymers with emphasis on controlling architectures and fundamentals. Later, application aspects are studied for these electroactive fluorinated materials. In general, fluorinated electroactive polymers present a broad range of attractive multiple material family uses. Thus, continued studies are going to lead eventually in almost a certainty manner in terms of discoveries of novel materials that show higher performances in many regards.

Keywords: Fluorinated electroactive polymers, fluorine, fuel cells, batteries

**Corresponding author*: soneycgeorge@amaljyothi.ac.in

Raneesh Balakrishnan, Sobi K. Chacko and Visakh P.M. (eds.) Electroactive Polymer-Based Smart Materials and Applications, (185–206) © 2026 Scrivener Publishing LLC

7.1 Introduction

7.1.1 What are Polymers?

Polymers are large molecules consisting of thousands of repeating subunits known as monomers. Polymers can be found everywhere—from the natural polymers that make up our bodies and the plants and animals we feed to the synthetic polymers that are used in everything from plastics, rubber, and textiles.

Polymers are divided into two, including synthetic polymers and natural polymers. Natural polymers are produced by living organisms including proteins, polysaccharides, and nucleic acids. Synthetic polymers that include plastics, rubber, and synthetic fibers are produced by humans. Polymers have various properties depending on the type of monomer used and how they are linked together. Some polymers are rigid, durable, and hard; other polymers are soft, brittle, and easily breakable. Some polymers have high electrical conductivity; some polymers are not very conductive.

The piezoelectric effect describes how electrical and mechanical energy are transferred in a material. Specifically, it is the linear coupling between the material's stress condition and electrical polarization [1]. This particular effect was established in the case of polymers too, making it more applicable to the field of electrochemical applications.

7.1.2 Electroactive Polymers

Electroactive polymers are a class of polymers that react to an electrical stimulation by changing their shape or size. EAPs are rapidly developing classes of materials with wide-ranging potential applications. While the research continues, new EAPs with developed properties and functionalities are in development. Thus, new and innovative devices and applications are emerging. Many kinds of EAPs are vastly different in their characteristics. Some of the most common categories of EAPs include:

a) Piezoelectric polymers: When mechanically distorted, piezoelectric polymers generate an electric field.
b) Pyroelectric polymers: When heated or cooled, pyroelectric polymers generate an electric field.
c) Ferroelectric polymers: polarized by an electric field and retain their polarization even after the field has been removed.

d) Ionic EAPs: change shape or size as ions migrate within the polymer.
e) LCEs (liquid crystal elastomers): An EAP that can change shape or size in response to an electric field or light. The electroactive polymers are classified into mainly 3 categories;
   1. Π-conjugated
   2. Electronically conducting polymers
   3. Ion-exchange polymers

Electroactive polymer technology is employed in various mechatronic applications, mostly as actuators. Indeed, electroactive polymers are lightweight and flexible, making them ideal for replacing electromagnetic actuators or motors in aerospace, robotics, and medical applications [2]. Electroactive polymers (EAPs) are considered to be a class of polymers that change dimensions or shape upon an application of electrical stimulus. Thus, EAPs have innumerable applications in fields as diverse as possible. These can be used as sensors to measure physical characteristics, such as pressure force or strain. EAPs are also used to make actuation devices that either perform a movement or deformation, under the influence of electrical fields. They can be used in the design of devices which exploit either mechanical oscillations or thermal differentials. They are also utilized to

Table 7.1 Specific applications of different electroactive polymers.

Electroactive polymer	Applications
Poly (vinylidenefluoride) [PVDF]	Osteogenic material applications [3]
Poly (Lactic acid) [PLA]	Biomaterial Applications [3]
PVDF- trifluoro ethylene [PVDF-TrFE]	Pressure Sensor and energy harvesting applications [4]
Polypropylene [PP]	Fuel cells, corrosion protection, biosensors, drug delivery and tissue engineering applications [3]
Polyaniline [PANI]	Biosensors, neural prostheses, drug delivery and tissue engineering applications [3]
PEDOT [Polythiophene derivative]	Tissue engineering applications [3]

create new forms of electronic devices, such as flexible screens and transistors, and are currently under investigation for being researched for a wide range of biomedical applications (Table 7.1).

7.2 Fluorinated Electroactive Polymers

7.2.1 Origin of Fluorinated Electroactive Polymers

The history of fluorinated electroactive polymers can be traced back into the early 1950s when American chemist Dr. Roy W. Plunkett accidentally discovered by mistake the polymer polytetrafluoroethylene abbreviated as Teflon, which is a very highly fluorinated polymer that revealed excellent properties such as chemical inertness, thermal stability, and electrical insulation. Scientific investigation with the possible use of fluorinated polymers for electroactive applications started in the late 1950s until early 1960. One of the first fluorinated polymers investigated for this application was the semi-crystalline polymer poly (vinylidene fluoride) [PVDF]. This polymer possesses a unique dipole moment, giving it piezoelectric and pyroelectric properties. This implies that on application of mechanical stress or temperature changes, PVDF will have an electrical charge.

Japanese researchers discovered in 1969 that copolymerizing PVDF with other monomers such as trifluoroethylene (TrFE) and chlorotrifluoroethylene (CTFE) might make it even more electroactive. These PVDF copolymers have greater piezoelectric and pyroelectric coefficients than PVDF homopolymers. Researchers proceeded to produce novel fluorinated polymers with improved electroactive characteristics during the next few decades.

One of the most impressive developments is poly (vinylidene fluoride-hexafluoropropylene), a copolymer, outstanding for its piezoelectric properties and high thermostability. Then there is poly(3,4-ethylenedioxythiophene) or PEDOT, which is a conductive polymer that may be doped further by ions. Finally, there is poly (fluorinated ethylene propylene) or FEP, and it is a fluoridated copolymer that demonstrates supreme chemical and electrical insulation properties.

Fluorinated electroactive polymers are a rapidly expanding area of research with many potential applications. Fluorinated electroactive polymers are going to play an important role in our lives as new materials and technologies are developed.

7.2.2 FEPs—Properties and Characteristics

Fluorinated electroactive polymers, or FEPs, are a type of polymer that has special features from fluorine and can conduct electricity. It is a very electronegative element, thus giving FEPs good qualities, such as having a high dielectric constant, resistance to chemicals, stability in heat, strength in mechanics, and the ability to repel water.

When electrical stimulation exists, the electroactive polymers are said to alter in size or shape. Ferroelectricity, pyroelectricity, and piezoelectricity are a few of the electroactive properties found in FEPs. The capability of generating an electric field by mechanically deforming matter is known as piezoelectricity. The ability of heat or cold to produce an electric field is recognised as pyroelectricity. The ability of an object to become orientated by an electric field and maintain that polarisation after the field is eliminated is known as ferroelectricity.

One of the most noteworthy insulators for organic electrical devices is FEP. They are now adaptable to a wide range of electrical applications due to the alterations in how they react to an applied electric field. However, it is still challenging to add functionality to FEPs because of the chemical stillness of these polymers and the very sophisticated synthetic methods needed to manufacture them. Kallitsis suggested a simple method for the implementation of fundamental functionality into the FEPs [5]. FEPs have a vast potential application in many fields, such as sensors and actuators that measure physical quantities such as pressure, force, and strain. They can also be used in building actuators, which move or deform in response to an electric field. In addition, FEPs can be used in creating vibrational or thermal gradient-powered devices that could be used as new kinds of electrical devices and flexible screens or transistors. Also, FEPs are studied for certain biomedical applications like artificial muscles and drug delivery systems.

A distinct class of material in fluorine polymers has a number of intriguing characteristics due to its chemical structure, which was produced by strong carbon-carbon bonds (340 kJ min^{-1}) and stable carbon-fluorine bonds (485 kJ min^{-1}). Moreover, fluorine's electric structure has an extremely short van der Waals radius of 1.32A°, which is unbroken, high electronegativity, and impoverished polarisability [6]. Some of the limitations of Fluorinated homopolymers include high crystallinity that necessitates expensive processing, poor solubility in common solvents that limit characterization, and Cross-linking/functionalization is delicate. Functional comonomer(s) could be incorporated into the backbone of

Fluorinated homopolymers that could possibly solve these problems [7]. In other words, the electron-directing influence of the F group has relevance to the crosslinking mechanisms linked to these fluorinated electroactive polymers. Han *et al.* addressed the negative consequences regarding the electron-directing influence of the F group on linkage locations between the neighboring repeat units during polymer backbone formation [8]. Here are some specific applications for FEPs:

i. PVDF is found in many piezoelectric sensors and actuators, including ultrasonic transducers, microphones, and artificial muscles.
ii. In pyroelectric sensors for fire alarms and thermostats, PCTFE is employed.
iii. In ferroelectric transistors and memory devices, PTFE is employed.
iv. Flexible displays and other electrical gadgets make use of FEP.
v. PHBV is being studied for potential applications in artificial muscles and medicine delivery systems.

FEPs are one of the emerging flexible and promising classes of material that can transform a multitude of different industries. Therefore, we could conclude that the FEPs will definitely play major roles in further ventures.

7.3 PVDF and Its Copolymers—The Next-Generation FEP

Polyvinylidene fluoride is a semi-crystalline thermoplastic fluoropolymer prepared by polymerizing vinylidene difluoride. It is a versatile material with many applications because of its special properties, such as chemical resistance to solvents, acids, and bases, thermal stability and flame retardance, low density, excellent mechanical strength, and properties of piezoelectric and pyroelectric materials.

PVDF is used widely in various industries, with applications in chemical processing to allow the production of pipes, valves and other equipment that withstand strong chemical solutions. It also finds other applications in electrical and electronic hardware to make insulation, coatings among other parts, manufacture implantable devices like artificial valves for the heart and vascular implants, construction, and architectural applications to produce roofing membrane, siding, among other roofing materials. The high electrical dipole moment of the PVDF monomer unit ($5–8 \times 10^{-30}$ C m) is caused by the electronegativity of fluorine atoms in comparison to carbon atoms, and it is linked to many of the intriguing properties of PVDF, particularly those related to its use as a sensor or actuator [9]. Besides this, it is used in the manufacturing of membranes and other components for water and wastewater treatment systems. It is also being used nowadays in the manufacture of batteries, fuel cells, and other components for energy storage and transportation applications.

PVDF is a useful material with numerous applications. Because of its unique qualities and versatility, it is projected to expand in popularity in the next few years. Here are some specific PVDF application examples:

1. PVDF pipes: They are used in the chemical processing industry to transport corrosive substances such as acids and bases. The flow of these substances is controlled using PVDF valves. Corrosive chemicals are transferred using PVDF pumps.
2. PVDF insulation: It is used in the electrical and electronics industries to insulate electrical wires and cables against moisture, chemicals, and abrasion. PVDF coatings are applied to electronic components to protect them against corrosion and other environmental threats. PVDF films are utilized in the production of capacitors and other electrical devices.
3. PVDF artificial heart valves: They are long-lasting and resistant to blood clots. PVDF vascular grafts are used to restore blood vessels that have been damaged.
4. PVDF roofing membranes: They are extremely robust and resistant to UV radiation, chemicals, and fire. PVDF siding requires little upkeep and is resistant to fading and cracking.
5. Water and wastewater treatment: PVDF membranes are used to filter water and wastewater in reverse osmosis and ultrafiltration systems.
6. PVDF batteries: They are lightweight and have a high energy density, making them ideal for transportation. PVDF fuel cells are highly efficient and generate clean energy. PVDF components are employed in electric vehicles and other forms of transportation.

However, two significant issues limit PVDF membrane applications, which are fouling in water treatment and wetting in membrane contactors. The membrane's efficiency will be reduced as a result of fouling and wetness, which in turn impair the performance, resulting in an increased operation costing money and possibly creating failure. PVDF membranes with appropriate hydrophilic or hydrophobic modifications are effective in enhancing fouling or wetting resistance and have gained a large number of customers' attention [10]. On increasing research into the area of PVDF, it was found that PVDF can be used as a membrane functionality in various applications concerning the area of electroactive polymers. PVDF recently became one of the more prominent membrane materials. PVDF is more hydrophobic than other materials, such as polysulfone (PS), polyethersulfone (PES), and polyimide (PI), yet the hydrophobicity may vary [11].

7.3.1 Polar Properties of PVDF

In addition to comprising, PVDF, a semi-crystalline polymer with four polymorphs—α, β, γ, and δ—has unique significance. At degrees lower than 160°C, melt crystallization develops the α-phase more easily. Because of the nonpiezoelectric shape and the antiparallel arrangement of fluoride atoms along the carbon backbone, its chain conformation is Trans (T)-gauche (G)-Transgauche (TG + TG) and it has a net dipole moment of zero [12].

7.3.2 Physical Properties of PVDF

With the chemical formula $(CH_2CF_2)_n$, PVDF is a well-known pure fluoropolymer that is nonreactive. The tensile strength and impact toughness of PVDF illustrate its durability as well as strength due to its simplistic chemical structure. PVDF is exceptionally resistant to fatigue and creep. Additionally, to its weather resistance, chemical resistance, UV resistance, and nuclear radiation resistance, PVDF also boasts a high dielectric constant, immense strength, and high thermal stability. Depending on how the sample is processed, PVDF can acquire a variety of molecular crystal forms that can both exist and change. It permits the synthesis of polymers to take on many different forms. At room temperature, PVDF has a density of 1.77 g/cm^3 and a melting point of roughly 177°C [13]. Figure 7.1 depicts the molecular structure of polyvinylidene fluoride.

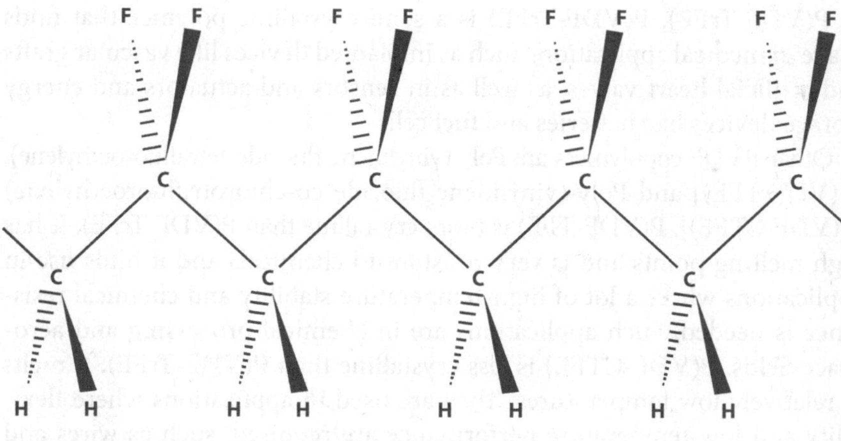

Figure 7.1 Molecular structure of polyvinylidene fluoride [1].

7.3.3 Chemical Properties of PVDF

PVDF is lighter than polytetrafluoroethylene [PTFE], fluorinated ethylene propylene [FEP], and perfluoroalkoxy polymer [PFA] because it has the least amount of fluorine. Compared to PTFE, PFA, and FEP, PVDF has a more limited chemical inertness spectrum. However, it can tolerate a significant number of stresses due to its extraordinary inertness [14]. Due to its amazing mechanical and chemical properties, PVDF is now identified as the chemical-resistant material of choice [13]. The incorporation of fluorine atoms within the polymer backbone provides the chemical properties of PVDF. Since fluorine is the most electronegative atom within the periodic table, fluorine has a strong pull-on electron; therefore, the carbon–fluorine bond becomes quite stable and resistant to any kind of chemical attack. PVDF has many divergent chemical characteristics as a material. It is also a significant resource to quite a few industrial sectors. It is expected to experience growth in utilization in the following years.

7.3.4 PVDF and Its Copolymers

The copolymerization of PVDF with other monomers, such as trifluoroethylene (TrFE), tetrafluoroethylene (TFE), and chlorotrifluoroethylene (CTFE), gives PVDF copolymers. PVDF copolymer properties can be modified based on the nature and the extent of the other monomers involved. The most well-known of them is Poly (vinylidene-fluoride-trifluoro ethylene), or P(VDF-TrFE). P(VDF-TrFE) is a semi-crystalline polymer that finds usage in medical applications such as implanted devices like vascular grafts and artificial heart valves, as well as in sensors and actuators and energy storage devices like batteries and fuel cells.

Other PVDF copolymers are Poly (vinylidene fluoride-tetrafluoroethylene), [P(VDF-TFE)] and Poly (vinylidene fluoride-co-chlorotrifluoroethylene) [P(VDF-CTFE)]. P(VDF-TFE) is more crystalline than P(VDF-TrFE). It has high melting points and is very resistant to chemicals and it finds use in applications where a lot of high-temperature stability and chemical resistance is needed. Such applications are in chemical processing and aerospace fields. P(VDF-CTFE) is less crystalline than P(VDF-TrFE). It melts at relatively low temperatures. They are used in applications where flexibility and low-temperature performance are required, such as wires and cables.

Important polymers with flexibility and electroactive properties like ferroelectricity, pyroelectricity, and piezoelectricity are PVDF and its copolymers. In the past few years, technology has been used in a wide variety of applications. Memory, transducers, actuators, and energy harvesting are a few of these promising technologies that are being investigated and brought to market [15]. Because of their exceptional electrical qualities and wide range of uses in electronics and electrical machinery, certain fluoropolymers, like PVDF, have been the focus of several studies. With copolymers, it is now the material of choice for a wide range of data communications applications, including insulation, sensors, energy harvesting devices, actuators, and front-end processors. High polarity, thermal and mechanical properties, wettability by organic solvents, chemical inertness, cathodic stability environment, and tailored porosity by binary and ternary solvent/non-solvent interaction systems represent a few of PVDF-CTFE's exceptional features and characteristics that are used in the synthesis of battery separators. These are semi-crystalline, partly fluorinated polymers with amorphous phase spherulites in between crystalline lamellae. It crystallizes in a variety of crystalline phases based on processing conditions and temperature. The two most significant crystalline phases of PVDF and its copolymers are the α-phase, which is stable when the material is taken straight from the melt, and the β-phase, which possesses ferroelectric, piezoelectric, and pyroelectric attributes [16].

P(VDFTrFE), also known as poly (vinylidene fluoride-trifluoroethylene), is one of the copolymers that has been explored the most. Regardless of the processing method, the insertion of the third fluoride in the TrFE monomer structure with a notable steric barrier favours the all-trans conformation and introduces the ferroelectric-phase, whereas PVDF always has the ferroelectric crystalline phase at specific molar ratios. The conclusion is that PVDF, its copolymers, and composites continue to be the foundation of an extremely dynamic, fruitful, and captivating research field that will surely support some of the challenging applications of the near future, primarily due to their superior electrical properties over those of any other polymer [9].

7.4 Polymer Membranes as Electrolytes

7.4.1 Polymer Membranes

A membrane is an interphase that acts as a selective barrier between two adjacent phases and controls the transportation of materials from one of the two compartments [17]. Polymer membranes are thin films of semi-permeable polymeric material. It means it permits certain chemical materials to pass while restraining others. The main applications include usage in water filtration, gas separation, and fuel cells. Rubbery or glassy polymers can be used to produce polymeric membranes. An amorphous polymer transitions from a glassy to a rubbery form when heated above the rubbery transition temperature, or Tg. Glassy polymers, which demonstrate outstanding purity, have great selectivity but poor permeability, whereas rubbery polymers frequently display high gas permeabilities but poor selectivity [18].

Polymer membranes are frequently made from synthetic polymers, such as polysulfone, polyethersulfone, and polyamide. Depending upon the application, they might be porous or non-porous. Non-porous polymer membranes are dense and merely allow certain molecules to flow through by diffusion, whilst porous polymer membranes consist of small pores that allow certain molecules to pass through. Polymer membranes serve a purpose in water filtering processes such as reverse osmosis, nanofiltration, and ultrafiltration. Reverse osmosis membranes remove dissolved salts and other contaminants from water, whereas the use of nanofiltration and ultrafiltration membranes remove larger molecules like bacteria and viruses. The separation of hydrocarbon vapors from air is expected to be one of the most significant applications of membranes [19]. Polymer membranes are also used in gas separation applications, for example in isolating CO_2 from natural gas or nitrogen from air. In general, gas separation membranes are prepared from porous polymers that provide higher permeability to specific gases than others (Figure 7.2).

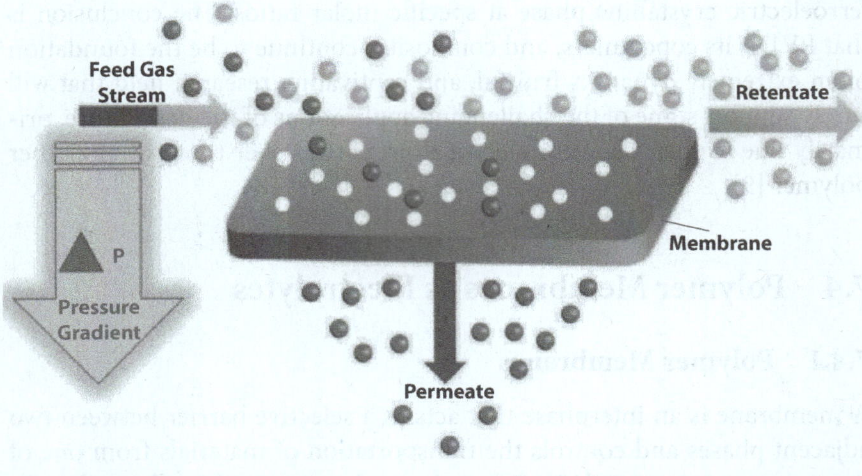

Figure 7.2 The membrane gas separation process's theoretical framework [20].

A thorough investigation into the ion exchange processes involved in water purification as well as the ideal membrane properties was conducted by Kamcev *et al.* [21]. Fuel cells require polymer membranes. Chemical energy from fuel sources like hydrogen or methanol may be converted into electrical energy using fuel cells. In fuel cells, the polymer membranes serve as electrolytes, permitting ions to pass through from the anode electrode to the cathode electrode. The various facets of how nano-clay affects the ionic conductivity and electrochemical characteristics of poly (vinylidene fluoride)-based porous polymer membranes and their use as polymer electrolytes in lithium-ion batteries were investigated by Prasanth *et al.* [22].

7.4.2 Applications for Polymer Membranes

Polymer membranes are a robust yet vital material class with numerous applications. New and innovative applications for polymer membranes are expected to be discovered as research continues. In addition to the applications outlined above, polymer membranes are utilized in a variety of other domains. Polymer membranes are utilized to distinguish blood components and oxygenate blood in artificial organs such as hemodialyzers and artificial lungs. They are used in the food and beverage industries to concentrate and eliminate items such as juices, syrups, and wine, in the drug manufacturing industry to produce and refine drugs, for environmental applications to remove pollutants from water and air. They also serve the purpose to separate gases like carbon dioxide from methane or nitrogen from oxygen. Gas permeation is the process that accomplishes this. Fuel cells and batteries are two distinct kinds of devices that use polymer membranes for energy production and storage. They are also employed for absorption of carbon dioxide from fuel gas. A novel class of materials, polymer membranes have numerous potential applications spanning numerous industries.

Further applications of polymer membranes include their utilization for making artificial organs such as kidney dialysis equipment and artificial lungs. They can deliver medications to the body in a controlled and targeted manner. Moreover, they can be used to generate sensors that detect a broad spectrum of substances, notably gases, ions, and biomolecules. Polymer membranes are a flexible yet vital class of materials with multiple uses. As research in this area continues, we are likely to experience even more inventive and advantageous applications of polymer membranes in the future.

7.4.3 Electrolytes Composed of Polymer Membranes

Polymer membranes can be hired as electrolytes in an extensive variety of electrochemical devices that include fuel cells, batteries, and water electrolyzers. Polymer electrolytes have multiple benefits over conventional liquid electrolytes. They have additional advantages as they are solid-state materials, which makes them less prone to leakage as well as easier to transport and store. They possess high ionic conductivity, enabling electrochemical devices to function with greater efficacy. In addition to this, they are mechanically stable making them as a pioneer in applications that require durability.

Nafion, a perfluorosulfonic acid (PFSA) membrane, is undoubtedly the most used type of polymer electrolyte. Proton exchange membrane fuel cells are the ideal option for PFSA membranes because they are proton-conducting membranes that move protons from the anode to the cathode inside an electrochemical device. Another kind of polymer electrolyte would be an AEM. Anion-conductive AEMs have tremendous promise since they can move anions from the anode to the cathode of an electrochemical device *via* the membrane. Although AEMs are still in their infancy, they have the potential to support a wide range of applications, from metal-air batteries to alkaline fuel cells. For example, polymer electrolytes are also considered promising material for water electrolyzers. Water electrolyzers split water into hydrogen and oxygen using electricity. Polymer electrolytes may enable water electrolyzers to operate more efficientlyand for longer times. As a class, polymer membranes are an attractive material to be used as electrolytes in electrochemical devices. They have several advantages over traditional liquid electrolytes and are being explored for application in many contexts.

When using electrolytes with different component ratios in KOH electrolyte, Wu *et al.* talked about the intriguing characteristics of the PVA/PAA polymer membrane, additionally referred to as Poly (vinyl alcohol) [PVA]/ Poly (Acrylic Acid) [PAA]. The alkaline PVA/PAA polymer membrane electrolyte has an ionic conductivity of 0.301 S cm-1 at room temperature. The polymer PVA/PAA electrolyte membrane in this study revealed a suitable balance between improved ionic conductivity and appropriate mechanical strengths, both of which were vital to present the solid polymer electrolyte at an optimal composition ratio of 10:5 [23].

Polymer membranes are employed as electrolytes in electrochemical devices. PEM fuel cells are used in portable applications such as electric cars. The essential part of proton exchange membrane fuel cells (PEMFCs) is the polymer electrolyte membrane (PEM). In addition to providing an electron and fuel barrier, the PEM would function as an electrolyte to move protons from the anode to the cathode. Because of this, many researchers have concentrated on creating dependable, high-performing poly electrolyte membranes that have low fuel permeability, good oxidative stability, strong mechanical stability, high proton conductivity, and inexpensive manufacturing and assembly costs. Nafion has been selected as a standard for polymeric electrolyte fuel cells due to its strong ionic conductivity (0.1 S/cm when fully hydrated) and thermal and chemical stability [24]. The most prevalent form of polymer electrolyte utilized in PEM fuel cells is perfluorosulfonic acid (PFSA) membranes. Moreover, alkaline fuel cells are more impervious to contaminants than PEM fuel cells, making them appropriate for usage with biofuels. Alkaline fuel cells are being developed to use Anion Exchange Membranes (AEMs). Metal-air batteries have a high energy density, but they are tough to fabricate due to the formation of dendrites on the metal electrode. AEMs are being evaluated as a potential remedy to this problem. Polymer electrolytes can help boost water electrolyzer productivity and lifespan. AEMs are being researched for implementation in water electrolyzers as they can work at elevated levels of pH, which is desirable for the oxygen evolution reaction.

A thorough look into the potential for ion transport in composite polymer gel electrolytes was conducted by Osinska *et al.*; this work will have numerous uses in lithium battery technology. The idea of electrolyte membranes has garnered a lot of attention lately. This relies on microporous membranes made of PVDF/HFP (vinylidene fluoride-hexafluoropropylene) that have been ballooned by an appropriate liquid electrolyte. According to Osińska *et al.*, this concept is one of the few examples of effective "non-liquid" electrolytes for Li-ion battery systems since its electrolytes have good processability [25].

7.4.4 Electrospun Polymer Membranes as Electrolytes

Electrospun polymer membranes (EPMs) are a type of polymer membrane made by the electrospinning method. Electrospinning is a flexible means of creating highly porous and vast-area nanofiber mats. In light of these features, EPMs have been optimized for use as electrolytes in electrochemical devices. Considering the membranes have a large porosity, they have a fairly high electrolyte absorption, resulting in elevated ionic conductivity. The use of a Polyethylene, PE that is based on the activation of an electro spun P (VDF-HFP) membrane integrated with 1-butyl-3-methylimidazolium (BMI)-based room temperature ionic liquids (RTILs) was carefully investigated by Cheruvally *et al.* [26]. EPMs are preferable to conventional polymer membranes in plenty of ways, including ionic conductivity, mechanical stability, and tunable characteristics.

The potential use of EPMs as electrolytes in a variety of electrochemical devices has been researched. In addition to increasing the efficiency and durability of water electrolysers, EPMs have been observed to improve the power density and durability of PEM fuel cells, which are implemented in alkaline fuel cells where they have the potential to improve performance while also cutting costs. Further to this, EPMs are being researched for their use in metal-air batteries, where they could help solve dendrite development and methanol crossover difficulties.

Overall, electrochemical polymers are a promising technology that can be used as an electrolyte in electrochemical devices. They have their distinguishing features that make them a good fit for virtually every application and are already attracting the attention of investigators across the globe.

7.4.5 Polymer Membranes as Electrolytes

Polymer membranes are one sort of solid-state electrolyte of numerous kinds of devices connected with energy storage as well as conversion, batteries, fuel cells, and membrane electrolysis. Polymer-based ionic conducting electrolytes consist of polymer matrices that host ionic species. The embedded ionic species give ionic conductivities while the polymer host or matrix gives structural and mechanical rigidity. Polymer-based membrane exhibits several advantages when in comparison to other varieties like liquid and solid-state inorganic-based electrolytes. Polymer membranes are usually non-flammable, non-corrosive, and can be operated over a wide temperature range. They are also relatively easy to manufacture and assemble in various sizes and designs.

7.4.5.1 *Polymer Membranes as Fuel Cell Electrolytes*

In a particular sort of fuel cell, PEMFC, protons are transferred from the anode to the cathode through the polymer membrane electrolyte. Based on its great efficiency and ability to operate at very low temperatures, PEMFCs have a wide range of possibilities in portable electric power and transportation.

The Polymer Electrolyte Membrane (PEM) fuel cell is the most promising possible power source for vehicles and other uses. The polymer electrolyte membranes in PEMFCs are essential for them to efficiently separate the fuel and oxidant from other components and to prevent ions from travelling from the cathode to the anode to wrap up the redox process.

Proton transport *via* the hydrated polymeric matrix occurs mostly through two mechanisms: "hopping" and "vehicular." A proton in the membrane initiates "hopping" as it transitions from one protonic species—H^+, H_3O^+, and even NH_4^+—to another. When an electrochemical force moves a hydrated proton through an aqueous medium or another fluid, the vehicular process takes place. When the protons attach to a vehicle site, like water, they get carried by the medium's diffusion. The surface and chemical characteristics of the interface between the inorganic and organic phases govern the proton transport mechanism in the hybrid system [24].

Several ways are being developed by researchers to improve proton conduction in PEMs. One strategy would be to create new polymer materials with high-water absorption and retention capability. This can assist in keeping the PEM moisturized even in hot temperatures and low humidity. Another option would be to create novel polymer materials with functional groups that enable proton transport. Some researchers, for example, are working on polymer membranes with sulfonic acid or imidazole groups. These functional groups can aid in the movement of protons across the membrane. Finally, some researchers are working on composite PEMs that include nanomaterials like carbon nanotubes or graphene. These nanoparticles can help to increase the water retention capacity of the PEM by providing more routes for proton transport.

Researchers can make PEM fuel cells and other energy storage and conversion devices more efficient and durable by enhancing proton conduction in PEMs.

7.4.5.2 Polymer Membranes as Battery Electrolytes

Another place where one can find polymer membrane electrolytes is in batteries. For example, lithium-ion and lithium-polymer batteries' safety and efficiency can be enhanced through the employment of polymer membrane electrolytes. Polymer membrane electrolytes, for instance, can help eliminate the occurrence of lithium dendrite growth, which tends to create short circuits within the battery and causes them to malfunction. Polymer membrane electrolytes can also make the battery's ionic conductivity more efficient, so that its power density is raised and its charging periods shorter. Figure 7.3 lists the many types of polymer-based composite solid electrolytes now in use.

Due to their excellent ionic conductivity at room temperature and the fact that there is no chance of electrolyte leaks, gel polymer electrolytes, or GPEs, have garnered a lot of attention lately. The application of PVB as a polymer membrane and as a polymer gel electrolyte for lithium-ion batteries was examined by Lian *et al.* [27]. An electrostatic field would be created throughout the electrolyte by the crosslinking polymer membranes with

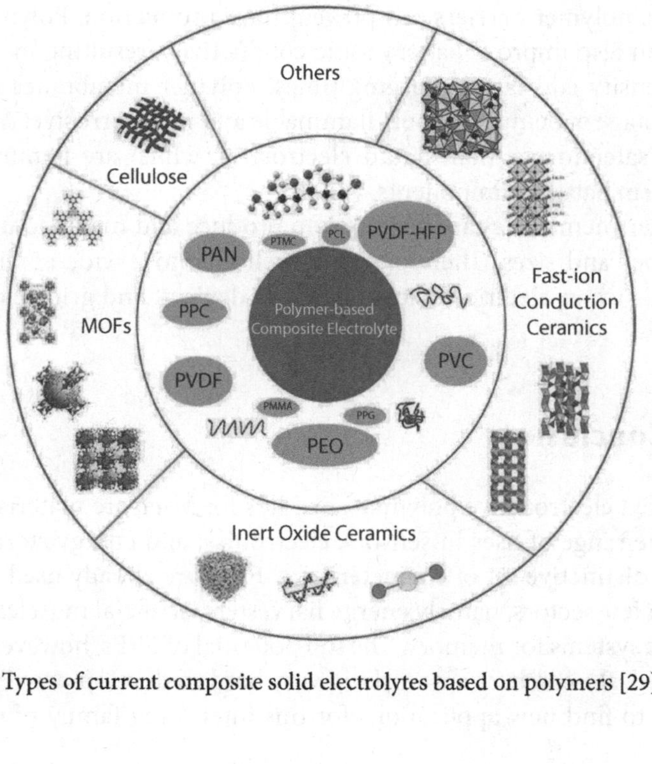

Figure 7.3 Types of current composite solid electrolytes based on polymers [29].

pendant sulfonate isomers, as demonstrated by Stalin *et al.* They creatively expanded this to follow the idea of several immobilized capabilities in an electrochemical cell. They have developed a simple method for lithium-ion battery safety using the aforementioned changes [28].

Researchers are currently collaborating on novel polymer membrane electrolytes for batteries that will perform even better. The development of polymer membrane electrolytes compatible with high-voltage electrode materials is one interesting area of research. The creation of batteries with greater power density and longer lifespans may result from this. In general, polymer membranes represent a promising class of materials that could be employed as electrolytes in a variety of energy conversion and storage systems. Compared to other types of electrolytes, including liquid and solid-state inorganic electrolytes, polymer membranes have several advantages. Researchers are working to create new polymer membrane electrolytes with improved durability, affordability, and performance.

One of the primary benefits of employing polymer membranes as electrolytes is that they can aid in the prevention of lithium dendrite formation. Dendrites of lithium are needle-like structures that can form on the anode and cathode of lithium-ion batteries. Short circuits and battery failure can be caused by dendrites. By inhibiting the growth routes of lithium dendrites, polymer barriers can prevent their production. Polymer membranes can also improve battery ionic conductivity, resulting in improved power density and faster charging times. Polymer membranes also have the advantage of being both non-flammable and non-corrosive. As a result, they are safer to use than liquid electrolytes, which are flammable and might harm battery components.

Polymer membranes are very easy to produce and manufacture in various shapes and sizes. Therefore, they well fit into a wide application of batteries, such as electric vehicles, portable devices, and grid-scale energy storage.

7.5 Conclusion

Fluorinated electroactive polymers, or FEPs for brief, are materials featuring a wide range of uses in sensors, electronics, and energy storage. FEPs exhibit a distinctive set of characteristics. FEPs are already used commercially in a few sectors, namely energy harvesters, artificial muscles, and ferroelectric systems for memory. The full potential of FEPs, however, is yet to be realized. Researchers are working very hard to develop new FEP materials and to find new applications for this interesting family of materials.

Some of the major challenges and opportunities for FEPs are that their synthesis is more complicated and costly, will get degraded in specific situations, and long-term reliability is still being explored. Despite the challenges above, FEPs are going to revolutionize many industries. For instance, electronics, energy, and healthcare industries are constantly developing. The cost of producing FEP is going to come down with advancing technology.

The outlook for FEPs appears bright in the future. FEPs have the potential to allow numerous new technologies and applications. Researchers are working hard to address the hurdles that FEPs face, and the commercialization of FEP-based devices is gaining momentum. FEPs, in our opinion, have the potential to become one of the most important groups of materials in the twenty-first century. FEPs are expected to make radical changes in a huge sector and also change many lives in one way or the other.

Acknowledgments

We are grateful to ANERT, KERALA (SRI-51/2020-21) for financial assistance and APJ Abdul Kalam Technological University (KTU), Thiruvananthapuram (KTU/RESEARCH5/2712/2023) for the research fellowship.

AI disclosure statement: The author acknowledges the use of AI-based tools for minor language editing or grammatical correction. However, all intellectual contributions, analyses, and conclusions presented in this chapter are entirely the author's own.

References

1. Smith, M. and Kar-Narayan, S., Piezoelectric polymers: theory, challenges and opportunities. *Int. Mater. Rev.*, 67, 1, 65–88, Jan. 2022, doi: 10.1080/09506608.2021.1915935.
2. Jean-Mistral, C., Basrour, S., Chaillout, J.-J., Comparison of electroactive polymers for energy scavenging applications. *Smart Mater. Struct.*, 19, 8, 085012, Aug. 2010, doi: 10.1088/0964-1726/19/8/085012.
3. Ning, C., Zhou, Z., Tan, G., Zhu, Y., Mao, C., Electroactive polymers for tissue regeneration: Developments and perspectives. *Prog. Polym. Sci.*, 81, 144–162, Jun. 2018, doi: 10.1016/j.progpolymsci.2018.01.001.
4. Wang, T., *et al.*, Electroactive polymers for sensing. *Interface Focus*, 6, 4, 20160026, Aug. 2016, doi: 10.1098/rsfs.2016.0026.

5. Kallitsis, K., *et al.*, Introducing Functionality to Fluorinated Electroactive Polymers. *Macromolecules*, 52, 21, 8503–8513, Nov. 2019, doi: 10.1021/acs.macromol.9b01942.
6. Cardoso, V., Correia, D., Ribeiro, C., Fernandes, M., Lanceros-Méndez, S., Fluorinated Polymers as Smart Materials for Advanced Biomedical Applications. *Polymers*, 10, 2, Feb. 2018, doi: 10.3390/polym10020161.
7. Dolui, S., Kumar, D., Banerjee, S., Ameduri, B., Well-Defined Fluorinated Copolymers: Current Status and Future Perspectives. *Acc. Mater. Res.*, 2, 4, 242–251, Apr. 2021, doi: 10.1021/accountsmr.1c00015.
8. Han, C.-C. and Chen, H.-Y., Highly Conductive and Electroactive Fluorine-Functionalized Polyanilines. *Macromolecules*, 40, 25, 8969–8973, Dec. 2007, doi: 10.1021/ma071745t.
9. Martins, P., Lopes, A.C., Lanceros-Mendez, S., Electroactive phases of poly(vinylidene fluoride): Determination, processing and applications. *Prog. Polym. Sci.*, 39, 4, 683–706, Apr. 2014, doi: 10.1016/j.progpolymsci.2013.07.006.
10. Kang, G. and Cao, Y., Application and modification of poly(vinylidene fluoride) (PVDF) membranes – A review. *J. Membr. Sci.*, 463, 145–165, Aug. 2014, doi: 10.1016/j.memsci.2014.03.055.
11. Liu, F., Hashim, N.A., Liu, Y., Abed, M.R.M., Li, K., Progress in the production and modification of PVDF membranes. *J. Membr. Sci.*, 375, 1–27, Jun. 2011, doi: 10.1016/j.memsci.2011.03.014.
12. Mokhtari, F., Latifi, M., Shamshirsaz, M., Electrospinning/electrospray of polyvinylidene fluoride (PVDF): piezoelectric nanofibers. *J. Text. Inst.*, 107, 1037–1055, Sep. 2015, doi: 10.1080/00405000.2015.1083300.
13. Saxena, P. and Shukla, P., A comprehensive review on fundamental properties and applications of poly(vinylidene fluoride) (PVDF). *Adv. Compos. Hybrid Mater.*, 4, 1, 8–26, Mar. 2021, doi: 10.1007/s42114-021-00217-0.
14. Inderherbergh, J., Polyvinylidene Fluoride (PVDF) Appearance, General Properties and Processing. *Ferroelectrics*, 115, 4, 295–302, Mar. 1991, doi: 10.1080/00150193.1991.11876614.
15. Yin, Z., Tian, B., Zhu, Q., Duan, C., Characterization and Application of PVDF and Its Copolymer Films Prepared by Spin-Coating and Langmuir–Blodgett Method. *Polymers*, 11, 12, 2033, Dec. 2019, doi: 10.3390/polym11122033.
16. Barbosa, J., Dias, J., Lanceros-Méndez, S., Costa, C., Recent Advances in Poly(vinylidene fluoride) and Its Copolymers for Lithium-Ion Battery Separators. *Membranes*, 8, 3, 45, Jul. 2018, doi: 10.3390/membranes8030045.
17. Ulbricht, M., Advanced functional polymer membranes. *Polymer*, 47, 7, 2217–2262, Mar. 2006, doi: 10.1016/j.polymer.2006.01.084.
18. Ozdemir, S., Buonomenna, M., Drioli, E., Catalytic polymeric membranes: Preparation and application. *Appl. Catal. Gen.*, 307, 2, 167–183, Jul. 2006, doi: 10.1016/j.apcata.2006.03.058.
19. Koros, W.J., Coleman, M.R., Walker, D.R.B., Controlled Permeability Polymer Membranes.

20. Elhenawy, S., Khraisheh, M., AlMomani, F., Hassan, M., Key Applications and Potential Limitations of Ionic Liquid Membranes in the Gas Separation Process of CO2, CH4, N2, H2 or Mixtures of These Gases from Various Gas Streams. *Molecules*, 25, 18, 4274, Sep. 2020, doi: 10.3390/molecules25184274.
21. Kamcev, J. and Freeman, B.D., Charged Polymer Membranes for Environmental/Energy Applications. *Annu. Rev. Chem. Biomol. Eng.*, 7, 1, 4.1–4.23, Jun. 2016, doi: 10.1146/annurev-chembioeng-080615-033533.
22. Prasanth, R., Shubha, N., Hng, H.H., Srinivasan, M., Effect of nano-clay on ionic conductivity and electrochemical properties of poly(vinylidene fluoride) based nanocomposite porous polymer membranes and their application as polymer electrolyte in lithium ion batteries. *Eur. Polym. J.*, 49, 2, 307–318, Feb. 2013, doi: 10.1016/j.eurpolymj.2012.10.033.
23. Wu, G.M., Lin, S.J., Yang, C.C., Preparation and characterization of PVA/PAA membranes for solid polymer electrolytes. *J. Membr. Sci.*, 275, 1–2, 127–133, Apr. 2006, doi: 10.1016/j.memsci.2005.09.012.
24. Kim, D.J., Jo, M.J., Nam, S.Y., A review of polymer–nanocomposite electrolyte membranes for fuel cell application. *J. Ind. Eng. Chem.*, 21, 36–52, Jan. 2015, doi: 10.1016/j.jiec.2014.04.030.
25. Osińska, M., Walkowiak, M., Zalewska, A., Jesionowski, T., Study of the role of ceramic filler in composite gel electrolytes based on microporous polymer membranes. *J. Membr. Sci.*, 326, 2, 582–588, Jan. 2009, doi: 10.1016/j.memsci.2008.10.036.
26. Cheruvally, G., *et al.*, Electrospun polymer membrane activated with room temperature ionic liquid: Novel polymer electrolytes for lithium batteries. *J. Power Sources*, 172, 2, 863–869, Oct. 2007, doi: 10.1016/j.jpowsour.2007.07.057.
27. Lian, F., Wen, Y., Ren, Y., Guan, H., A novel PVB based polymer membrane and its application in gel polymer electrolytes for lithium-ion batteries. *J. Membr. Sci.*, 456, 42–48, Apr. 2014, doi: 10.1016/j.memsci.2014.01.010.
28. Stalin, S., Choudhury, S., Zhang, K., Archer, L.A., Multifunctional Cross-Linked Polymeric Membranes for Safe, High-Performance Lithium Batteries. *Chem. Mater.*, 30, 6, 2058–2066, Mar. 2018, doi: 10.1021/acs.chemmater.7b05353.
29. Yao, P., *et al.*, Review on Polymer-Based Composite Electrolytes for Lithium Batteries. *Front. Chem.*, 7, 522, Aug. 2019, doi: 10.3389/fchem.2019.00522.

8

Electrospun Fiber Mats of Fluoropolymer Composites for Wound Healing and Drug Delivery Applications

Merin Tomy and Xavier T.S.*

Center for Advanced Materials Research, Department of Physics, Government College for Women, Thiruvananthapuram, Kerala, India

Abstract
This chapter explores the innovative applications of fluoropolymer composite electrospun fiber mats in medication administration and wound healing. Electrospun fiber mats have garnered significant attention due to their flexibility and potential for enhancing wound healing outcomes. The incorporation of fluoropolymers into these mats has opened new avenues for developing highly efficient and functionalized drug delivery systems and wound dressings. We delve into the unique properties of fluoropolymers, the electrospinning process, and how these materials can be tailored to meet the needs of medication delivery and wound care. This chapter aims to provide insights into the current state-of-the-art and future prospects of fluoropolymer-based electrospun fiber mats in the healthcare industry.

Keywords: Electrospun fiber mats, fluoropolymers, wound healing, medication administration, drug delivery systems, bioactive substances

8.1 Introduction

One essential biological process that is important to tissue regeneration and repair is wound healing. Over the years, significant advancements have been made in the disciplines of biomaterials and nanotechnology to

**Corresponding author*: xavierkattukulam@gmail.com; ORCID: https://orcid.org/0000-0001-8955-0743

Raneesh Balakrishnan, Sobi K. Chacko and Visakh P.M. (eds.) Electroactive Polymer-Based Smart Materials and Applications, (207–234) © 2026 Scrivener Publishing LLC

provide innovative approaches to enhance the outcomes of wound healing. In recent years, electrospun fibre mats—a flexible and promising family of materials—have received a lot of attention due to their potential uses in medication administration and wound healing. Specifically, the incorporation of fluoropolymers into these electrospun fibre mats has created novel and exciting prospects for the development of highly efficient and functionalized drug delivery systems and wound dressings.

This chapter delves into the intriguing realm of fluoropolymer composite electrospun fiber mats and their uses in medication administration and wound healing. We will explore the special qualities of fluoropolymers, the electrospinning method, and the ways in which these materials can be customized to fulfil the demands of medication administration and wound care. We will also go over the several approaches of adding bioactive substances to these fibre mats to increase their medicinal potential. We hope that this investigation will shed light on the state-of-the-art and future potential applications of electrospun fiber mats based on fluoropolymers in the healthcare industry.

8.1.1 Electrospun Fiber Mats

Electrospinning, an electrostatic technique, produces continuous nanofibers with high surface area-to-volume ratios and oriented structures. These nanofibers are promising candidates for a wide range of applications, including energy storage, active packaging, adsorption, bioactive delivery, and enzyme immobilization. For a very long time, continuous ultrafine fibers with adjustable widths were created using an antiquated method called electrostatic spinning, or electrospinning, which produced fibers. This allowed for the construction of nonwoven or aligned structures. This method's primary discovery was made by Formhals, who also published its initial form in 1934. A typical electrospinning system includes a metal collector, a high voltage power supply, and a polymer container connected by a capillary or needle, as shown in Figure 8.1. In order to create a strong electrostatic field across a conductive needle that joins the metal collector and the polymer solution container, a high voltage power supply is utilized in the electrospinning process.

When the electrostatic field strength reaches a certain value, charges on the surface of the polymer drop cause the shape to become unstable and transition from hemispherical to the conical Taylor cone shape. When the electrostatic field intensity exceeds a critical value, surpassing the surface tension of the polymer, a charged polymer jet is expelled from the tip of the cone. The jet of polymer gets quite long and narrow due to the

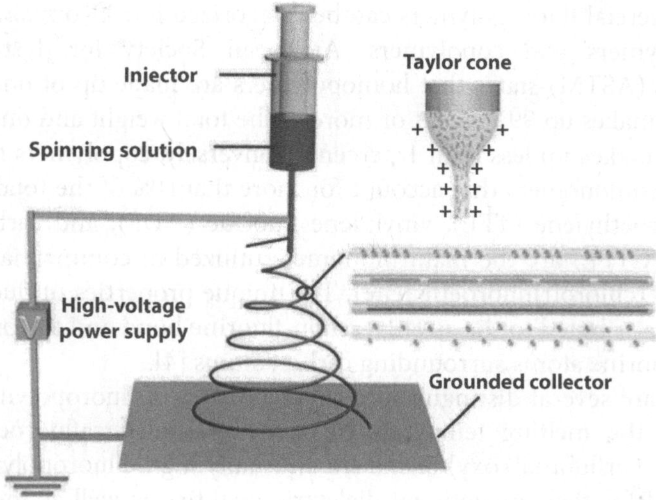

Figure 8.1 Schematic diagram of Electro spinning instrument [1].

elongation and instability phase it goes through. The ultrafine fibers on the metal collector are the consequence of the solvent evaporating and the fibers hardening as the polymer jet moves. Different working parameters in electrospinning have an impact on the fiber output and structure. These parameters can be categorized into three groups, in theory: ambient, process, and system parameters. Thus, by varying these parameters, the morphologies, diameters, and characteristics of the electrospun fibers can be changed.

8.1.2 Fluoropolymers in Biomaterials

Fluoropolymers, a class of polymers, consist of wholly or partially fluorinated olefinic monomers. Among these monomers, vinylidene fluoride and tetrafluoroethylene are notable examples. When Roy Plunkett discovered polytetrafluoroethylene (PTFE) in 1938 while working on refrigerant research at E. I. DuPont De Nemours & Co., it marked the beginning of the development of fluoropolymers. Perfluoropolymer research and commercialization were made possible by this discovery [2]. Many polymers containing fluorine have been created over time, and the last 20 years have seen a notable rise in this manufacturing. These polymers comprise silicon, nitrogen, and chlorine, as well as variants of poly(tetrafluoroethylene) (PTFE). They are among a wide range of compounds that have numerous uses in different sectors of the economy [3].

Commercial fluoropolymers can be categorized into two main groups: homopolymers and copolymers. American Society for Testing and Materials (ASTM) states that homopolymers are made up of one monomer that makes up 99 percent or more of the total weight and one monomer that makes up less than 1 percent. Conversely, copolymers have one or more comonomers that account for more than 1% of the total weight. Tetrafluoroethylene (TFE), vinylidene fluoride (VDF), and carboxylene fluoride (CTFE) are the main monomers utilized in commercial fluoropolymers (chlorotrifluoroethylene). The unique properties of fluoropolymers are attributed to the strong carbon-fluorine bond and the protective role of fluorine atoms surrounding carbon atoms [4].

There are several distinguishing characteristics of fluoropolymers. For example, the melting temperatures of PTFE (polytetrafluoroethylene) and PFA(Perfluoroalkoxy) resins are unusually high. Fluoropolymers are renowned for their exceptional dielectric qualities as well as their resistance to strong chemicals and ordinary solvents. Other than PTFE and PVF (polyvinyl fluoride), a range of commercial fluoropolymers are melted and processed to produce sheets, moldings, films, and profiles using certain manufacturing techniques. A polymer's photostability, flame resistance, and resistance to chemicals and solvents all improve when fluorine content rises in it. Additionally, the polymer helps to improve the melting point, thermal stability, dielectric constant, friction coefficient, and electrical characteristics. But when the concentration of fluorine rises, the mechanical qualities usually decrease and the polymer's solvent solubility decreases as well [5, 6]. It should be noted that exhaust ventilation is required during high-temperature manufacturing operations since fluoropolymers might release dangerous air pollutants when subjected to temperatures that are marginally higher or lower than their permitted processing temperatures. In recent decades, a variety of fluorinated polymers with unique and remarkable sets of properties have been produced to get around these restrictions. Many industries, such as automotive, aircraft, building, chemistry, aerospace, microelectronics, textile, energy, and biological applications, use high molecular weight fluorinated polymers extensively [7, 8].

The intriguing super-wetting properties of fluorinated polymers can be attributed to the formation of micro-textured surfaces resulting from the interaction at a stagnant air-liquid interface [9]. This feature makes it possible to create a variety of surfaces, including multifunctional surfaces with adhesion, superwettability, superhydrophilicity, superphobicity, and directed surfaces for liquid transfer [10]. These varied wetting properties give the material improved properties that make it appropriate for a variety of uses. The use of fluorinated compounds to modify silicon and copper micro- and nanostructured materials is one approach to creating superhydrophobic surfaces. These surfaces can find applications in anti-icing systems and beyond [11]. Currently, more than 20% of all pharmaceutical products available on the market include fluorine [12]. By producing additional dipole/charged dipole interactions, fluorination of medicines improves binding interactions with targets and boosts membrane mobility [13]. The addition of fluorine atoms to proteins and peptides gives these biomolecules interesting properties. When compared to their non-fluorinated counterparts, proteins, and peptides containing fluorinated amino acids, such as leucine and phenylalanine, show noticeably higher thermal and chemical stability [14]. Fluorine has the ability to efficiently alter the mechanical characteristics, bioactivities, assembly behaviors, and structures of proteins and peptides [15]. Moreover, nucleic acids are more resistant to nucleases when fluorine atoms are added to nucleobases or deoxyribose. Furthermore, fluorinated chemicals have been used in the development of probes for magnetic resonance imaging (MRI). The fluorine nucleus (^{19}F) has ½ (half) spin and a gyromagnetic ratio of 40.08/42.58 MHz/T, which is substantially larger than that of the ^{1}H nucleus (about 350 vs 210 ppm) [16].

A subclass of fluorinated chemicals known as perfluorocarbons has low toxicity and is bioinert, which makes it useful in a wide range of medical applications [17]. Because of their low polarisation and weak van der Waals interactions, its viscosity is low, high vapor pressure, and a low coefficient of friction. Because of its high affinity for gases such as oxygen and carbon monoxide, PFCs are useful as gas delivery carriers. They have been used in place of blood to reduce ischemia and hypoxia, which has enhanced photodynamic cancer therapy [18]. PFCs are widely used as contrasting agents in imaging modalities [19], lubricating solvents for self-cleaning and non-fouling applications [20], and propellant-based inhalers for respiratory treatments [21, 22] due to their low intermolecular forces and greater vapor pressures. PFCs can be expelled from the lungs thanks to the reticuloendothelial system, which removes them from the body. Additionally, the chemical structures of renowned fluoropolymers depicted in Figure 8.2 illustrate that polymers incorporating fluorine yield high-performance materials such as Teflon, PTFE, and PVDF.

With more than 60% of the fluoropolymer industry going to PTFE, it is clear that this material is the most widely used [24]. The strong repulsion between fluorine atoms in its structure gives it a helical configuration that makes PTFE chemically inert to nearly all substances. Because of this characteristic, medical instruments such as catheters and vascular grafts frequently have coverings made of PTFE [25]. But PTFE's helical chains

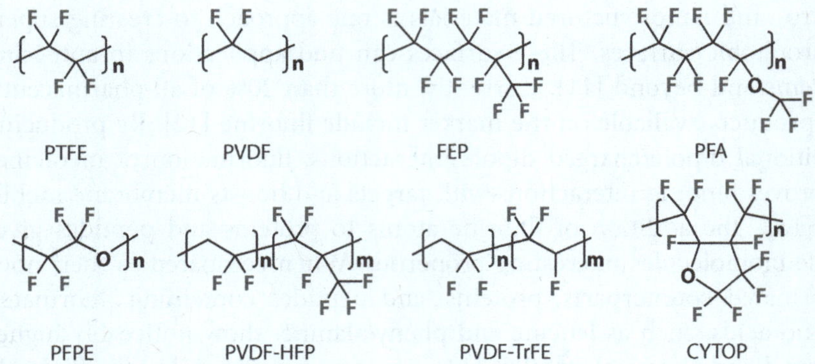

Figure 8.2 Chemical structures of well-known fluoropolymers [23].

make movement simple, which lowers fatigue resistance. Its impenetrable aspect is further enhanced by the way that light is scattered by its crystalline nanostructures. Expanded variants such as PFA and FEP were created to get around these restrictions; in many cases, this involved adding a trifluoromethyl chain length to decrease crystallinity while improving mechanical strength, optical clarity, and thermal characteristics.

The high viscosity and melt temperature of PTFE present difficulties for 3D modeling and other production techniques. As a result, PFPE, a fluoropolymer with superior optical clarity, was developed especially for devices and coatings that would withstand chemical exposure [26]. PVDF is a well-known fluoropolymer in the field of materials science and chemistry that is distinguished by its superior tensile strength, manufacturing ease, and dielectric constants when compared to PTFE [27]. Piezoelectric characteristics are exhibited by PVDF and its copolymers, including PVDF-TrFE and PVDF-HFP. These characteristics have been used to stimulate stem cells for neural regeneration without the need for external electrical input and to imitate sinoatrial node cells in cardiac pacing [28].

Applications for PVDF and related copolymers can be found in a variety of industries, including tissue engineering, and user-friendly medical electronics [29–31]. They can be used to wrap sensors, biological tissues, and optoelectronic devices because of their resistance to chemicals [32]. The purpose of this study is to examine the latest developments in the specialized use of fluoropolymers, including their production, processes of assembly, and benefits in certain biomedical applications.

A class of materials known as fluoropolymers has shown great promise for use in a range of biological applications. Their exceptional chemical resistance, low surface energy, and biocompatibility are just a few of the special qualities that make them desirable for use in the creation of cutting-edge biomaterials. This section will explore the world of fluoropolymers, highlighting their unique qualities and explaining why they have emerged as the preferred material for medication delivery and wound healing. In order to solve important healthcare issues, we will reveal the workings of these polymers' interactions with biological systems and lay the groundwork for their incorporation into electrospun fiber mats.

8.2 Electrospinning Technique

Electrospinning is a versatile and widely used method for creating nanofibrous structures from polymer solutions or melts. It has found use in a variety of fields, including medicine, tissue engineering, filtration, and electronics. An electric field is used to draw and stretch polymer fibres out of a liquid solution or melt. Understanding the fundamentals of electrospinning is critical for optimizing the production of nanofibers with the desired properties for specific applications.

Here's a more detailed explanation of the electrospinning process and the main variables that influence nanofiber creation:

Polymer Solution/Melt Preparation:
For solution-based electrospinning, a polymer solution or melt is prepared, typically with the use of a solvent. The polymer solution should have suitable rheological properties to allow the formation of continuous fibers during the electrospinning process.

Spinneret and Electrospinning Setup:
The prepared polymer solution is loaded into a syringe equipped with a metallic needle or spinner for controlled ejection. Subsequently, a high voltage is applied to the spinneret, generating a strong electric field between the spinneret and a grounded collector. This process facilitates the electrospinning of polymer fibers.

Formation of Taylor Cone:
As the electric field strength increases, the droplet at the tip of the spinneret elongates and forms a cone-like structure known as the Taylor cone.

Fiber Formation:
When the electric field reaches a critical point, a jet of polymer solution is ejected from the Taylor cone and directed toward the grounded collector.

During flight, solvents evaporate or polymers solidify, resulting in the formation of nanofibers.

Collector:
The grounded collector, which can be a rotating drum or a stationary plate, collects the electrospun nanofibers and forms a nonwoven mat from them.

Key Variables Affecting Electrospinning:

Voltage:
Higher voltage produces a stronger electric field, which influences the diameter and morphology of the nanofibers.

Solution Flow Rate:
The rate of dispensing the polymer solution influences fibre diameter and uniformity.

Spinneret-to-Collector Distance: The distance between the spinneret and collector determines the polymer jet's flight time, which affects fibre alignment and mat density.

Polymer Concentration:
Higher polymer concentrations can result in increased fibre diameter and mat thickness.

Solvent Type and Evaporation Rate:
The choice of solvent influences the polymer solutions viscosity as well as the rate of evaporation of solvent during electrospinning.

Temperature and Humidity:
Environmental factors can influence the drying kinetics of the polymer solution, influencing fibre morphology. Understanding and optimising these variables allows researchers and engineers to tailor the electrospinning process to specific applications, such as creating custom electrospun fibre mats for medication delivery and wound healing in the medical field.

Advantageous for a range of biomedical applications. Here are some notable bio-applications of electrospun fiber mats:

a. Drug Delivery Systems:
Pharmaceutical Encapsulation: Electrospinning can be used to encapsulate drugs, proteins, and other therapeutic agents within nanofibers. The mats'

high surface area and porosity enable controlled release of the loaded substances, making them excellent choice for drug delivery systems.
Targeted Delivery: Surface modification of nanofibers can improve drug delivery to specific cells or tissues.

b. Wound Healing and Tissue Engineering:
Electrospun fiber mats can mimic the structure of the extracellular matrix (ECM), fostering a supportive environment that enhances cell adhesion, promotes cell proliferation, and supports tissue regeneration. By replicating the ECM's structure, these fiber mats provide a scaffold for healthy tissue growth and repair, making them a valuable tool in the field of tissue engineering and regenerative medicine. This biomimetic approach has the potential to revolutionize the way we approach tissue repair and replacement, enabling the creation of functional tissue substitutes that can replace or repair damaged or diseased tissues.

Scaffold for Tissue Engineering: Nanofiber mats are excellent scaffolds for tissue engineering applications, as they provide mechanical support and guide cell growth.
Incorporation of Growth Factors: Growth factors and bioactive molecules can be incorporated into fibers to improve tissue regeneration.

c. Biodegradable Implants:
Temporary Support Structures: Electrospun nanofibers, particularly those made of biodegradable polymers, can act as temporary support structures for damaged tissues. As the fibers degrade, they are replaced by regenerated tissue.

d. Anti-microbial Applications:
Infection Prevention: Electrospun fibre mats infused with antimicrobial agents can be used to prevent infections in wound dressings and other medical devices.
Filters and Membranes: Nanofiber mats can act as effective filters in medical applications, removing bacteria and other contaminants.

e. Diagnostic Tools:
Electrospun fibers can be functionalized for sensing applications, such as detecting specific biomolecules or pathogens for diagnostic use.

f. Dental Applications:
Sensing Applications: Electrospun fibers can be functionalized to detect specific biomolecules or pathogens for diagnostic purposes.

g. Neural Tissue Engineering:
Nerve Regeneration: Electrospun fibers can be designed to promote nerve regeneration, which could have applications in treating nerve injuries or disorders.

h. Vascular Grafts:
Electrospun nanofiber mats can be used to create scaffolds for vascular tissue engineering, which helps to develop artificial blood vessels.

8.2.1 Characterization of Electrospun Fiber Mats

Because of their distinct structural and functional properties, electrospun fiber mats have received a lot of attention and applications in a variety of bio-related fields. These nanofiber mats have been demonstrated to be highly characterizing electrospun fiber mats is an important step in understanding their properties and optimizing their performance for particular applications. Electrospun nanofiber mats are evaluated using a variety of analytical techniques, including morphology, structure, composition, and mechanical properties. Here are some common methods used to characterize electrospun fiber mats:

a. Scanning Electron Microscopy (SEM):
Purpose: SEM is widely used to examine the surface morphology and structure of electrospun fiber mats.
Information Obtained: Provides high-resolution images, allowing visualization of fiber diameter, morphology, and distribution.

b. Transmission Electron Microscopy (TEM):
The purpose of TEM is to provide detailed information on the internal structure of electrospun fibers.
Information obtained: Allows for the study of nanoscale features such as the core-shell structure and nanoparticle distribution within the fibers.

c. Atomic Force Microscopy (AFM):
Purpose: AFM is employed to study the topography and mechanical properties of electrospun fibers at the nanoscale.

Information Obtained: Measures surface roughness, fiber diameter, and mechanical properties such as stiffness and elasticity.

d. Fourier Transform Infrared Spectroscopy (FTIR):
The FTIR technique is used to identify and analyze the chemical composition of electrospun fibers.
Information obtained includes functional groups, polymer structure, and potential chemical interactions within the fibers.

e. X-ray Diffraction (XRD):
The purpose of using XRD is to investigate the crystallinity and orientation of electrospun fibers.
Information obtained: Detects crystal phases, degree of crystallinity, and molecular orientation.

f. Thermogravimetric Analysis (TGA):
TGA is used to investigate the thermal stability and degradation of electrospun fibers.
Information obtained: Data is provided on weight loss, decomposition temperature, and polymer degradation.

g. Mechanical Testing:
Mechanical testing measures the tensile strength, elasticity, and other mechanical properties of electrospun fiber mats.
Information obtained: Quantifies the mats' mechanical behavior, which is critical for applications such as tissue engineering and filtration.

h. Contact Angle Measurement:
Purpose: Contact angle measurement assesses the wettability and surface energy of electrospun fiber mats.
Information Obtained: Influences the interaction of fibers with liquids and helps understand their potential for applications like drug delivery.

i. Porosity and Pore Size Analysis:
Contact angle measurement determines the wettability and surface energy of electrospun fiber mats.
Information obtained: Influences how fibers interact with liquids and aids in understanding their potential for applications such as drug delivery.

j. Rheological Analysis:
The purpose of rheological studies is to assess the flow behavior of the polymer solutions used in electro spinning.
Information obtained: aids in the optimization of solution properties for successful electrospinning.

Medication distribution and wound healing are two critical components of healthcare that have received a great deal of attention and research over the years. These professions can significantly improve the quality of life for many people and are critical in the treatment of a wide range of medical conditions, from minor injuries to chronic diseases. Understanding the history and significance of medication delivery and wound healing is essential for recognizing the contribution of novel materials and technologies, including fluorinated polymer electrospun fiber mats of fluoropolymer composites, to the advancement of contemporary healthcare.

8.2.2 Drug Delivery: Precision Medicine at its Core

In contrast, drug delivery is a fundamental aspect of contemporary medicine that aims to maximize the therapeutic effectiveness of pharmacological drugs while reducing their adverse effects. From the earliest days of medicine, drug design and administration have advanced significantly, from straightforward herbal cures to intricate pharmaceutical formulas. Precise control over the release of medicinal substances is the goal of drug delivery systems, which guarantees that the right amount and at the right time reach the intended spot in the body.

Drug distribution is important because it has the potential to completely change how many different medical problems are treated. It makes possible the advancement of personalized medicine, in which treatments are customized based on each patient's own genetic composition and medical background. Furthermore, by offering convenient and user-friendly dosage forms, it significantly contributes to increased patient compliance. However, difficulties like some medications' low bioavailability, the requirement for prolonged release, and the desire to prevent harmful side effects have prompted scientists to look at cutting-edge materials and delivery methods.

8.2.3 Wound Healing: A Vital Biological Process

Our bodies constantly go through the complex and dynamic biological process of wound healing when we are hurt. The body starts a sequence of carefully controlled processes to regenerate and repair damaged tissue, regardless of the severity of the injury—whether it's a minor cut, surgical incision, or more complicated wound. To restore the afflicted area's structural and functional integrity, these processes entail the coordination of cells, growth factors, cytokines, and extracellular matrix components.

The importance of wound healing goes beyond the simple idea of tissue regeneration; it includes the avoidance of infections, the restoration of organ function, and the maintenance of the general health and well-being of the patient. Impaired tissue regeneration, chronic wounds, and heightened susceptibility to infections are just a few of the issues that can arise from compromised wound healing. These consequences frequently lead to longer hospital stays, a lower quality of life, and higher healthcare expenses. Consequently, the hunt for novel approaches to wound healing has propelled medical research, with an emphasis on creating materials and technologies that accelerate the healing process and enhance results.

8.3 The Mergence of Medication Administration and Wound Healing

The intersection of medication distribution and wound healing has become a novel area of study in healthcare in recent years. In the context of persistent wounds, tissue regeneration, and targeted disease treatment, in particular, this confluence holds promise for addressing complex medical issues. An intriguing direction at this intersection is the application of fluoropolymer composite electrospun fiber mats.

A flexible platform for the targeted administration of medications and bioactive chemicals to the site of an injury or illness is provided by electrospun fiber mats. Due to their high surface area-to-volume ratio, these mats made of nanofibers allow for effective drug loading and long-term release. Fluoropolymers are a great option for these applications due to their special qualities, which include hydrophobicity, chemical stability, and biocompatibility.

We shall examine the complex realm of electrospun fiber mats of fluoropolymer composites for drug delivery and wound healing applications as we go deeper into this chapter. We will finally contribute to the continuous search for better healthcare solutions that improve patients' lives all across the world by learning more about the mechanisms, design considerations, and therapeutic potential of these cutting-edge biomaterials.

8.3.1 Wound Healing Applications

The search for novel approaches to wound healing has resulted in the creation of electrospun fiber mats with specialized uses. We will take a tour of the field of wound healing applications in this section, looking at the ways in which electrospun fiber mats—especially those that use fluoropolymers—can be engineered to encourage tissue regeneration, quicken the healing process, and act as an infection-prevention barrier. We'll learn how these biomaterials could transform wound care, from design concerns to mechanisms of action.

Human health relies heavily on effective wound management, as wounds - defined as any injury or surgical incision that disrupts tissue continuity - require proper care to facilitate the healing process and achieve optimal functional and aesthetic outcomes. Polymer materials have garnered significant interest in biomedical applications due to their unique properties, such as biocompatibility, biodegradability, and versatility, making them ideal for wound dressing applications. Contrary to the long-held belief that wounds heal faster when kept dry and uncovered, research has shown that maintaining a moist environment enhances the healing process by preventing tissue desiccation, promoting enzymatic activity that removes debris during the early stages of wound healing, and fostering a conducive environment for cell proliferation and tissue regeneration. This understanding has led to the development of advanced wound dressings that maintain a moist environment, protect the wound from further damage, and promote optimal healing outcomes. By leveraging the unique properties of polymer materials and the benefits of a moist environment, healthcare professionals can provide more effective wound care, improving patient outcomes and quality of life.The body's natural procedure for returning wounded tissues to a state that closely resembles their pre-injury state is known as wound healing. Dermal and epidermal tissue regeneration is required for this restoration, and it happens in a sequence of precise steps. The many processes that comprise wound healing can be divided into four stages: hemostasis (i) inflammation (ii), granulation tissue development (iii), and remodelling (iv). Individual differences exist in the length and severity of various phases related to wounds, including age, nutritional state, behaviour, systemic disorders, and medicine, in addition to the size, depth, aetiology, and cause of the wounds.

The three types of wound dressings that are commonly used are interactive, bioactive, and passive. Gauze and other passive wound dressings generally function as a covering for the injured region. Interactive treatments for wounds, like polymer dressings, have permeability. Bioactive wound dressings promote tissue growth at the wound sites by containing antibacterial agents, nanosilver, growth-promoting chemicals, and other bioactive compounds [33, 34]. In 1962, Winter's study examined the effects of polymer film dressings on pig wounds. The results showed that, in comparison to wounds exposed to air, wounds treated with these dressings healed about twice as quickly and showed significant proliferation of epithelial tissues. Wound care is currently one of the most lucrative and quickly growing medical industries for both retailers and distributors.

Wet and dry wound dressing techniques have been used historically. The dry technique was once thought to be the best course of action for healing wounds. Prior to the 1970s, nonwoven rayon, blends of polyester and cotton, and woven cotton gauze were among the frequently used commercial wound dressing materials. These traditional dressing techniques performed a number of functions, including absorbing exudates, cushioning, facilitating drying, hiding the wound, and preventing infection. To hasten the process of further drying the injured areas, these dressings were easily removed.

Winter's further studies in 1965 [32] showed that wounds exposed to air formed scabs that impeded the process of epithelialization, hence delaying the healing process. On the other hand, wounds that did not create a scab healed more quickly in a damp atmosphere. This happens because moist wound dressings help to create new skin during the healing process without causing scab development since they match the skin's natural permeability and ideal water content of 85% [35].

Fluorinated polymer nanofibers are undoubtedly a good alternative for healing burns or wounds on a variety of human skin types. These tiny polymer nanofibers, which are made from polymers and spun using an electric field, can be applied straight to burnt or wounded skin areas to create a fibrous membrane or mat treatment [36]. By directly applying a fibrous membrane dressing to the site of injury, scar tissue that is typically formed by conventional wound treatments is prevented and natural skin or tissue growth is facilitated.

Nonwoven nanofiber membranes used for wound dressing have pore diameters that are remarkably tiny, ranging from 500 nm to 1 μm. Aerosol particle trapping systems help to prevent bacterial penetration of wounds due to their size. Enhanced surface area, between 5 and 100 m^2/g, is essential for cutaneous delivery and efficient fluid absorption [37]. In 2008, Powell and associates [38] carried out a comparative analysis of collagen nano-scaffolds made by an Electrospinning process vs freeze-drying.

According to the results of their investigation, scaffolds made by the freeze-drying process can be substituted with electrospun nano-scaffolds.

Choosing the right polymer has a big impact on the quality of electrospun nanofibers, which are important for applications that involve wound healing. To aid in wound healing, polysaccharides like cellulose, chitosan, and hyaluronic acid, as well as proteins like collagen and silk, are frequently electrospun. Renowned for its hemostatic and antibacterial qualities, chitosan is a great natural polymer to use for creating scaffolds for wound healing applications.

Concurrently, artificial polymers like PVA, PCL, PLA, and PEO are extensively used in wound-healing applications. Synthetic polymers are more suitable for the electrospinning process because they are soluble in a variety of solvents and have a higher mechanical strength than their natural counterparts. By combining natural and synthetic polymers and modifying their morphological, mechanical, and degrading qualities to suit the unique needs of each patient, wound-healing electrospun nanoscaffolds (NS) can be created [39, 40].

Electrospun nanofibers offer a multifunctional approach to wound management, serving as a breathable and protective wound dressing that permits oxygen permeation, promotes healthy tissue regeneration, acts as a barrier against contamination and infection, and prevents dehydration, maintaining a moist environment conducive to healing. Various synthetic and natural polymers have been successfully electrospun for wound treatment and dressing applications, including collagen, chitosan, carboxymethyl chitosan, silk fibroin, polyurethane (PU), and polyethylene glycol (PEG). These polymers bring unique properties to the table, such as enhanced tissue regeneration, antimicrobial and hemostatic effects, biocompatibility, biodegradability, flexibility, and drug release capabilities. By leveraging these electrospun polymer nanofibers, researchers and clinicians can develop advanced wound dressings that promote wound healing, tissue repair, and regeneration, ultimately improving patient outcomes and quality of life [41–44].

Chronic wounds are defined as wounds that take longer than eight weeks to heal naturally, indicating a delay or disruption in the healing process. The two main causes of the difficulty in healing chronic wounds are: (1) an imbalance between the processes of tissue creation and degradation in the injured area; and (2) insufficient or inadequate production of a functional Extracellular Matrix (ECM). The healing process is greatly aided, directed, and organised by the typical ECM. Consequently, the sensible approach to treating non-healing/chronic wounds is the development of sophisticated wound-dressing materials and techniques that perfectly imitate the absent tissues.

Electro spin nanofibers provide an effective way to imitate the extracellular matrix (ECM). Applying NFs to wounds causes cells to identify them ECM, which promotes adhesion, differentiation, re-epithelialization, growth, and the deposition of temporary matrices. There are various benefits to including NFs into contemporary wound dressings [45]. NFs have the ability to replace natural elastin and collagen because of their size, shape, and mechanical characteristics. Furthermore, because of their gas permeability, NFs effectively eliminate wound exudates and guard against wound infection. Moreover, NFs composed of hydrophilic materials promote the migration of epithelial cells from the wound's periphery into its centre by creating a moist wound environment, which improves healing [45]. Furthermore, the biomimetic property of nanofibrous membranes helps to prevent the formation of scars and promote wound healing [46, 47].

NFs can be loaded with active materials by chemical bonding on the surface or physical absorption. Physical entrapment is now the most widely used method of loading drugs into nanofluidizers (NFs) because it may protect the medication from unfavorable environmental factors and regulate drug release over an extended period of time [45, 47–49]. Furthermore, the drug's integration into the fibre matrix is a somewhat simple and useful process.

Using this technique, the medicine to be loaded is evenly distributed throughout the polymer solution prior to the Electrospinning process starting. Since there is less time for drug recrystallization during fibre creation, amorphous medicines are frequently preferred. A nearly linear and continuous release pattern follows an initial burst effect in the ideal drug release profile from such NFs.Core-shell NFs provide a way to construct a drug reservoir system that controls the pace of drug diffusion while protecting the loaded medication. The first medication put into NFs was the antibiotic tetracycline, which was encased in polyethylene and vinyl acetate copolymers and PLA mixes. Afterwards, a wide variety of compounds, such as biomolecules, anti-inflammatory agents, antiseptics, antibiotics, antifungals, non-steroidal and anticancer medications, and nucleic acids, have been incorporated into or adsorbed onto NFs.

An innovative method used to treat bacterial infections in long-term wounds is to incorporate silver nanoparticles into NFs, which have strong antibacterial qualities [50]. Electrospun Nanofibers are characterized by their high surface area and fiber holes, which stimulate cell signaling pathways to rapidly trigger fibroblastic cell responses. These NFs have a great deal of promise for creating cosmetic masks that are intended to cleanse and repair skin. Electrospinning skin masks have a very large surface area, which makes it easier for the ingredients to be applied to the skin. Because of their great flexibility, these masks are simple to apply and remove from the skin without causing any pain [51–53]. Adding more components to the electrospun NF matrix is necessary to improve skin therapy [54]. Furthermore, these scaffolds have the potential to draw cells into the dermal layer, which encourages the release of components of the Extracellular Matrix (ECM), such as growth factors, collagen, and cytokines, which are critical for healing wounds. Nonwoven NFs show up as excellent options for materials used in wound dressings.

8.4 Drug Delivery Applications

The precise dispensing of medications is among the most crucial elements of contemporary medical care. Electrospun fibre mats provide a high surface area and configurable drug release, which can help with innovative drug delivery systems. This section will cover the kinetics and controlled release mechanisms that govern drug delivery, as well as the inclusion of pharmaceuticals and bioactive materials into fluoropolymer composites electrospun mats. By the time you're done reading this section, you'll know how these materials, which provide a customized and continuous release of medicinal substances, might drastically alter how medications are administered.

Evaluating the extent of drug release is essential, and most release methods require the drug and delivery vehicle to be kept apart. The drug stacking capability of nanoparticles is clearly defined by the amount of drug bound for each polymer mass, or, to put it another way, by the moles of the drug for every milligram of polymer or by mg drug/mg polymer. A percentile in relation to the polymer can also be used to quantify it. These variables, which include gel filtration, centrifugal ultrafiltration, ultracentrifugation, and ultrafiltration, would be determined using a range of methods, including UV spectroscopy and HPLC. Similar methods are used in drug release studies and drug-loading assays, which look at the drug release mechanism over time [55].

According to the concept behind their use, nanoparticles could deliver a concentrated dose of medication close to tumour targets, hence enhancing retention and permeability or permitting active targeting by ligands on the surface of the nanoparticles [56]. To minimize drug exposure to healthy tissues, nanoparticles restrict drug distribution to the target organ. For successful medication delivery, nanoparticles must also be capable of targeting malignancies beyond MPS-rich areas. Over the past ten years, much time and work have gone into creating PEGylated nanoparticles, commonly referred to as "stealth" particles, that are imperceptible to macrophages and phagocytes. The blood-brain barrier (BBB) is the main barrier preventing the entrance of new drugs for the central nervous system (CNS).

The theory behind utilizing nanoparticles to target tumours is that they could deliver a concentrated dose of medication close to the target, enhancing permeability and retention or permitting ligands on the surface of the nanoparticles to actively target the tumor [56]. By restricting drug distribution to the intended organ, nanoparticles lessen the amount of drug exposure to healthy tissues. The ability of nanoparticles to target tumors outside of MPS-rich regions is necessary for their efficacy as a medication delivery mechanism. The creation of PEGylated nanoparticles, commonly referred to as "stealth" particles, that are imperceptible to phagocytes and macrophages has required a great deal of time and work over the past ten years. The blood-brain barrier is by far the biggest barrier to the introduction of novel drugs for the central nervous system.

Currently, the main focus in medicine is on how to give therapeutic medications to specific patients in a way that is most compatible with their bodies. The key to improving drug absorption in human bodies is to reduce the geometry of the drug and the coating material that is intended for encapsulation. The use of particulate nanofibers in drug delivery applications stems from a basic idea: raising the surface area of the drug and its carrier both accelerates the pace at which the drug particles dissolve [35].

While research on fluoropolymer composites of electrospun mats for medication administration is still in its infancy, developments in this field must focus on developing a more intelligent and useful delivery method in order to boost output and effectiveness [35, 57]. The capacity to regulate the drug's release rate is a crucial prerequisite for drug delivery applications. The carrier containing the medicine or active ingredient is released at a consistent pace in a controlled-release drug delivery system. Because of their high surface-area-to-volume ratio, electrospun nanofibers provide an effective drug delivery system. Controlling the release profile in electrospun nanofibers requires adroit manipulation of the NFs' shape, porosity, and content. Therapeutic chemicals can be encapsulated within PNFs due to the ease of use of the electrospinning process.

There are several ways to load pharmaceuticals into electrospun NFs: coaxial or emulsion ES techniques can be used for coating, encapsulating, or embedding the drugs. Drug molecules are physically or chemically incorporated into or cross-linked to the surface of electrospun NFs in the coating process [58]. Another technique is to electrospin a polymer solution that contains the medicinal substance. To ensure that drug particles in a solid state disperse well into the host polymer solution, this approach requires compatibility between the drug and the polymer solution. It is particularly difficult to incorporate bioactive medications utilizing the embedding approach into electrospun NFs. The method of electrospinning that is most often used to encapsulate medicines into PNFs is coaxial and emulsion electrospinning.

By far the most promising method for developing sustained or controlled drug release systems at the moment is coaxial electrospinning. The drug can be encapsulated or encased at the core of the nanofibers, shielded by the polymeric shell, thanks to this technique, which keeps the polymer and drug solution apart until they exit the nozzles. Since drug release through this mechanism only happens by diffusion across a barrier, burst release is completely avoided [59].

The drug reservoirs that make up the shell part are mostly made of biodegradable polymers. By employing both drug-driven and degradation-driven pathways, these polymers enable the control of drug-release kinetics [59]. Although electrospun NFs are good drug transporters, there are still a number of serious limitations. Large-volume medicines added to PNFs using conventional electrospinning processes frequently lead to poor control over drug release because of drug agglomeration and adsorption on the NFs' surface. NFs with less initial burst or bulk release have been made using the coaxial and emulsion electrospinning techniques, nevertheless [60].

The application of surfactants helps reduce surface tensions and NF diameter, which promotes medication homogeneity and lessens the burst impact, according to several studies [35, 61–63] Electrospun nanofibers have been successfully used to load a variety of medicines, such as proteins, antibiotics, and anticancer agents, for drug delivery applications. The electrospinning method not only monitors regulated drug release but also greatly reduces toxicity and improves therapeutic efficacy. Because electrospun nanofibers may hold many medicines, they are an excellent choice for a drug delivery carrier [54].

A comprehensive analysis has been conducted on the effects of biodegradable and non-biodegradable polymers on obtaining regulated and sustained drug release utilizing electrospun nanofibers. It is possible to create drug-loaded particulate nanofiber membranes using synthetic and natural polymers. In tissue engineering, natural polymers are ideal for simulating the Extracellular Matrix, while synthetic polymers added to medications are advantageous for drug delivery applications [64–67]. When it comes to drug release characteristics, electrospun nanofibers outperform traditional casting methods [54].

Drug loading, localization, and release are all affected by the physicochemical properties of both the carrier matrix and the drug itself. As a result, the solubility of the medication in the polymer solution, as well as the interaction of the polymer and the drug in their solid phases, determine the nanofiber loading mechanism. Because the electrospinning process takes so little time, the ideal drug would be more soluble and amorphous. Several characterization studies, such as differential scanning calorimetry (DSC), FTIR spectroscopy, X-ray diffraction (XRD), and proton nuclear magnetic resonance (H-NMR), have confirmed that the high voltage (HV) used during ES does not affect the chemical, biological, or structural integrity of the loaded drug.

Nanofibers from fluoropolymer composites are extremely useful for targeted therapeutics and medicine delivery. The drug release profile of NFs loaded with pharmaceuticals is based on the idea that a larger NF surface area leads to a faster rate of drug release. The ideal polymers for drug delivery should be able to hold onto the drug of choice and release it only into the intended tissues; they should also be able to maintain maximum drug loading capacity for sustained release over an extended period of time; and finally, they should be able to ensure consistent drug release over an extended period of time in the intended areas.

Electrospun nanofibers of fluoropolymer composites offer several distinct advantages when used as drug delivery systems. Because of their high surface-to-volume ratio, PNFs are especially useful for delivering many insoluble medicines. Solid dispersion is a useful method for increasing the bioactivity and dissolution rates of weakly water-soluble medications. Traditional methods of manufacturing solid dispersion systems, however, have practical drawbacks such as complexity, unpredictable physicochemical properties, poor dosage formation, and difficulty scaling up production.

Electrospun nanofibers have been improved to speed up the dissolution of poorly miscible chemicals. Second, the composition, shape, processing, and microstructure of NF mats can be easily changed to alter drug release profiles. Third, different dosage forms, also known as mid-dosage forms, can be dynamically constructed using NFs to maximise the bioavailability of drug moieties for various delivery routes.

Furthermore, electrospun nanofibers are more efficient at drug encapsulation than other nanotechnologies.

Drug-loaded NFs also have a shorter diffusion channel due to their small NF diameter, effective drug release, and high NF surface area, which aids in mass transfer. The one-dimensional structure of electrospun PNFs containing pharmaceuticals has numerous advantages, including the ability to modify the drug molecule's biopharmaceutical and biopharmacokinetic properties to achieve desired therapeutic outcomes [68, 69].

8.5 Biocompatibility and Toxicity Assessment

When using biomaterials in healthcare, it is critical to ensure their safety and biocompatibility. In this section, we will look at the stringent evaluation procedures used to see how electrospun fibre mats, particularly those containing fluoropolymers, interact with biological systems. We will look at *in-vivo* and *in vitro* studies that provide information about the materials' toxicity, biocompatibility, and clinical translation potential.

8.6 Future Trends and Challenges

The field of fluoropolymer composite electrospun fibre mats is dynamic, with shifting trends and persistent challenges. We will look ahead in this area, discussing new research directions, emerging trends, and potential applications of these materials in healthcare. In order to advance the field, we will acknowledge the challenges that researchers face while also considering potential solutions.

8.7 Conclusion

As this chapter concludes, let us summarise the main findings and lessons learned from our investigation into electrospun fiber mats of fluoropolymer composites for drug delivery and wound healing. We'll look at this field's significant contributions to healthcare, as well as its potential to address unmet medical needs. The knowledge synthesised in the chapter's conclusion will inspire professionals and researchers to continue pushing the boundaries of biomaterial science in search of better healthcare solutions.

AI disclosure statement: The author acknowledges the use of AI-based tools for minor language editing or grammatical correction. However, all intellectual contributions, analyses, and conclusions presented in this chapter are entirely the author's own.

References

1. Yang, X., Chen, Y., Zhang, C., Duan, G., Jiang, S., Electrospun carbon nanofibers and their reinforced composites: Preparation, modification, applications, and perspectives. *Composites, Part B*, 249, 110386, 2023.
2. Plunkett, R.J., The history of polytetrafluoroethylene: discovery and development of high performance polymers, their Orig. Dev., 261–266, 1986.
3. Ameduri, B. and Boutevin, B., *Well-Architectured Fluoropolymers: Synthesis, Properties and Applications*, Technology & Engineering, Elsevier, 2004.
4. Drobny, J.G., *Technology of Fluoropolymers*, CRC Press, Boca Raton, 2008.
5. Ebnesajjad, S., *Introduction to Fluoropolymers: Materials, Technology, and Applications*, Plastics Design Library, William Andrew, 2020.
6. Cui, Z., Drioli, E., Lee, Y.M., Recent progress in fluoropolymers for membranes. *Prog. Polym. Sci.*, 39, 1, 164–198, 2014.
7. Gardiner, J., Fluoropolymers: origin, production, and industrial and commercial applications. *Aust. J. Chem.*, 68, 1, 13–22, 2014.
8. Liu, M., Wang, S., Jiang, L., Nature-inspired superwettability systems. *Nat. Rev. Mater.*, 2, 7, 117, 2017.
9. Yabu, H. and Shimomura, M., Single-step fabrication of transparent superhydrophobic porous polymer films. *Chem. Mater.*, 17, 21, 5231–5234, 2005.
10. Tourkine, P., Le Merrer, M., Quéré, D., Delayed freezing on water repellent materials. *Langmuir*, 25, 13, 7214–7216, 2009.
11. Muller, K., Faeh, C., Diederich, F., Fluorine in pharmaceuticals: looking beyond intuition. *Science*, 317, 5846, 1881–1886, 2007.
12. Ojima, I., Exploration of fluorine chemistry at the multidisciplinary interface of chemistry and biology. *J. Org. Chem.*, 78, 13, 6358–6383, 2013.
13. Marsh, E.N.G., Fluorinated proteins: from design and synthesis to structure and stability. *Acc. Chem. Res.*, 47, 10, 2878–2886, 2014.
14. Loureiro, J.A., Crespo, R., Börner, H., Martins, P.M., Rocha, F.A., Coelho, M., *et al.*, Fluorinated beta-sheet breaker peptides. *J. Mater. Chem. B*, 2, 16, 2259–2264, 2014.
15. Tirotta, I., Dichiarante, V., Pigliacelli, C., Cavallo, G., Terraneo, G., Bombelli, F.B., *et al.*, 19F magnetic resonance imaging (MRI): from design of materials to clinical applications. *Chem. Rev.*, 115, 2, 1106–1129, 2015.
16. Zhang, T., Zhang, Q., Tian, J.H., Xing, J.F., Guo, W., Liang, X.J., Perfluorocarbon-based nanomedicine: emerging strategy for diagnosis and treatment of diseases. *MRS Commun.*, 8, 2, 303–313, 2018.

17. Cheng, Y., Cheng, H., Jiang, C., Qiu, X., Wang, K., Huan, W., *et al.*, Perfluorocarbon nanoparticles enhance reactive oxygen levels and tumour growth inhibition in photodynamic therapy. *Nat. Commun.*, 6, 1, 18, 2015.
18. Houvenagel, S., Moine, L., Picheth, G., Dejean, C., Brûlet, A., Chennevière, A., *et al.*, Comb-like fluorophilic-lipophilic-hydrophilic polymers for nanocapsules as ultrasound contrast agents. *Biomacromolecules*, 19, 8, 3244–3256, 2018.
19. Gao, D., Xu, M., Cao, Z., Gao, J., Chen, Y., Li, Y., *et al.*, Ultrasound-triggered phase-transition cationic nanodroplets for enhanced gene delivery. *ACS Appl. Mater. Interfaces*, 7, 24, 13524–13537, 2015.
20. Leach, C.L., Kuehl, P.J., Chand, R., McDonald, J.D., Nasal deposition of HFA-beclomethasone, aqueous fluticasone propionate and aqueous mometasone furoate in allergic rhinitis patients. *J. Aerosol Med. Pulm. Drug Deliv.*, 28, 5, 334–340, 2015.
21. Epstein, A.K., Wong, T.S., Belisle, R.A., Boggs, E.M., Aizenberg, J., Liquid-infused structured surfaces with exceptional anti-biofouling performance. *Proc. Natl. Acad. Sci.*, 109, 33, 13182–13187, 2012.
22. Lv, J. and Cheng, Y., Fluoropolymers in biomedical applications: state-of-the-art and future perspectives. *Chem. Soc. Rev.*, 50, 9, 5435–5467, 2021.
23. Emdadi, B., Yusifov, R., Moradi, R., Synthesis methods and characterization techniques of fluoropolymers, in: *Advanced Fluoropolymer Nanocomposites*, pp. 29–77, Woodhead Publishing Series in Composites Science and Engineering, 2023.
24. Wong, T.S., Kang, S.H., Tang, S.K., Smythe, E.J., Hatton, B.D., Grinthal, A., *et al.*, Bioinspired self-repairing slippery surfaces with pressure-stable omniphobicity. *Nature*, 477, 7365, 443–447, 2011.
25. Bharatwaj, B., Wu, L., Whittum-Hudson, J.A., da Rocha, S.R., The potential for the non-invasive delivery of polymeric nanocarriers using propellant-based inhalers in the treatment of Chlamydial respiratory infections. *Biomaterials*, 31, 28, 7376–7385, 2010.
26. Conti, D.S., Bharatwaj, B., Brewer, D., da Rocha, S.R., Propellant-based inhalers for the non-invasive delivery of genes *via* oral inhalation. *J. Control. Release*, 157, 3, 406–417, 2012.
27. Royo-Gascon, N., Wininger, M., Scheinbeim, J.I., Firestein, B.L., Craelius, W., Piezoelectric substrates promote neurite growth in rat spinal cord neurons. *Ann. Biomed. Eng.*, 41, 1, 112–122, 2013.
28. Cardoso, V.F., Correia, D.M., Ribeiro, C., Fernandes, M.M., Lanceros-Méndez, S., Fluorinated polymers as smart materials for advanced biomedical applications. *Polymers*, 10, 2, 161, 2018.
29. Zhang, Y., An, Q., Tong, W., Li, H., Ma, Z., Zhou, Y., *et al.*, A new way to promote molecular drug release during medical treatment: a polyelectrolyte matrix on a piezo-electricdielectric energy conversion substrate. *Small*, 14, 37, 1802136, 2018.

30. Kim, S.H. and Ko, H.C., Double-sided printed circuit textiles based on stencil-type layer-by-layer coating with PEDOT: PSS: Ag nanowires and chitosan for electrothermochromic displays. *J. Mater. Chem. C*, 7, 46, 14525–14534, 2019.
31. Karlicky, F., Kumara Ramanatha Datta, K., Otyepka, M., Zboril, R., Halogenated graphenes: rapidly growing family of graphene derivatives. *ACS Nano*, 7, 8, 6434–6464, 2013.
32. Gupta, B., Agarwal, R., Alam, M.S., Textile based smart wound dressings. *Indian J. Fibre Text. Res.*, 35, 2, 174–187, 2010.
33. Winter, G.D., Formation of the scab and the rate of epithelization of superficial wounds in the skin of the young domestic pig. *Nature*, 193, 293–294, 1962.
34. Winter, G.D., A note on wound healing under dressings with reference to perforated—film dressings. *J. Invest. Dermatol.*, 45, 4, 299–302, 1965.
35. Huang, Z.M., Zhang, Y.Z., Kotaki, M., Ramakrishna, S., A review on polymer nanofibers by electrospinning and their applications in nanocomposites. *Compos. Sci. Technol.*, 63, 15, 2223–2253, 2003.
36. Hasan, M.M., Alam, A.K.M.M., Nayem, K.A., Application of electrospinning techniques for the production of tissue engineering scaffolds: a review. *Eur. Sci. J.*, 10, 15, 1857-7431, 2014.
37. Powell, H.M., Supp, D.M., Boyce, S.T., Influence of electrospun collagen on wound contraction of engineered skin substitutes. *Biomaterials*, 29, 7, 834–843, 2008.
38. Ahmad, J., Deshmukh, K., Habib, M., Hagg, M.B., Influence of TiO2 nanoparticles on morphological thermal and solution properties of PVA/TiO2 nanocomposite membranes. *Arabian J. Sci. Eng.*, 39, 6805–6814, 2014.
39. Sadeghi-Avalshahr, A., Nokhasteh, S., Molavi, A.M., Khorsand-Ghayeni, M., Mahdavi-Shahri, M., Synthesis and characterization of collagen/PLGA biodegradable skin scaffold fibers. *Regener. Biomater.*, 4, 5, 309–314, 2017.
40. Abrigo, M., McArthur, S.L., Kingshott, P.L., Electrospun nanofibres as dressings for chronic wound care: advance, challenges and future prospects. *Macromol. Biosci.*, 14, 6, 772–792, 2014.
41. Pawde, S.M. and Deshmukh, K., Influence of c irradiation on the properties of polyacrylonitrile films. *J. Appl. Polym. Sci.*, 110, 2569–2578, 2008.
42. Uttayarat, P., Jetawattana, S., Suwanmala, P., Eamsiri, J., Tangthong, T., Pongpat, S., Antimicrobial electrospun silk fibroin mats with silver nanoparticles for wound dressing application. *Fibers Polym.*, 13, 8, 999–1006, 2012.
43. Deshmukh, K., Ahmad, J., Hagg, M.B., Fabrication and characterization of polymer blends consisting of cationic polyallylamine and anionic polyvinyl alcohol. *Ionics*, 20, 957–967, 2014.
44. Gao, Y., Bach, T.Y., Zhu, Y., Kyratzis, I.L., Electrospun antibacterial nanofibers: production, activity and *in vivo* applications. *J. Appl. Polym. Sci.*, 131, 18, 40797–40810, 2014.

45. Rosic, R., Kocbek, P., Pelipenko, J., Krist, J., Baumgartner, S., Nanofibers and their biomedical use. *Acta Pharm.*, 63, 295–304, 2013.
46. Mohanapriya, M.K., Deshmukh, K., Ahamed, M.B., Chidambaram, K., Pasha, S.K.K., Zeolite 4A filled poly (3, 4-ethylenedioxythiophene): (polystyrenesulfonate) and polyvinyl alcohol blend nanocomposites as high-k dielectric materials for embedded capacitor applications. *Adv. Mater. Lett.*, 7, 996–1002, 2016.
47. Rath, G., Hussain, T., Chauhan, G., Garg, T., Goyal, A.K., Collagen nanofiber containing silver nanoparticles for improved wound-healing applications. *J. Drug Target.*, 24, 6, 520–529, 2016.
48. Deshmukh, K., Ahamed, M.B., Polu, A.R., Sadasivuni, K.K., Pasha, S.K.K., Ponnamma, D., AlMaadeed, M.A.A., Deshmukh, R.R., Chidambaram, K., Impedance spectroscopy, ionic conductivity and dielectric studies of new Li+ ion conducting polymer blend electrolytes based on biodegradable polymers for solid state battery applications. *J. Mater. Sci.: Mater. Electron.*, 27, 11410–11424, 2016.
49. Aytimur, A. and Uslu, I., Promising materials for wound dressing: PVA/PAA/PVP electrospun nanofibers. *Polym. Plast. Technol. Eng.*, 53, 7, 655–660, 2014.
50. Kim, S.E., Heo, D.N., Lee, J.B., Kim, J.R., Park, S.H., Jeon, S.H., Kwon, I.K., Electrospun gelatin/polyurethane blended nanofibers for wound healing. *Biomed. Mater.*, 4, 4, 044106, 2009.
51. Deshmukh, K., Ahamed, M.B., Deshmukh, R.R., Pasha, S.K.K., Bhagat, P.R., Chidambaram, K., Biopolymer composites with high dielectric performance: interface engineering, in: *Biopolymer composites in electronics*, vol. 1, pp. 27–128, Elsevier Publications, Amsterdam, 2017.
52. Pillai, C.K.S. and Sharma, C.P., Electrospinning of chitin and chitosan nanofibres. *Trends Biomater. Artif. Organs*, 22, 3, 179–201, 2009.
53. Haider, A., Haider, S., Kang, I.K., A comprehensive review summarizing the effect of electrospinning parameters and potential applications of nanofibers in biomedical and biotechnology. *Arabian J. Chem.*, 11, 8, 1165–1188, 2015.
54. Magenheim, B., Levy, M.Y., Benita, S., A new *in vitro* technique for the evaluation of drug release profile from colloidal carriers-ultrafiltration technique at low pressure. *Int. J. Pharm.*, 94, 1–3, 115–123, 1993.
55. Pathak, Y. and Thassu, D., Drug delivery nanoparticles formulation and characterization. *Drugs Pharm. Sci.*, 191, 239251, 2009.
56. Yu, D.G., Zhu, L.M., White, K., White, C.B., Electrospun nanofiber-based drug delivery systems. *Health*, 1, 2, 67–75, 2009.
57. Elahi, F., Lu, W., Guoping, G., Khan, F., Core-shell fibers for biomedical applications—a review. *J. Bioeng. Biomed. Sci.*, 3, 1, 1–14, 2013.
58. Focarete, M.L. and Gualandi, C., Potentialities of electrospun polymeric nanofibers in the biomedical field. *J. Tissue Sci. Eng.*, 4, 1, 1–3, 2013.

59. Gupta, K.C., Haider, A., Choi, Y., Kang, I.K., Nanofibrous scaffolds in biomedical applications. *Biomater. Res.*, 18, 2, 27–38, 2014.
60. Deshmukh, K., Ahamed, M.B., Deshmukh, R.R., Pasha, S.K.K., Chidambaram, K., Sadasivuni, K.K., Ponnamma, D., AlMaadeed, M.A.A., Eco-friendly synthesis of graphene oxide reinforced hydroxypropyl methyl cellulose/polyvinyl alcohol blend nanocomposites filled with zinc oxide nanoparticles for high-k capacitor applications. *Polym. Plast. Technol. Eng.*, 12, 1240–1253, 2016.
61. Zeng, J., Xu, X., Chen, X., Liang, Q., Bian, X., Yang, L., Jing, X., Biodegradable electrospun fibers for drug delivery. *J. Control. Release*, 92, 3, 227–231, 2003.
62. Sill, T.J. and Recum, H.A.V., Electrospinning: applications in drug delivery and tissue engineering. *Biomaterials*, 29, 13, 1989–2006, 2008.
63. Sharma, J., Lizu, M., Stewart, M., Zygula, K., Lu, Y., Chauhan, R., Yan, X., Guo, Z., Wujcik, E.K., Wei, S., Multifunctional nanofibers towards active biomedical therapeutics. *Polymers*, 7, 2, 186–219, 2015.
64. Ma, Z., Kotaki, M., Ramakrishna, S., Surface modified non-woven Polysulphone (PSU) fiber mesh by electrospinning: a novel affinity membrane. *J. Membr. Sci.*, 272, 1–2, 179–187, 2006.
65. Deshmukh, K., Ahamed, M.B., Deshmukh, R.R., Pasha, S.K.K., Sadasivuni, K.K., Polu, A.R., Ponnamma, D., AlMaadeed, M.A.A., Chidambaram, K., Newly developed biodegradable polymer nanocomposites of cellulose acetate and Al2O3 nanoparticles with enhanced dielectric performance for embedded passive applications. *J. Mater. Sci.: Mater. Electron.*, 28, 973–986, 2017.
66. Dhandayuthapani, B., Yoshida, Y., Maekawa, T., Kumar, D.S., Polymeric scaffolds in tissue engineering application: a review. *Int. J. Polym. Sci.*, 2011, 290602, 19, 2011.
67. Sathapathy, K.D., Deshmukh, K., Ahamed, M.B., Sadasivuni, K.K., Ponnamma, D., Pasha, S.K.K., AlMaadeed, M.A.A., Ahmad, J., High-quality factor poly (vinylidene fluoride) based novel nanocomposites filled with graphene nanoplatelets and vanadium pentoxide for high-Q capacitor applications. *Adv. Mater. Lett.*, 8, 288–294, 2017.
68. Rosic, R., Pelipenko, J., Kristl, J., Kocbek, P., Baumgartner, S., Properties, engineering and applications of polymeric nanofibers: current research and future advances. *Chem. Biochem. Eng. Q.*, 26, 4, 417–425, 2012.
69. Illa, M.P., Khandelwal, M., Sharma, C.S., Bacterial cellulose – derived carbon nanofibers as anode for lithium – ion batteries. *Emergent Mater.*, 1, 3–4, 1–6, 2018.

9

Exploring the Potential of PEDOT:PSS: From Fundamental Properties to Diverse Applications

Rishi Mohanan, Raneesh Balakrishnan* and Karthika S.

Department of Physics, Catholicate College, Pathanamthitta, Kerala, India

Abstract

PEDOT:PSS (poly(3,4-ethylenedioxythiophene): polystyrene sulfonate) has become a vital conducting polymer due to its electrical conductivity, mechanical deformability, and chemical stability. Herein, we present an overview of the fundamental properties of PEDOT:PSS, from synthesis and structural aspects to its mode of conductivity. PEDOT:PSS is a versatile material and can be used in diverse areas: one of the most important is bioelectronics, where PEDOT:PSS plays as a suitable material for fabricating devices such as biosensors, implants, and flexible electrodes. The chapter further elaborates on recent work to tailor the properties of PEDOT:PSS to improve its performance within a given application, for example, biocompatibility and stability in aqueous environments. Various functionalizations of PEDOT:PSS with additives and composite materials are discussed, demonstrating a wide use in flexible electronic applications, including organic solar cells and wearable electronics. This chapter discusses the potential properties of PEDOT:PSS by reviewing various approaches to optimizing its attributes for different applications, along with possible future research directions regarding conducting polymers.

***Keywords*:** PEDOT:PSS, conductive polymer, thin films, sensors, EMI, flexible electronics

**Corresponding author*: raneesh.b@catholicatecollege.ac.in; ORCID: 0000-0003-0056-559X

Raneesh Balakrishnan, Sobi K. Chacko and Visakh P.M. (eds.) Electroactive Polymer-Based Smart Materials and Applications, (235–264) © 2026 Scrivener Publishing LLC

9.1 Introduction

Following the identification of the first conductive polymer, polyacetylene, other synthetic variants were created and thoroughly investigated [1]. Due to their many benefits, including their unique blend of mechanical and optoelectronic properties, chemical diversity, low processing costs, easy handling, tuneability, and biocompatibility, conductive polymers have become increasingly appealing for various electronic applications. Due to the alternating single and double bonds between the carbon atoms in their structure, conductive polymers allow electricity to pass through, unlike normal polymers, mostly used for packing and insulation in the plastics sector. These artificial materials are chained organic macromolecules whose conjugated backbones supply delocalized π electrons by sp^2 hybridization, which results in the molecular framework generating true electrical conductivity [2]. These low-density organic molecules can be used as conductors to replace the device's metal components or as semiconductors for various electronic device applications, depending on how much electricity they can conduct. Compared to traditional electronic materials like metals, polymers offer significant advantages including low weight and inert, non-corrosive characteristics [3, 4].

Much research has been put on the conductive polymer PEDOT:PSS as a possible material platform for upcoming wearable device technology. Its mechanical compliance, conductivity, and active optoelectronic versatility have enabled various applications, such as bioelectronic devices, wearable displays, and upcoming augmented/virtual reality (AR/VR) systems. By altering the conjugated polymer backbones oxidation state, conducting polymers, such as PEDOT:PSS, can have their optical and electronic characteristics electronically modified [5]. Because of PEDOT:PSS excellent electrical conductivity, hygroscopicity, and flexibility make it an ideal choice for detecting devices for gas, biological, electrochemical, and physical sensing applications [6]. It also has the ability to function as a scaffold for cell proliferation and differentiation because it is nontoxic and biocompatible [7]. Furthermore, in hostile conditions, including acidic and alkaline ones, as well as high temperatures, PEDOT:PSS exhibits exceptional environmental resilience. PEDOT:PSS-based hydrogels are anticipated to find extensive use in the domains of environmental pollution management, bioelectronics, drug delivery, tissue engineering, interface engineering, sealing, wound dressing, and energy storage [8–10].

9.2 Structure of PEDOT:PSS

One of the polythiophene derivatives, poly(3,4-ethylenedioxythiophene) (PEDOT), is regarded as the widely studied conducting polymer, drawing special attention. In addition, PEDOT offers several benefits, chief among them being its high electrical conductivity (about 400–600 S/cm) and environmental solid stability in conducting state at room temperature. It is also noted that PEDOT can be treated in an aqueous solution with poly(styrene sulfonate) (PSS) in proper ratio, resulting in the formation of a homogeneously dispersed polymer complex called PEDOT:PSS [11].

The positively charged PEDOT chains in the PEDOT:PSS structure form a polyionic complex with the negatively charged PSS chains through an electrostatic bond. Figure 9.1 displays the chemical structures of the PSS and the PEDOT. The PSS structure's –OH groups dissociated to H^+ in the water. The H^+ actively engaged the thiophene double bond in the PEDOT, causing hydrogen bonding. Oxygen greater electronegativity, however, drew electrons from the C–O shared electron pairs. The occurrence caused the C to take on a positive charge in the C–O bond. Ultimately, the C^+ of the PEDOT structure and the O^- of the PSS structure established an ionic connection [12].

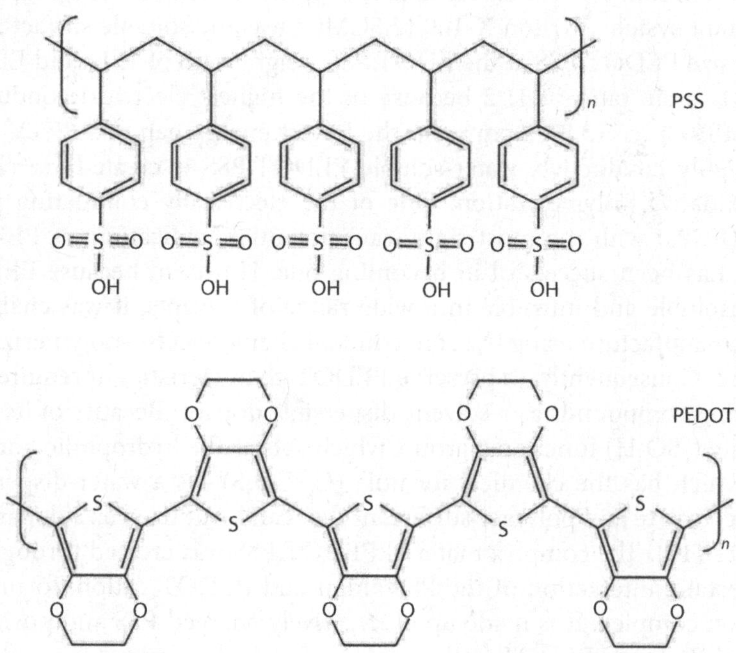

Figure 9.1 Chemical structure of PEDOT:PSS [12] (CC BY).

9.3 Synthesis of PEDOT:PSS

Aniline, pyrrole, phenols, diphenyl sulfide, thiophenols, and thiophene are examples of aromatic chemicals from which polymers (oligomers) are synthesized by chemical oxidative polymerization (COP). Strong electron donor characteristics and a strong oxidation sensitivity define the monomers employed in COP. An applied potential (i.e., electrochemical oxidative polymerization) or an inorganic oxidizing agent can oxidize monomers. COP causes the monomer molecules to produce cation or cation radical sites, which starts the polymer synthesis process [13]. Powdered PEDOT:PSS is a viable preparation option. Research has examined the impact of varying the weight ratio between 1:2.5 and 1:7.5 on the electrical conductivity of the powder, with variations ranging from 0.3 to 1.3 S cm^{-1}. According to Wichiansee *et al.* produced PEDOT:PSS powder by COP at room temperature, employing ferric sulfate ($Fe_2(SO_4)_3$) as the catalyst and sodium persulfate ($Na_2S_2O_8$) as the oxidizing agent. The electrical conductivity of the synthesized PEDOT:PSS powder was 27.5 ± 0.6 S cm^{-1} [14]. Chanthaanont *et al.* produced PEDOT:PSS powder in an aqueous solution at room temperature with $Na_2S_2O_8$ as the oxidizing agent and $Fe_2(SO_4)_3$ as the catalyst its electrical conductivity was between 11.69 ± 0.06 S cm^{-1} [15]. But in some special cases, Sakunpongpitiporn *et al.* investigated three surfactant systems, Triton X-100 (2.5CMC) was the suitable surfactant to synthesize PEDOT:PSS at the EDOT:PSS weight ratio of 1: 11 and EDOT: $Na_2S_2O_8$ mole ratio of 1: 2 because of the highest electrical conductivity of 1879.49 ± 13.87 S cm^{-1} and the lowest energy gap of 1.80 eV [16]. The highly conductive, water-soluble PEDOT:PSS is created *via* chemical oxidative polymerization. One of the electrically conducting polymers (ECPs) with the most significant potential applications is PEDOT, which has been successful in becoming one. However, because PEDOT was insoluble and infusible in a wide range of solvents, it was challenging to manufacture using the conventional chemo/electro-polymerization method. Consequently, to preserve PEDOT characteristics, it requires the following compounds as a solvent-dispersible dopant: Because of its ester sulphate ($-SO_3H$) functional group, which acts as the hydrophilic portion, PSS, which has the chemical formula $(C_8H_8O_3S)_n$, is a water-dispersible polyelectrolyte and polymer surfactant that can be utilized as a dopant for PEDOT [17]. The complex material PEDOT:PSS was created through the electrostatic interaction of the PSS anion and PEDOT cation, forming a poly-ion complex. It is made up of negatively charged PSS and positively charged P- doped PEDOT [18].

9.4 Properties of PEDOT:PSS

PEDOT:PSS exhibits a range of significant properties, including high electrical conductivity, tunable work function, excellent electrochemical stability, notable bio compatibility and morphological adaptability. The key properties are summarized in Figure 9.2 and are discussed in detail in the following sections.

9.4.1 Conductivity

Commercially accessible water-based dispersion of PEDOT:PSS is easily produced into thin films on substrates, which can be rigid or flexible by coating or printing. PEDOT:PSS is also mechanically flexible and exhibits great transparency in the visible spectrum. On the other hand, as-prepared PEDOT films deposited from an aqueous solution show less than 1 S/cm of conductivity. Compared to ITO, its conductivity is noticeably lower. ITO on plastic has a conductivity of roughly 2000 S/cm, but ITO on glass has a 3000-6000 S/cm conductivity. Polar solvents can enhance the conductivity of PEDOT:PSS significantly as the transparent electrode [19, 20]. Organic solvents, including sorbitol, glycerol, meso-erythritol, xylitol, diethylene glycol, dimethyl sulfoxide (DMSO), tetrahydrofuran (THF), N,N-dimethyl formamide (DMF), and methoxy ethanol. While acetone belongs in the non-OH group, other organic solvents like ethanol (EtOH) and methanol (MeOH) can be categorized within the OH group.

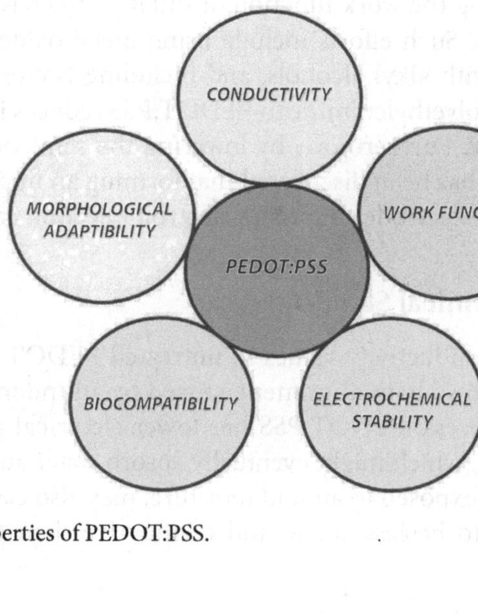

Figure 9.2 Properties of PEDOT:PSS.

According to the procedure and the solvent, a notable improvement in conductivity ranging from 0.5 S cm^{-1} to 1233 S cm^{-1} is obtained. This procedure, known as the solvent additive method, is typically applied before the PEDOT:PSS dispersion deposition using a high boiling solvent. This kind of strategy is straightforward, adaptable, and simple to use, according to an earlier study on the solvent addition method to increase the conductivity of PEDOT:PSS. Researchers find this method more practical for assessing how different grades of PEDOT:PSS would produce varying conductivities following the addition of the solvent. For instance, the conductivity of the PEDOT:PSS has greatly increased by adding ethylene glycol or DMSO [21].

9.4.2 Work Function

Perovskite solar cells (PSCs), organic solar cells (OSCs), organic light-emitting diodes (OLEDs), and quantum dot (QD)-LEDs are examples of high-performance thin-film optoelectronic devices that require at least one electrode with a low enough work function (WF) to inject electrons into the lowest unoccupied molecular orbitals of a particular semiconductor.

Polymer electrolytes are ideal for batteries and supercapacitors, especially in high-performance solid-state applications, because of their excellent ionic conductivity and low voltage polarization. Unfortunately, existing polymer electrolytes frequently have poor electrical conductivity and adequate electronic capabilities despite having strong ionic conductivity [22]. Improving the work function of PEDOT:PSS has been a focus of numerous efforts. Such efforts include using metal oxides as dopants, surface treatment with alkyl alcohols, and including Nafion. It has been found that adding polyethylenimine to PEDOT:PSS reduces its work function to about 4.0 eV. Furthermore, by lowering the semiconducting layer's vacuum level, it has been discovered that forming an interfacial dipole between conductive electrodes improves electron injection capability [23].

9.4.3 Electrochemical Stability

The low electrical conductivity values in untreated PEDOT:PSS films are usually caused by low chain alignment caused by abundant PSS. When PSS is present in excess, PEDOT:PSS has lower electrical performance. The presence of PSS, which might eventually absorb water and shorten the lifetime of the films exposed to air and moisture, may also cause untreated PEDOT:PSS films to be less stable and conductive than treated films.

The use of specific additives, such as crosslinking agents DVS and DMSO like solvents, to prevent polymer redispersion or post-process delamination is necessary to meet the specifications for wearable electrodes and films based on PEDOT:PSS. Treatments such as methanol (MeOH) combined with 1-ethyl-3-methylimidazolium bis(trifluoromethyl sulfonyl)imide (EMITFSI) and ethylene glycol (EG) combined with EMITFSI significantly enhance both electronic and ionic conductivity, thereby improving the electrochemical stability of PEDOT:PSS. These treatments enhance the charge storage and charge transfer capacity, resulting in lower total impedance and faster charging of PEDOT:PSS-based devices. Research indicates that post treatments of this polymer can substantially increase ionic conductivity, facilitating accelerated electrochemical reactions [24, 25].

9.4.4 Biocompatibility

PEDOT:PSS, is now recognized in bioelectronics because of its exceptional biocompatibility. It can be used in several different biological applications, including drug delivery systems, sensors, and scaffolds for tissue engineering. Based on current studies, there are various ways to improve PEDOT:PSS biocompatibility.According to Yang *et al.* PEDOT:PSS hydrogels that were generated without the use of potentially dangerous reagents had remarkable conductivity (up to 3265 S m^{-1}) and good biocompatibility, which made them perfect for real-time electrochemical sensing involving living cells [26]. The non-toxic character of PEDOT:PSS has also been confirmed by its successful use as a drug carrier in silk fibroin hydrogels, which exhibited high cell viability (≥80%) for human skin fibroblast cells. Moreover, it has been proven that incorporating PEDOT:PSS into soft conductive electrodes can improve cell endurance and attachment while reducing foreign body reactions in biological conditions. Applications involving brain tissue engineering benefit substantially from this capability. In addition, novel post-treatment techniques involving ionic liquids increase conductivity and PEDOT:PSS's biocompatibility. These developments make PEDOT:PSS a flexible option for applications in regenerative medicine and wearable medical technology [27].

9.4.5 Morphological Adaptability

PEDOT:PSS has been attributed to phase-separated grain formation; each grain is composed of a thin PSS-rich shell around a PEDOT-rich core, where excess PSS functions as an insulator instead of a dopant, as shown by photoelectron spectroscopy. Organic additives, such as sorbitol, ethylene glycol, and dimethyl sulfoxide, can be added to PEDOT:PSS aqueous dispersions before casting or spin coating and are effective in enhancing the conductivity of polymer films [28]. To process PEDOT:PSS film, a complex must develop with PSS. This complex causes the PEDOT segments to disperse in an aqueous solution and acts as a charge-balancing counterion. The conductivity of the films is doubled when treated with organic solvents like glycerol and dimethyl sulfoxide (DMSO). Initially, this phenomenon was thought to be caused by either a solvent-induced conformational change in the PEDOT chain, which increases the interchain interaction, or the screening effects of polar solvents on the Coulomb interaction between positively charged PEDOT grains and negatively charged PSS dopant [20]. According to recent studies, morphological alterations in the film structure and a phase-segregation process are the cause for this conductivity improvement. It is known that PEDOT:PSS combines to generate a phase-segregated material with excess weakly ionic conducting PSS surrounding highly conducting PEDOT grain. In the phase-segregation scenario, a recent correlation was found between the anisotropic conductivity of PEDOT:PSS and the nanoscopic morphology. It is well known that the polar solvent increases the process and degree of phase separation. Because the surplus PSS is partially phase split, it can effectively insulate the conducting PEDOT:PSS particles to a large extent, improving conduction routes and increasing the overall connections in the PEDOT:PSS network as shown in the Figure 9.3 [29].

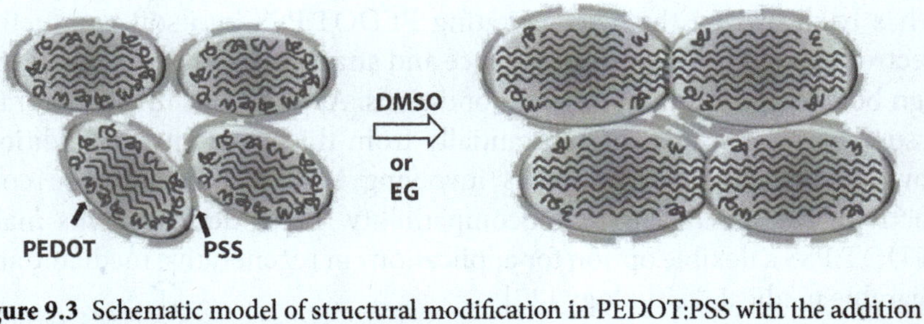

Figure 9.3 Schematic model of structural modification in PEDOT:PSS with the addition of co-solvents. Reproduced from [30] (CC BY).

9.5 Applications of PEDOT:PSS

The diverse applications of PEDOT:PSS are introduced in Figure 9.4 and further elaborated in the subsequent sections.

9.5.1 Energy Harvesting and Storage

A flexible conductive polymer, PEDOT:PSS has been thoroughly investigated for a range of energy-related applications, such as energy harvesting and conversion applications. Due to its special qualities, which include high conductivity, transparency, flexibility, and solution processability, it is a material that is desirable for utilizing a variety of energy sources [31].

9.5.1.1 Solar Cells

The conductive polymer blend PEDOT:PSS is used in solar cells because of its inherent absorber qualities. The conductive PEDOT part increases absorptance in the solar spectrum and reflectance in the infrared. High absorptance and low emittance can be produced by optimizing the thickness of the PEDOT:PSS coating, which qualifies it for solar energy harvesting applications. According to research, PEDOT:PSS coatings have

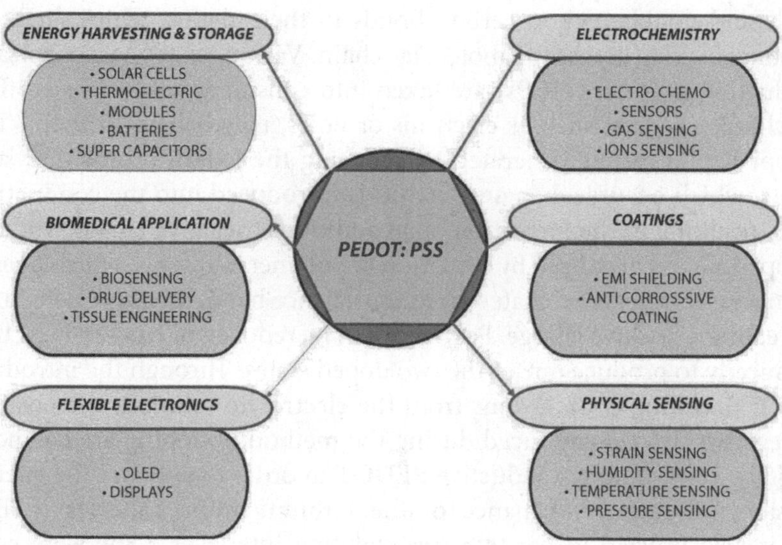

Figure 9.4 Applications of PEDOT:PSS.

excellent reflectance in the infrared region and good intrinsic absorption in the solar spectrum, indicating their potential for use efficiently in solar cell technology [32].

9.5.1.2 *Thermoelectric Modules*

Flexible thermoelectric energy generating systems use PEDOT:PSS, a promising thermoelectric material with low heat conductivity and strong electrical conductivity. Applying an alternating electric field during film production improves the thermoelectric characteristics of PEDOT:PSS films, which are likely due to increased carrier mobility and enhance crystallinity. This approach presents a viable means of enhancing the efficiency of thermoelectric modules based on PEDOT:PSS for energy harvesting purposes. Chonan Y. *et al.* examined the impact of an AC electric field on PEDOT:PSS thermoelectric characteristics during film formation. Three different electrical conductivity values were found for the films created under DC, square wave and sine wave electric fields: 1079 S/cm, 1373 S/cm, and 1244 S/cm. The Seebeck coefficient remained nearly constant irrespective of the AC electric field frequency and waveform [33].

9.5.1.3 *Batteries*

Organic conducting polymers exhibit conductivity due to alternating single and double carbon-carbon bonds in their chains, which facilitate π-orbital overlap across the molecular chain. Various states of intrinsically conducting polymers (ICPs) are taken into consideration because different charge carriers, such as electrons or holes, may behave in them. The n-doped state can be generated by reducing the length of polymer segments, which occurs when an electron is introduced into the conduction band, resulting in the formation of an additional negative charge, and the p-doped state is produced by oxidation of polymer segments, which occurs when an electron is eliminated from the valence band, leaving a hole, such as creating a positive charge. Polymers can be reduced or oxidized electrochemically to produce one of the two doped states. Through the introduction of matching counter ions from the electrolyte solution, the positive or negative charges produced during the method of doping are balanced out [34]. For example, conductive PEDOT in order to enhance the energy density and rate performance of the Lithium-sulfur batteries (LSBs), PSS is also utilized to create a free-standing interlayer. LSBs were constructed with a lightweight, highly conductive PEDOT:PSS-CNT interlayer inserted into them. PEDOT:CNTs supplied porous transport channels for quick Li+ diffusion, while PSS served as the soft buffer in the polysulfide reservoir to prevent polysulfide migration. Consequently, the polysulfide migration to the lithium anode by chemical contact was significantly impeded by the free-standing composite interlayer. The PEDOT:PSS-CNT interlayer-assembled LSB demonstrated exceptional cycle durability [35]. On the other hand, studies on the electrochemical stability of PEDOT:PSS in battery electrolytes have shown that this material is suitable for use as a cathodic binder in lithium-ion batteries. After 200 cycles, there are no discernible changes in impedance in PEDOT:PSS, which demonstrates a stable response with repeated cycling over the operational potential range of 4.2V. This indicates that this polymer can be used as a viable cathode binder in lithium-ion batteries [36, 37].

9.5.1.4 *Supercapacitors*

PEDOT:PSS is a conductive polymer that is favored for improving the performance of supercapacitors because of its excellent thermal stability, low electrical resistance, and simplicity of application. PEDOT:PSS exhibits notable properties that could lead to the creation of solid-state polymer capacitors that are superior to their liquid counterparts in terms of low equivalent series resistance (ESR), high capacitance, and improved operating conditions. A significant development in energy storage technology, the use of PEDOT:PSS in supercapacitors closes the gap between batteries and traditional capacitors and opens the door to creative and sustainable energy storage solutions [38]. To build a complete capacitor structure, valve metals such as tantalum, niobium, or aluminum must be present as the negative electrode counterparts to PEDOT:PSS, which plays a crucial function as the positive electrode. Because of its lightweight and ease of processing, aluminum is especially preferred. However, to maximize its surface area and create an oxide layer that is necessary for good capacitor operation, aluminum must go through key surface treatments called etching and anodization [39, 40].

9.5.2 Electrochemistry

9.5.2.1 *Gas Sensing*

The theory behind PEDOT:PSS gas sensors is based on the special characteristics of this conducting polymer. Composites like meldola blue (MDB) and prussian blue (PB) are incorporated to improve sensitivity for hydrogen peroxide (H_2O_2) detection by facilitating electron transfer processes [41, 42]. Adding gold and silver nanoparticles into PEDOT: PSS composites enhances electron transfer efficiency, raising sensitivity to ammonia (NH_3) [43]. Blending PEDOT: PSS with metal complexes improves sensitivity and selectivity in the case of carbon monoxide (CO) and carbon dioxide (CO_2) detection [44]. The data gathered from these sensors, such as response times and detection limits, demonstrate how well PEDOT: PSS-based gas sensors work to detect particular gases precisely.

9.5.2.2 *Ionic Sensing*

Among these, PEDOT:PSS is particularly interesting with respect to ionic sensors for comprehensive ion detection due to its unique characteristics. As an ion-to-electron transducer layer, PEDOT:PSS improves the potentiometric stability and detection performance by controlling charge transport in the sensor. The sensor can only move certain ions while excluding interfering counter ion by interfacing PEDOT:PSS with an Ion-Selective Membrane (ISM). This can detect anions like Cl^- and cations like Na^+, K^+, $Ca^{2+}$and Mg^{2+}. These studies demonstrate the excellent ionic selectivity of PEDOT:PSS for a variety of ions and illustrate potential applications to develop sensor technology in various fields, including but not limited to biomedical, environmental, and analytical applications [45].

9.5.3 Biomedical Applications

PEDOT:PSS is a particularly promising material for biological and biomedical applications. This gives it a number of advantages over conventional metallic electrodes, with better electrochemical behavior that is ideal for recording and stimulating electrically excitable cells. PEDOT:PSS can be used in various bioelectronic applications due to its production, doping, and modification versatility. PEDOT based materials are perfect for next-generation bio electronic systems because of their bio stability, scalability, and reproducibility, which provide them with benefits over many traditional metals. Given these great properties, PEDOT:PSS is highly attractive and promising and can unleash new possibilities for the advancement of biomedical technologies, including cerebral stimulation [46].

9.5.3.1 *Biosensing*

Structural, Architectural, and Morphological Optimization of PEDOT:PSS plays a key role in enabling sensing applications across diverse analytes and environments. The work shows that PEDOT:PSS can combine well with a range of nanoscale catalytic materials such as metal nanoparticles and carbon nanomaterials, to boost electron transfer rates that are key for sensing. For instance, it has been evidenced that the combination of polymer with gold and silver nanoparticles enhances biosensors sensitivity and stability [47]. This illustrates the importance of nanostructuring and nanoscale engineering methods for effective sensor device realization by preparing PEDOT:PSS composite materials. These PEDOT:PSS-based sensors offer prospective to future investigations in flexible and wearable sensor technologies for environmental monitoring and healthcare [48]. PEDOT:PSS Organic Electrochemical Transistors (OECT) have also been explored in Chemical and Biological Sensing applications. Gate electrode modifications can be used to enhance OECT sensing power and specifically focus on types of nanomaterials such as Platinum nanoparticle or carbon-based materials that have been developed for signal amplification leading directly towards improvement of enzyme sensor sensitivity. The introduction of OECTs to photoelectrochemical technology offers new prospects in biosensing. In addition, potential applications for OECTs as sensitive sensors in detecting specific biomarkers, such as human influenza A virus and prostate-specific antigen, show the flexibility and usefulness provided by OECT-based immunosensors. OECTs are extremely sensitive transducers that can convert biological reactions into electronic signals. It is highly potent in enhancing the selectivity and performance of sensors across numerous world-class biosensing applications [49].

9.5.3.2 *Drug Delivery*

Several methods have been developed for the incorporation of drugs into PEDOT:PSS matrices with electrochemical control over drug release. One way is to drug-load in PEDOT:PSS during the electrochemical polymerization, such as dexamethasone in the polymer matrix [6]. However, PEDOT:PSS films can be loaded with drugs post-fabrication by soaking them in drug solutions, allowing independent tuning of the polymer and drug loading. The electrical stimulation causes redox cycling, which leads to swelling and de-swelling of the polymer matrix.

This helps in increasing drug diffusion and consequently initiates a sustained release mechanism for loaded pharmaceuticals out from PEDOT:PSS. In particular, PEDOT:PSS can be combined with other materials (e.g., hydrogels) and constitute composite drug delivery systems; it can be used alongside the hydrogel to provide direct electrical control over drug release [50]. There are still challenges ahead to improve drug loading into PEDOT:PSS and reduce burst release, ensuring sustained delivery. The development of PEDOT:PSS-based systems designed for localized drug delivery to specific tissues or cells, which, again combined with other stimuli-responsive materials, can further improve drug delivery application [51].

9.5.3.3 *Tissue Engineering*

According to Pange *et al.*, the vascular endothelium functions as a barrier to the movement of fluids and solutes between blood and tissues. scientists use tissue engineering, electrical and physical simulations, and other techniques to replace sick or insufficient endothelium cells and produce healthy structures [52, 53]. Ingvar (1920) was the first to show that electrical current substantially affected neuronal cells. Subsequent research has been conducted on the responses of many cellular and tissues, including neurons, epithelium, stem cells, and bone cells, to electrical and electromagnetic fields, both *in vitro* and *in vivo*. It is well known that during several biological processes, including cell division, a weak current can produce electrical fields. Although it has been demonstrated that electrical stimulation causes considerable angiogenesis *in vivo* and that an electric field temporarily increases the amount of filamentous actin in cultured bovine aortic endothelial cells by 80%, the exact mechanisms by which an electric field affects vascular endothelial cells remain to be fully explained [54]. Selecting appropriate materials is necessary for tissue engineering to electrically stimulate the intended tissue. Since conductive scaffolds are made of conducting polymers, these materials are appealing for application as biomedical implants [55], due to their good conductivity, stability, and redox characteristics. Additionally, they can fuse the electrical properties of metal with the properties of plastic, and it has recently been demonstrated that they have an impact on the *in vitro* functioning of mammalian cells [56, 57]. These conjugated systems have an intrinsically low conductivity, and doping is the procedure that turns these materials into conductors. To regulate and optimize the electrochemical properties of the material, it is crucial to regulate the size, charge, and stability of the counter ions. This significantly impacts the *in vitro* or *in vivo* performance of these conductive scaffolds.

The ionomer mixture of PEDOT:PSS is very intriguing in this scenario because of its conductivity, cytocompatibility, and stability. This system is very adaptable for study since the electrical characteristics can be further enhanced by adding additional dopants or treating it with a solvent [10].

9.5.4 Coating

The application of several coating processes, such as drop, spray, and dip coating, for the fabrication of PEDOT:PSS films has also been observed. Dip coating and drop casting are the two simplest but scalable ways to prepare a PEDOT:PSS film on free-form substrates. The substrates are dipped in liquid PEDOT for a set duration at a controlled dipping speed. Droplets of the polymer are deposited onto the substrate surface. After that, the material is dried, and the solvents in the solution evaporate. It is important to remember that both strategies have difficulties in achieving uniform coating dispersion, and exact control over film thickness is difficult. Compared to other types of layering methods, spin-coating is one of the most common because it permits quick creation and reliable reproduction at nanoscale thin films. The process involves dropping liquid PEDOT:PSS in the center of a substrate that is placed on top of a rotating platform with controlled speed and initial acceleration. Centrifugal forces then cause the material to spread outward radially until a thin film forms. The main issue with this is the extra material that is wasted and the uneven thickness around the edges [58]. Another economical method is spray coating, especially when covering uneven surfaces with large areas of thin film. Using a nozzle that is precisely placed away from the substrate, the solution is sprayed onto the substrate surface in this procedure. This produces a smooth coating, but adherence to the substrate may result in problems with surface moisture absorption [59].

9.5.4.1 EMI Shielding

As schematically depicted in Figure 9.5, EMI shields primarily cover the absorption and/or reflection of electromagnetically transmitted waves. The electromagnetic radiation is reflected off the EMI shielding surface due to its reflective properties. In light of the potential harm that the reflected radiation could cause to the surrounding environment, particularly to living things, the absorption process is preferred for shielding efficiency (SE) materials. Reflection of electromagnetic radiation is likewise undesirable for low visibility, especially as it could reach the radar and cause

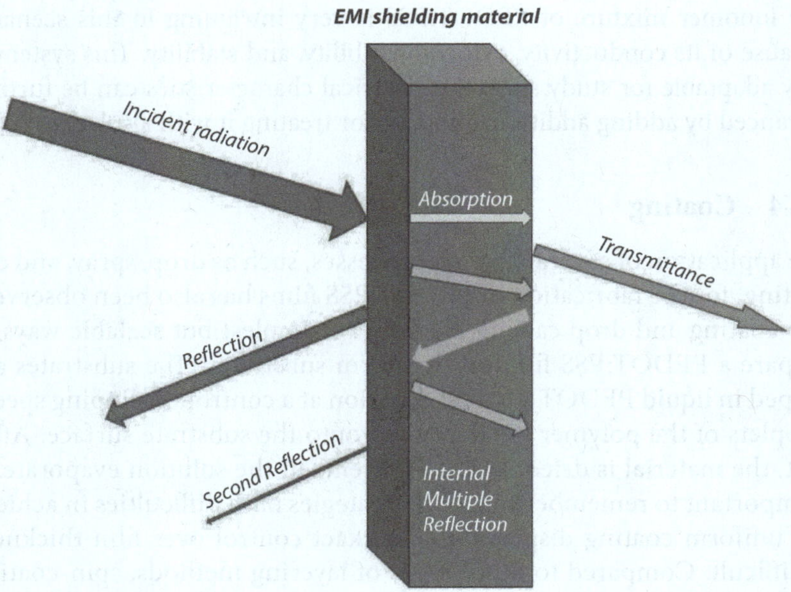

Figure 9.5 EMI shielding schematic diagram.

the object or system to be discovered. Reducing reflection can be achieved by decreasing the interfacial resistance between the fabric and the near medium. The interfacial resistance has an important ability to impact the conductivity of the fabrics, even though EMI shields are usually made of conductive materials [60].

The PEDOT:PSS film has an exceptionally strong mechanical structure along with a highly conductive layer structure. According to Xianrui Wu *et al.*, the combination of the added cellulose nanofibers (CNFs) and the phosphoric acid post-treatment results in high conductivity. PEDOT grains generate highly ordered long-range conducting routes by growing and stacking in an orderly manner along the CNF. Consequently, the film's mechanical strength and conductivity are enhanced. Additionally, the elimination of the insulating PSS and CNF immersing in phosphoric acid can significantly increase conductivity. As a result, when incoming electromagnetic waves interface with the high-density carriers and penetrate the highly conductive film, they will experience enormous ohmic losses and attenuation of electromagnetic wave energy. Additionally, the many internal reflections between the neighboring layers made possible by the densely stacked structure will further dissipate and attenuate the electromagnetic waves [61]. Because of their extreme moisture sensitivity, stand-alone PEDOT:PSS thin films are unsuitable for real-time EMI shielding applications. A crosslinking technique to create moisture-resistant PEDOT:PSS free-standing films was recently disclosed by Mantione *et al.* [24]. Crosslinked conducting polymer (PEDOT:PSS) films have specific remarkable characteristics, including increased durability and conductivity, particularly because they transition from hydrophilic to hydrophobic [62].

9.5.4.2 *Anticorrosive Coating*

Due to their non-toxicity, great environmental sustainability, and easy synthesis ability, intrinsically conducting polymers (CPs) are among the most trusted materials to protect metals against corrosion. Given PEDOT:PSS has strong electrochemical activity at low solid content; hence, efficient paint spread with minimal porosity shows exceptional promise for anticorrosive coatings. The findings of this study illustrate that epoxy paints with PEDOT:PSS greatly enhance mechanical performance, corrosion-inhibiting abilities, and barrier capabilities; all of them are important characteristics of coatings for corrosion protection. PEDOT:PSS represents a relatively inexpensive and effective method to limit corrosion on metallic surfaces, functioning as a conducting additive in epoxy-based coatings, and will be a valuable additive for future innovation in coating technology. In addition, due to the strong electrochemical activity of PEDOT even at low solid content, it enables efficient paint spread, resulting in coatings with minimal porosity and enhanced protective properties [63].

9.5.5 Flexible Electronics

PEDOT:PSS-based materials have been thoroughly studied in recent years for utilization in flexible and stretchable electronic devices and in related fields that is illustrated in the Figure 9.6. However, the pristine PEDOT:PSS electrochromic layer's thick shape impedes ion movement, which results in a narrow optical adjusting range and poor cycling stability [64]. Numerous works are being conducted to optimize the versatility and electrochromic characteristics of PEDOT:PSS by morphological or molecular design. The structure and crystallinity of PEDOT:PSS can be improved by organic polar solvent doping, acid, and processing with alcohol based solvents, which are advantageous to the material's electrical characteristics, electrochemical stability, and homogeneity in film formation [65]. Blending PEDOT:PSS with inorganic nanoparticles can improve its electrochromic characteristics, such as optical contrast, adaptability, and cycling strength, by adding different polymers and additives [66].

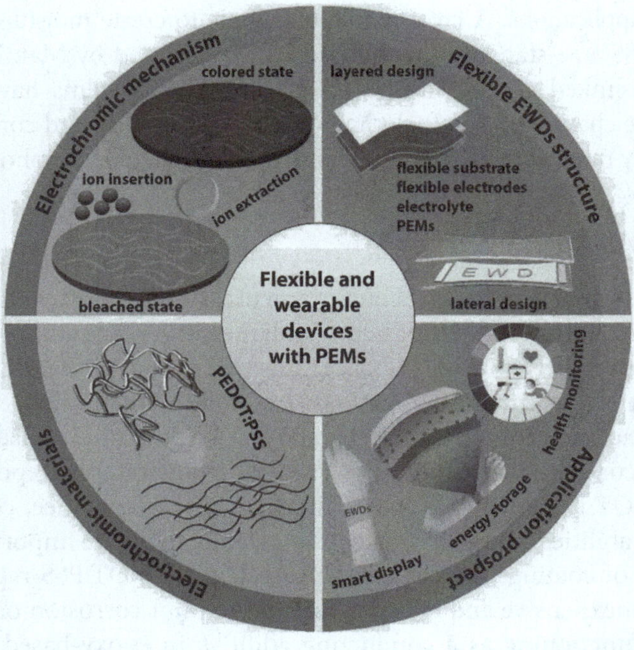

Figure 9.6 PEDOT:PSS-based electrochromic wearable devices (EWDs). Reproduced from [66], with permission from Elsevier.

9.5.5.1 OLED

Organic light-emitting diodes (OLEDs) are widely fabricated using conductive PEDOT:PSS polymer. Due to its specific abilities, it is a suitable candidate for OLED applications, mainly when used as a hole injection layer. It has been demonstrated that incorporating PEDOT:PSS to OLEDs greatly enhances performance. For example, using PEDOT:PSS/CNT films improves OLED functionality; adding CNTs to PEDOT:PSS electrodes has an enormous opportunity to further increase OLED effectiveness [67]. Furthermore, PEDOT:PSS could be combined with other substances, such as MWCNTs, to form nanocomposites that may be delivered as hole injection layers in OLEDs. This resulted in reduced turn-on voltages and improved current - voltage characteristics [68]. The material is utilized to produce transparent, metal-free, light-emitting electrochemical cells (LECs) [69].

9.5.5.2 Displays

Electrochromic display devices using PEDOT:PSS interact with transmitted or reflected light from external light sources. When an external voltage is applied, resulting in visible color shifts. When compared to existing flexible displays, the PEDOT:PSS-based electrochromic device features low driving voltage, low energy consumption, good optical contrast, biostability, and no blind angle or backlight. The benefit of using PEDOT:PSS as the electrochromic functional layer is that it is a fully recyclable, flexible biomass-based electrochromic display system with a rapidly changing speed and good optical brightness [70]. PEDOT:PSS and its composites require improvements in their electrochromic characteristics. Despite adding organic polymers or inorganic components to PEDOT:PSS improves cycling performance, environmental stability, optical contrast, and color change response time, these advancements remain insufficient for practical uses. A promising approach is to explore methods for synthesizing low-cost, high-performance PEDOT:PSS materials using inexpensive raw materials. Efficient PEDOT:PSS-derived compounds can be synthesized in this manner. Concurrently, techniques such as graft copolymerization and *in situ* polymerization can introduce additional reactive groups into the PEDOT:PSS structure, potentially reducing the overall cost of PEDOT:PSS composites while enhancing their microscopic morphology, chemical reactivity, and electrochromic properties. Furthermore, several flexible electrochromic wearable devices based on PEDOT:PSS and its composites are currently under development. Most devices can only achieve limited bending or stretching due to insufficient adhesion between surfaces during fabrication and the low stretchability of polyelectrolyte multilayers (PEM). Phase separation among PEDOT and PSS should be enhanced through an array of modification techniques in order to improve PEDOT structure and arrangement. Moreover, promoting PEDOT clustering is necessary. Active functional groups can be included in PEM surfaces, either through chemical synthesis or polymerization modifications, for better coating of the composite interface [71].

9.5.6 Physical Sensing

In physical sensing applications, PEDOT:PSS has become an important material, especially for sensors of strain, temperature, pressure, and humidity. The material is a good fit for the creation of sensors because of its exceptional electrical conductivity, hygroscopicity, and flexibility. According to studies, PEDOT:PSS-based sensors offer immediate response and great sensitivity, which makes them perfect for a variety of sensing applications [6].

9.5.6.1 *Strain Sensing*

The electrical resistance for a strain sensing application should vary in response to applied strain, and this tunability should be observed over a wide range of strains. In actuality, the strains that are imposed can be small or large. Fatigue research in materials, morphological characterization, and damage detection are some of the applications where tiny strains play a major role [72]. For example, PEDOT:PSS forms hydrogen bonds with the surface carboxylates of the cellulose nanofibers (CNFs) after the protonation of carboxyl groups. Modifications in the PEDOT benzenoid structure contributed to a significant improvement in the strength of the highly conductive aerogel. The conductivity of the aerogel can be doubled through annealing in ethylene glycol vapor, and the strain sensor created by incorporating polydimethylsiloxane (PDMS) into the aerogel exhibited an extended range and enhanced sensitivity [73]. Furthermore, PEDOT:PSS nanowires have a capillary action technique that allows them to cover a wide area. These nanowires were used in a nanomorphology-based strain sensor that showed significantly increased responsiveness. It responded quickly (230 ms) to even the smallest bending variations up to 200 μm and responded negatively or positively to inward or outward bending, respectively. This may be the reason why the in-bending of the sensor alters the arrangement of adjacent PEDOT:PSS nanocrystals, which can lead to a decrease in the impedance of the sensor and an increase in charge exchange efficiency. Outward bending produces the opposite effect. The sensor also shows anisotropic behavior, where parallel-aligned nanowires detect bending strain along their axis but remain unresponsive to strain applied perpendicularly. Since PEDOT:PSS is sensitive to humidity, the environment will have an impact on strain sensors according to PEDOT:PSS that have contact to the air. Injecting PEDOT:PSS into a microchannel as part of a processing technique can solve the issues of connection failure brought on by repeated bending and humidity influence [74].

9.5.6.2 *Humidity Sensing*

When there are high levels of humidity, the hydrophilic and insulating PSS shell absorbs water and expands. This increases the space between neighboring hydrophobic and conductive PEDOT cores, increasing the resistance of PEDOT:PSS for resistive sensing devices. Low humidity, on the other hand, causes water to desorb out of the PSS, which lowers the PSS's volume and the space between neighboring PEDOTs, which lowers the resistivity. In the case of PEDOT: PSS nanowires, an extremely high surface-to-volume ratio allowed them to be easily placed in the environment and speed up the rate at which water was absorbed and released. A gadget of the same kind has been used for examining human breath. Moreover, PEDOT:PSS has been combined with various moisture-sensitive materials to mitigate the limitations associated with the relatively narrow dynamic response range of a single material [75]. There have also been reports of capacitive sensors for humidity. It displays a common humidity sensor that combines interdigital capacitors with PEDOT:PSS film. The dielectric layer of the sensor, made of PEDOT:PSS film, showed good sensitivity to humidity. The humidity-sensitive part performs better for the PEDOT:PSS layer, which acts as a dielectric of this sensor. When there is high humidity, the effective dielectric constant of the PEDOT:PSS based layer will be altered due to expelling absorbed water vapor, which is used as a sensing principle [76]. On the other hand, a PEDOT:PSS thin film-coated piezoelectric resonator will inherently respond to humidity depending on the adsorption/desorption of water molecules onto/from its surface, introducing a mass loading effect. The change of the sensor's resonance frequency and transmission coefficient was reversible with humidity, maintaining good repeatability over the whole humidity range [77, 78].

9.5.6.3 Temperature Sensing

One of the most significant physical parameters for environmental and object monitoring is temperature. It is the most commonly measured parameter in patients by flexible temperature sensors. Ideally, temperature sensors should be placed as close to the body as possible; therefore, flexibility within devices is necessary for the rapid and facile acquisition of temperature information. In this respect, the main key factors that are relevant are excellent reliability and adaptability for temperature sensors [79, 80]. The temperature-sensitive behavior of PEDOT:PSS-based temperature sensors is primarily attributed to the amplification of charge carrier transport and the generation of charge carriers in response to thermal stimuli. As thermal energy increases, the limited carrier hopping and tunneling within individual grains and between adjacent nanosheets improve. As a result, the sensing film resistance exhibits a negative temperature coefficient (NTC), decreasing as the temperature rises. The sensitivity of the temperature sensor is characterized by its temperature coefficient of resistance (TCR) [81]. The hydrophilic PSS forms hydrogen bonds with water molecules at ambient temperature, allowing it to absorb moisture from the surrounding air. Electrons cannot hop between PEDOT:PSS nanoparticles because of the boundaries formed by the insulating PSS chains and non-conductive water. Water loss *via* PSS reduces the apparent dimension of particle barriers at high temperatures and the overall number of particle limits. In order to enhance the resistance as the temperature drops, efforts were made recently to boost the thermal sensitivity of these polymer-based devices. This is because the electrons may not have sufficient energy to get past the hurdles created by the particle limits. It is also discovered that utilizing materials with enhanced thermal conductivity such as GO, rGO, MWCNTs, and silver nanoparticles (AgNPs) for secondary doping of PEDOT:PSS is a rather simple technique [82–85].

9.5.6.4 Pressure Sensing

Flexible pressure sensors seek much attention because of their substantial results for several promising applications, such as rollable touch displays, artificial electronic skin, robotics, and healthcare monitoring. These sensors can convert the physical sense of pressure into meaningful information through a capacitive, resistive, or piezoelectric medium. The past few years have shown promising paths to enhance sensor functionality in terms of robustness, high reaction time, high sensitivity, and strong mechanical stability with recently developed improved functional materials. Resistive-type pressure sensors are widely reported to offer additional features to different sensing processes [86]. PEDOT:PSS is used as a pressure sensor because its unique characteristics make it a suitable candidate for pressure sensing applications. The material can also be mixed with other elements, like carbon nanotubes, for more enhanced performance. PEDOT:PSS is one of the key materials in the sensor technology area, as wearable devices can be realized using this material to identify and respond to changes in pressure with high precision [84].

9.6 Conclusions

This chapter describes PEDOT:PSS, one of the most conductive polymer blends with excellent electrical conductivity, mechanical flexibility, and environmental stability. Due to its unique properties, the electrical property shows a conductivity range of 0.1 to 1000 S/cm, which permits application into different areas such as organic electronics, sensors, and energy devices. Its excellent conductivity and tunable properties make it an ideal material for developing highly sensitive, and selective sensors in fields including biosensing and environmental monitoring. Its application in energy devices such as organic photovoltaics and supercapacitors demonstrates its promising potential for efficient sustainable deployments of emerging base-metal electrode materials. New doping strategies and composite formulations are being actively explored to enhance the stability and performance of PEDOT: PSS. As research into PEDOT:PSS progresses, it is clear that this material has great potential for shaping technological advancements. Achieving the full exploitation of PEDOT: PSS will require a multi-disciplinary approach with expert knowledge of its fundamental intrinsic material properties and innovation on new methodologies.

AI disclosure statement: The author acknowledges the use of AI-based tools for minor language editing or grammatical correction. However, all intellectual contributions, analyses, and conclusions presented in this chapter are entirely the author's own.

References

1. Shirakawa, H., McDiarmid, A., Heeger, A., Focus Article: Twenty-five years of conducting polymers. *Chem. Commun.*, 2003, 1, 1–4, dez. 2003, doi: 10.1039/b210718j.
2. Huseynova, G., Hyun Kim, Y., Lee, J.-H., Lee, J., Rising advancements in the application of PEDOT:PSS as a prosperous transparent and flexible electrode material for solution-processed organic electronics. *J. Inf. Disp.*, 21, 2, 71–91, abr. 2020, doi: 10.1080/15980316.2019.1707311.
3. Fichou, D. and Horowitz, G., Molecular and Polymer Semiconductors, Conductors, and Superconductors: Overview, in: *Encyclopedia of Materials: Science and Technology*, pp. 5748–5757, Elsevier, Amsterdam, Netherlands, 2001, doi: 10.1016/B0-08-043152-6/01000-7.
4. Lee, J., *et al.*, Room-Temperature Nanosoldering of a Very Long Metal Nanowire Network by Conducting-Polymer-Assisted Joining for a Flexible Touch-Panel Application. *Adv. Funct. Mater.*, 23, 34, 4171–4176, set. 2013, doi: 10.1002/adfm.201203802.
5. Doshi, S., *et al.*, Direct electron beam patterning of electro-optically active PEDOT:PSS. *Nanophotonics*, 13, 12, 2271–2280, jan. 2024, doi: 10.1515/nanoph-2023-0640.

6. Zhang, X., Yang, W., Zhang, H., Xie, M., Duan, X., PEDOT:PSS: From conductive polymers to sensors. *Nanotechnol. Precis. Eng.*, 4, 4, 045004, 1–19, dez. 2021, doi: 10.1063/10.0006866.
7. Panigrahy, S. and Kandasubramanian, B., Polymeric thermoelectric PEDOT:PSS & composites: Synthesis, progress, and applications. *Eur. Polym. J.*, 132, 109726, jun. 2020, doi: 10.1016/j.eurpolymj.2020.109726.
8. Abedi, A., Hasanzadeh, M., Tayebi, L., Conductive nanofibrous Chitosan/PEDOT:PSS tissue engineering scaffolds. *Mater. Chem. Phys.*, 237, 121882, nov. 2019, doi: 10.1016/j.matchemphys.2019.121882.
9. Sandu, G., *et al.*, Mechanochemical Synthesis of PEDOT:PSS Hydrogels for Aqueous Formulation of Li-Ion Battery Electrodes. *ACS Appl. Mater. Interfaces*, 9, 40, 34865–34874, out. 2017, doi: 10.1021/acsami.7b08937.
10. Guex, A.G., *et al.*, Highly porous scaffolds of PEDOT:PSS for bone tissue engineering. *Acta Biomater.*, 62, 91–101, out. 2017, doi: 10.1016/j.actbio.2017.08.045.
11. Feng-Xing, J., Jing-Kun, X., Bao-Yang, L., Yu, X., Rong-Jin, H., Lai-Feng, L., Thermoelectric Performance of Poly(3,4-ethylenedioxythiophene): Poly(styrenesulfonate). *Chin. Phys. Lett.*, 25, 6, 2202–2205, jun. 2008, doi: 10.1088/0256-307X/25/6/076.
12. Meen, T.-H., *et al.*, The Effects of Dilute Sulfuric Acid on Sheet Resistance and Transmittance in Poly(3,4-thylenedioxythiophene): Poly(styrenesulfonate) Films. *Int. J. Photoenergy*, 2013, 1–6, 2013, doi: 10.1155/2013/843410.
13. Abu-Thabit, N.Y., Chemical Oxidative Polymerization of Polyaniline: A Practical Approach for Preparation of Smart Conductive Textiles. *J. Chem. Educ.*, 93, 9, 1606–1611, set. 2016, doi: 10.1021/acs.jchemed.6b00060.
14. Wichiansee, W. and Sirivat, A., Electrorheological properties of poly(dimethylsiloxane) and poly(3,4-ethylenedioxy thiophene)/poly(stylene sulfonic acid)/ethylene glycol blends. *Mater. Sci. Eng., C*, 29, 1, 78–84, jan. 2009, doi: 10.1016/j.msec.2008.05.018.
15. Chanthaanont, P. and Sirivat, A., Effect of Transition Metal Ion-Exchanged into the Zeolite Y on Electrical Conductivity and Response of PEDOT-PSS/MY Composites toward SO 2. *Adv. Polym. Technol.*, 32, 4, 21367, dez. 2013, doi: 10.1002/adv.21367.
16. Sakunpongpitiporn, P., Phasuksom, K., Paradee, N., Sirivat, A., Facile synthesis of highly conductive PEDOT:PSS *via* surfactant templates. *RSC Adv.*, 9, 11, 6363–6378, 2019, doi: 10.1039/C8RA08801B.
17. Nurazizah, E.S., Aprilia, A., Risdiana, R., Safriani, L., Different Roles between PEDOT:PSS as Counter Electrode and PEDOT:Carrageenan as Electrolyte in Dye-Sensitized Solar Cell Applications: A Systematic Literature Review. *Polymers (Basel)*, 15, 12, 2725, jun. 2023, doi: 10.3390/polym15122725.
18. Jiang, H., *et al.*, *In situ* polymerization of PEDOT:PSS films based on EMI-TFSI and the analysis of electrochromic performance. *e-Polymers*, 21, 1, 722–733, set. 2021, doi: 10.1515/epoly-2021-0073.

19. Pei, Q., Zuccarello, G., Ahlskog, M., Inganäs, O., Electrochromic and highly stable poly(3,4-ethylenedioxythiophene) switches between opaque blue-black and transparent sky blue. *Polymer (Guildf)*, 35, 7, 1347–1351, mar. 1994, doi: 10.1016/0032-3861(94)90332-8.
20. Kim, J.Y., Jung, J.H., Lee, D.E., Joo, J., Enhancement of electrical conductivity of poly(3,4-ethylenedioxythiophene)/poly(4-styrenesulfonate) by a change of solvents. *Synth. Met.*, 126, 2–3, 311–316, fev. 2002, doi: 10.1016/S0379-6779(01)00576-8.
21. Shahrim, N.A., Ahmad, Z., Wong Azman, A., Fachmi Buys, Y., Sarifuddin, N., Mechanisms for doped PEDOT:PSS electrical conductivity improvement. *Mater. Adv.*, 2, 22, 7118–7138, 2021, doi: 10.1039/D1MA00290B.
22. Tong, B., *et al.*, A polymer electrolyte design enables ultralow-work-function electrode for high-performance optoelectronics. *Nat. Commun.*, 13, 1, 4987, 2022, doi: 10.1038/s41467-022-32651-z.
23. Woon, K.L., *et al.*, Work function modification of PEDOT:PSS by mixing with barium acetylacetonate. *RSC Adv.*, 10, 30, 17673–17680, 2020, doi: 10.1039/D0RA02575E.
24. Mantione, D., *et al.*, Low-Temperature Cross-Linking of PEDOT:PSS Films Using Divinylsulfone. *ACS Appl. Mater. Interfaces*, 9, 21, 18254–18262, maio 2017, doi: 10.1021/acsami.7b02296.
25. Hik, F., Taatizadeh, E., Takalloo, S.E., Madden, J.D.W., Fast electrochemical response of PEDOT:PSS electrodes through large combined increases to ionic and electronic conductivities. *Electrochim. Acta*, 468, 143136, nov. 2023, doi: 10.1016/j.electacta.2023.143136.
26. Yang, T., *et al.*, PEDOT:PSS hydrogels with high conductivity and biocompatibility for *in situ* cell sensing. *J. Mater. Chem. B*, 11, 14, 3226–3235, 2023, doi: 10.1039/D3TB00014A.
27. Liang, Y., Offenhäusser, A., Ingebrandt, S., Mayer, D., PEDOT:PSS-Based Bioelectronic Devices for Recording and Modulation of Electrophysiological and Biochemical Cell Signals. John Wiley and Sons Inc. *Adv. Healthcare Mater.*, 10, 11, 1 jun. 2021, doi: 10.1002/adhm.202100061.
28. Unuma, T., Yoshikawa, M., Nakamura, A., Kishida, H., Segmentation of conducting domains in PEDOT:PSS films induced by an additive for conductivity enhancement. *Appl. Phys. Express*, 9, 5, 051601, 2016, doi: 10.7567/APEX.9.051601.
29. Suchand Sangeeth, C.S., Jaiswal, M., Menon, R., Correlation of morphology and charge transport in poly(3,4-ethylenedioxythiophene)–polystyrenesulfonic acid (PEDOT–PSS) films. *J. Phys.: Condens. Matter*, 21, 7, 072101, fev. 2009, doi: 10.1088/0953-8984/21/7/072101.
30. Thomas, J.P., Zhao, L., McGillivray, D., Leung, K.T., High-efficiency hybrid solar cells by nanostructural modification in PEDOT:PSS with co-solvent addition. *J. Mater. Chem. A Mater.*, 2, 7, 2383, 2014, doi: 10.1039/c3ta14590e.

31. Fan, X., *et al.*, PEDOT:PSS materials for optoelectronics, thermoelectrics, and flexible and stretchable electronics. *J. Mater. Chem. A Mater.*, 11, 35, 18561–18591, 2023, doi: 10.1039/D3TA03213B.
32. Selvakumar, N., Karthik, G., Jayaraj, S., Barshilia, H.C., Sprayable PEDOT:PSS based spectrally selective coating for solar energy harvesting. *Sol. Energy Mater. Sol. Cells*, 221, 110906, mar. 2021, doi: 10.1016/j.solmat.2020.110906.
33. Chonan, Y., Sato, N., Komiyama, T., Kotani, K., Yamaguchi, H., Enhancement of Thermoelectric Properties of PEDOT:PSS Films by Applying an Alternating Electric Field During Preparation. *J. Electron. Mater.*, 48, 6, 3854–3858, jun. 2019, doi: 10.1007/s11664-019-07150-5.
34. Kondratiev, V. and Holze, R., Intrinsically Conducting Polymer Binders for Battery Electrodes. *Encyclopedia*, 2, 4, 1753–1762, out. 2022, doi: 10.3390/encyclopedia2040120.
35. Wang, A., *et al.*, Highly Conductive and Lightweight Composite Film as Polysulfide Reservoir for High-Performance Lithium–Sulfur Batteries. *ChemElectroChem*, 4, 2, 362–368, fev. 2017, doi: 10.1002/celc.201600579.
36. Das, P.R., Komsiyska, L., Osters, O., Wittstock, G., Electrochemical Stability of PEDOT:PSS As Cathodic Binder for Li-Ion Batteries. *ECS Trans.*, 68, 2, 45–58, nov. 2015, doi: 10.1149/06802.0045ecst.
37. Das, P.R., Komsiyska, L., Osters, O., Wittstock, G., PEDOT: PSS as a Functional Binder for Cathodes in Lithium Ion Batteries. *J. Electrochem. Soc.*, 162, 4, A674–A678, jan. 2015, doi: 10.1149/2.0581504jes.
38. Calabia Gascón, N., Terryn, H., Hubin, A., The role of PEDOT:PSS in (super) capacitors: A review. *Next Nanotechnol.*, 2, 100015, jun. 2023, doi: 10.1016/j.nxnano.2023.100015.
39. Wakabayashi, T., Katsunuma, M., Kudo, K., Okuzaki, H., pH-Tunable High-Performance PEDOT:PSS Aluminum Solid Electrolytic Capacitors. *ACS Appl. Energy Mater.*, 1, 5, 2157–2163, 2018, doi: 10.1021/acsaem.8b00210.
40. Sun, K., *et al.*, Review on application of PEDOTs and PEDOT:PSS in energy conversion and storage devices. *J. Mater. Sci.: Mater. Electron.*, 26, 7, 4438–4462, jul. 2015, doi: 10.1007/s10854-015-2895-5.
41. Siao, H.-W., Chen, S.-M., Lin, K.-C., Electrochemical study of PEDOT-PSS-MDB-modified electrode and its electrocatalytic sensing of hydrogen peroxide. *J. Solid State Electrochem.*, 15, 6, 1121–1128, jun. 2011, doi: 10.1007/s10008-010-1174-x.
42. Chen, Y.-C. and O'Hare, D., Exhaled breath condensate based breath analyser – a disposable hydrogen peroxide sensor and smart analyser. *Analyst*, 145, 10, 3549–3556, 2020, doi: 10.1039/C9AN02438G.
43. Lv, D., *et al.*, Enhanced flexible room temperature ammonia sensor based on PEDOT: PSS thin film with FeCl3 additives prepared by inkjet printing. *Sens. Actuators B Chem.*, 298, 126890, nov. 2019, doi: 10.1016/j.snb.2019.126890.
44. Memarzadeh, R., Noh, H.-B., Javadpour, S., Panahi, F., Feizpour, A., Shim, Y.-B., Carbon Monoxide Sensor Based on a B2HDDT-doped PEDOT:PSS

Layer. *Bull. Korean Chem. Soc.*, 34, 8, 2291–2296, 2013, doi: 10.5012/bkcs.2013.34.8.2291.

45. Urbanowicz, M., Pijanowska, D.G., Jasiński, A., Ekman, M., Bocheńska, M.K., A miniaturized solid-contact potentiometric multisensor platform for determination of ionic profiles in human saliva. *J. Solid State Electrochem.*, 23, 12, 3299–3308, dez. 2019, doi: 10.1007/s10008-019-04429-9.
46. Boehler, C., Aqrawe, Z., Asplund, M., Applications of PEDOT in Bioelectronic Medicine. *Bioelectron. Med. (Lond)*, 2, 2, 89–99, jun. 2019, doi: 10.2217/bem-2019-0014.
47. Xu, J., Peng, R., Ran, Q., Xian, Y., Tian, Y., Jin, L., A highly soluble poly(3,4-ethylenedioxythiophene)-poly(styrene sulfonic acid)/Au nanocomposite for horseradish peroxidase immobilization and biosensing. *Talanta*, 82, 4, 1511–1515, set. 2010, doi: 10.1016/j.talanta.2010.07.033.
48. Gao, N., *et al.*, Application of PEDOT:PSS and Its Composites in Electrochemical and Electronic Chemosensors. *Chemosensors*, 9, 4, 79, abr. 2021, doi: 10.3390/chemosensors9040079.
49. Liao, J., Si, H., Zhang, X., Lin, S., Functional Sensing Interfaces of PEDOT:PSS Organic Electrochemical Transistors for Chemical and Biological Sensors: A Mini Review. *Sensors*, 19, 2, 218, jan. 2019, doi: 10.3390/s19020218.
50. Hsiao, Y.-S., *et al.*, PEDOT:PSS-Based Bioelectrodes for Multifunctional Drug Release and Electric Cell-Substrate Impedance Sensing. *ACS Appl. Mater. Interfaces*, 15, 18, 21953–21964, maio 2023, doi: 10.1021/acsami.3c02769.
51. Boehler, C. and Asplund, M., A detailed insight into drug delivery from PEDOT based on analytical methods: Effects and side effects. *J. Biomed. Mater. Res. A*, 103, 3, 1200–1207, mar. 2015, doi: 10.1002/jbm.a.35252.
52. Pang, Z., Antonetti, D.A., Tarbell, J.M., Shear Stress Regulates HUVEC Hydraulic Conductivity by Occludin Phosphorylation. *Ann. Biomed. Eng.*, 33, 11, 1536–1545, nov. 2005, doi: 10.1007/s10439-005-7786-0.
53. Shinoka, T., *et al.*, Tissue engineering heart valves: Valve leaflet replacement study in a lamb model. *Ann. Thorac. Surg.*, 60, S513–S516, dez. 1995, doi: 10.1016/0003-4975(95)00733-4.
54. Hesam Mahmoudinezhad, M., Karkhaneh, A., Jadidi, K., Effect of PEDOT:PSS in tissue engineering composite scaffold on improvement and maintenance of endothelial cell function. *J. Biosci.*, 43, 2, 307–319, jun. 2018, doi: 10.1007/s12038-018-9748-3.
55. Khorshidi, S. and Karkhaneh, A., Formation of three-dimensionality in polyaniline-based nanofibers: A highly conductive permeable scaffold for stem cells residence. *Int. J. Polym. Mater. Polym. Biomater.*, 65, 12, 629–635, 2016, doi: 10.1080/00914037.2016.1157794.
56. Ahlborn, P., Schachner, M., Irintchev, A., One hour electrical stimulation accelerates functional recovery after femoral nerve repair. *Exp. Neurol.*, 208, 1, 137–144, nov. 2007, doi: 10.1016/j.expneurol.2007.08.005.

57. Molapo, K.M., *et al.*, Electronics of Conjugated Polymers (I): Polyaniline. *Int. J. Electrochem. Sci.*, 7, 12, 11859–11875, dez. 2012, doi: 10.1016/S1452-3981(23)16509-6.
58. Hu, L., Liu, Z., Hu, W., Wang, Y., Tan, J., Wu, F., Petri-net-based dynamic scheduling of flexible manufacturing system *via* deep reinforcement learning with graph convolutional network. *J. Manuf. Syst.*, 55, 1–14, abr. 2020, doi: 10.1016/j.jmsy.2020.02.004.
59. Seiti, M., Giuri, A., Corcione, C.E., Ferraris, E., Advancements in tailoring PEDOT: PSS properties for bioelectronic applications: A comprehensive review. *Biomater. Adv.*, 154, 213655, nov. 2023, doi: 10.1016/j.bioadv.2023.213655.
60. Orasugh, J.T. and Ray, S.S., Functional and Structural Facts of Effective Electromagnetic Interference Shielding Materials: A Review. *ACS Omega*, 8, 9, 8134–8158, mar. 2023, doi: 10.1021/acsomega.2c05815.
61. Wu, X., *et al.*, Facile Fabrication of PEDOT:PSS-Based Free-Standing Conducting Film for Highly Efficient Electromagnetic Interference Shielding. *Macromol. Mater. Eng.*, 308, 4, 2200554, 2023, doi: 10.1002/mame.202200554.
62. Bora, P.J., Anil, A.G., Ramamurthy, P.C., Tan, D.Q., MXene interlayered crosslinked conducting polymer film for highly specific absorption and electromagnetic interference shielding. *Mater. Adv.*, 1, 2, 177–183, 2020, doi: 10.1039/D0MA00005A.
63. Hou, J., Zhu, G., Xu, J., Liu, H., Anticorrosion Performance of Epoxy Coatings Containing Small Amount of Inherently Conducting PEDOT/PSS on Hull Steel in Seawater. *J. Mater. Sci. Technol.*, 29, 7, 678–684, jul. 2013, doi: 10.1016/j.jmst.2013.03.023.
64. Fu, K., *et al.*, Mixed ion-electron conducting PEO/PEDOT: PSS miscible blends with intense electrochromic response. *Polymer (Guildf)*, 184, 121900, dez. 2019, doi: 10.1016/j.polymer.2019.121900.
65. Tounakti, C., Decorse, P., Kouki, F., Lang, P., Relationship between enhancement of PEDOT : PSS conductivity by solvent treatment and PSS chain reorganization. *J. Polym Sci.*, 61, 7, 582–603, abr. 2023, doi: 10.1002/pol.20220605.
66. Wang, Z. and Liu, R., PEDOT:PSS-based electrochromic materials for flexible and stretchable devices. *Mater. Today Electron.*, 4, 100036, jun. 2023, doi: 10.1016/j.mtelec.2023.100036.
67. Tian, Y., *et al.*, High-Performance Transparent PEDOT: PSS/CNT Films for OLEDs. *Nanomaterials*, 11, 8, 2067, 2021, doi: 10.3390/nano11082067.
68. Wang, G., Tao, X., Wang, R., Fabrication and characterization of OLEDs using PEDOT:PSS and MWCNT nanocomposites. *Compos. Sci. Technol.*, 68, 14, 2837–2841, nov. 2008, doi: 10.1016/j.compscitech.2007.11.004.
69. Auroux, E., Huseynova, G., Ràfols-Ribé, J., Miranda La Hera, V., Edman, L., A metal-free and transparent light-emitting device by sequential spray-coating fabrication of all layers including PEDOT:PSS for both electrodes. *RSC Adv.*, 13, 25, 16943–16951, 2023, doi: 10.1039/D3RA02520A.

70. Li, J., *et al.*, Ultrathin Smart Energy-Storage Devices for Skin-Interfaced Wearable Electronics. *ACS Energy Lett.*, 8, 1, 1–8, jan. 2023, doi: 10.1021/acsenergylett.2c02029.
71. Xue, R., *et al.*, Fabrication of Flexible Electrochromic Devices with Degradable and Fully Recyclable Features. *ACS Biomater. Sci. Eng.*, 8, 3, 1320–1328, mar. 2022, doi: 10.1021/acsbiomaterials.2c00106.
72. Seyedin, M.Z., Razal, J.M., Innis, P.C., Wallace, G.G., Strain-Responsive Polyurethane/PEDOT:PSS Elastomeric Composite Fibers with High Electrical Conductivity. *Adv. Funct. Mater.*, 24, 20, 2957–2966, maio 2014, doi: 10.1002/adfm.201303905.
73. Zhou, J. and Hsieh, Y.-L., Conductive Polymer Protonated Nanocellulose Aerogels for Tunable and Linearly Responsive Strain Sensors. *ACS Appl. Mater. Interfaces*, 10, 33, 27902–27910, 2018, doi: 10.1021/acsami.8b10239.
74. Bhattacharjee, M., Soni, M., Dahiya, R., Microchannel based Flexible Dynamic Strain Sensor, in: *2019 IEEE International Conference on Flexible and Printable Sensors and Systems (FLEPS)*, jul. 2019, IEEE, pp. 1–3, doi: 10.1109/FLEPS.2019.8792317.
75. Zhou, C., Zhang, X., Tang, N., Fang, Y., Zhang, H., Duan, X., Rapid response flexible humidity sensor for respiration monitoring using nano-confined strategy. *Nanotechnology*, 31, 12, 125302, mar. 2020, doi: 10.1088/1361-6528/ab5cda.
76. Yao, X. and Cui, Y., A PEDOT:PSS functionalized capacitive sensor for humidity. *Measurement*, 160, 107782, 2020, doi: 10.1016/j.measurement.2020.107782.
77. Muckley, E.S., Lynch, J., Kumar, R., Sumpter, B., Ivanov, I.N., PEDOT:PSS/QCM-based multimodal humidity and pressure sensor. *Sens. Actuators B Chem.*, 236, 91–98, nov. 2016, doi: 10.1016/j.snb.2016.05.054.
78. Park, J.-K., Kang, T.-G., Kim, B.-H., Lee, H.-J., Choi, H.H., Yook, J.-G., Real-time Humidity Sensor Based on Microwave Resonator Coupled with PEDOT:PSS Conducting Polymer Film. *Sci. Rep.*, 8, 1, 439, jan. 2018, doi: 10.1038/s41598-017-18979-3.
79. Su, Y., *et al.*, Printable, Highly Sensitive Flexible Temperature Sensors for Human Body Temperature Monitoring: A Review. *Nanoscale Res. Lett.*, 15, 1, 200, dez. 2020, doi: 10.1186/s11671-020-03428-4.
80. Barmpakos, D. and Kaltsas, G., A Review on Humidity, Temperature and Strain Printed Sensors—Current Trends and Future Perspectives. *Sensors*, 21, 3, 739, jan. 2021, doi: 10.3390/s21030739.
81. Bießmann, L., Kreuzer, L.P., Widmann, T., Hohn, N., Moulin, J.-F., Müller-Buschbaum, P., Monitoring the Swelling Behavior of PEDOT:PSS Electrodes under High Humidity Conditions. *ACS Appl. Mater. Interfaces*, 10, 11, 9865–9872, mar. 2018, doi: 10.1021/acsami.8b00446.
82. Zhou, C., *et al.*, Simultaneously Optimize the Response Speed and Sensitivity of Low Dimension Conductive Polymers for Epidermal Temperature Sensing Applications. *Front. Chem.*, 8, 194, mar. 2020, doi: 10.3389/fchem.2020.00194.

83. Benchirouf, A., *et al.*, Electrical properties of multi-walled carbon nanotubes/PEDOT:PSS nanocomposites thin films under temperature and humidity effects. *Sens. Actuators B Chem.*, 224, 344–350, mar. 2016, doi: 10.1016/j.snb.2015.10.009.
84. Zhou, J., *et al.*, The temperature-dependent microstructure of PEDOT/PSS films: insights from morphological, mechanical and electrical analyses. *J. Mater. Chem. C*, 2, 46, 9903–9910, 2014, doi: 10.1039/C4TC01593B.
85. Zhang, F., *et al.*, Multi-modal strain and temperature sensor by hybridizing reduced graphene oxide and PEDOT:PSS. *Compos. Sci. Technol.*, 187, 107959, fev. 2020, doi: 10.1016/j.compscitech.2019.107959.
86. Lee, J.J., *et al.*, Alcohol-based highly conductive polymer for conformal nanocoatings on hydrophobic surfaces toward a highly sensitive and stable pressure sensor. *NPG Asia Mater.*, 12, 1, 65, dez. 2020, doi: 10.1038/s41427-020-00238-z.

10

EAPs and Environmental Impact

Anu Radha Pathania[1]* and Harman Deep Kour[2]

[1]*Department of Chemical Sciences, Parul Institute of Applied Sciences, Parul University, Limda, Vadodara, Gujarat, India*
[2]*Department of Chemistry, University Institute of Sciences, Chandigarh University, Gharuan, Mohali, Punjab, India*

Abstract

Electroactive polymers (EAPs) are fascinating materials that function as actuators, capable of changing their dimensions, shapes, or conductivity in response to electrical signals. This unique ability allows EAPs to be used in a wide range of applications, from artificial muscles and actuators to sensors and energy harvesting devices. However, it is essential to consider the environmental impacts associated with these materials. In this chapter, we take a closer look at the manufacturing processes, disposal options, and life cycle analyses related to EAPs, aiming to clarify their environmental footprint. We also discuss the potential of EAP-based technologies to address environmental challenges and promote sustainability, highlighting both their advantages and limitations. Our goal is to provide a comprehensive overview of electroactive polymers and their applications while addressing the environmental concerns that come with them. By fostering a deeper understanding of these issues, we hope to empower stakeholders—including researchers, policymakers, and industry leaders—to make informed and sustainable choices about the future use of these promising materials.

***Keywords*:** Electroactive polymers, piezoelectric effect, nanotechnology, biocompatibility, supercapacitors

**Corresponding author*: anuthakur2014@gmail.com

Raneesh Balakrishnan, Sobi K. Chacko and Visakh P.M. (eds.) Electroactive Polymer-Based Smart Materials and Applications, (265–296) © 2026 Scrivener Publishing LLC

10.1 Introduction

The phrase EAPs describes a class of polymers that change in size and shape in response to an electric field. The majority of the time, sensors and actuators employ this type of material. EAPs frequently go through substantial deformation while being subjected to severe stress. Rentgen researched it for the first time in 1880, and Sacerdote followed suit in 1899 [1]. As a result, Eguchi discovered the piezoelectric polymer Electret in 1925. Electret is an insulating substance that can hold an electric charge after being polarised in an electric field [2]. The output an electret produces under high voltage, however, restricts the device's use to sensors rather than actuators. Kawai discovered a significant piezoelectric effect in polyvinylidene fluoride in 1969 [3]. This work was carried out by Fukada, and it has been shown that some EAPs can display up to a 380% strain, making it applicable to actuators [4]. A system's actuators are the parts that turn energy or impulses into motion.

EAPs Robotic Arm vs. Human Challenge In 1999, Yoseph Bar-Cohen made the first Arm Wrestling Match suggestion [5]. The aim of this competition, which comprised research teams from all over the world, was a robotic arm built of EAP muscles that could beat a human in an arm-wrestling match. A symposium on electroactive polymer actuators and devices hosted the first competition in 2005. Since they are silent, strong, light, and compact, electroactive polymer-based artificial muscles, also known as soft actuators, are a desirable alternative to electrical motors for driving human-like robots [6]. A division of Bayer Material Science, Artificial Muscle was established in 2010.

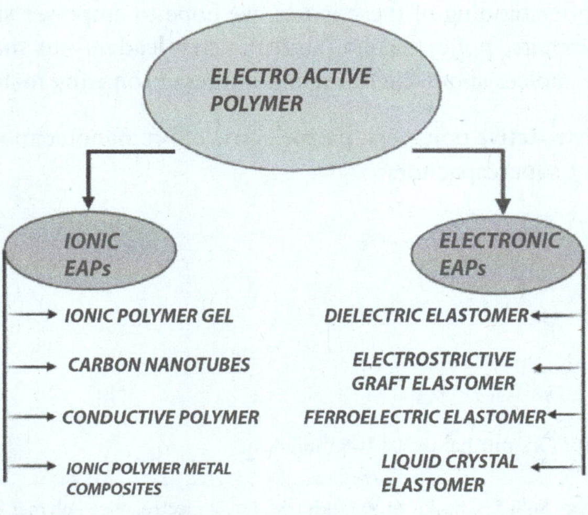

Figure 10.1 Types of electroactive polymer.

EAPs may be classified into two types

1) Ionic EAPs
2) Electronic EAPs

Ionic EAPs transfer energy *via* ions, as opposed to electronic EAPs, which do so through electric current. Each form of energy transfer mechanism has distinct benefits and drawbacks due to the diverse ways in which they operate, such as variable actuation force and operation voltage. As indicated in Figure 10.1, these two types are further separated into numerous groups. Each EAP type's working theory, subcategorization, benefits, and drawbacks are briefly discussed in this section.

10.1.1 Ionic EAPs

Smart materials known as ionic electroactive polymers (Figure 10.2) (IEAPs) display reversible mechanical deformation in response to an applied electrical voltage. Ion movement inside the polymer substance is the main cause of this deformation. Ions within the polymer travel towards one electrode when a voltage is applied, which causes the polymer to either expand or contract depending on the type of ions and the direction of the voltage. This phenomenon, often referred to as electrostriction or ion migration, can be used in actuators, sensors, artificial muscles, and other devices. Ion diffusion, electrostatic force, and polymer structure interact to create the mechanism [8]. In Table 10.1, the major benefits and drawbacks of ionic EAP are enumerated.

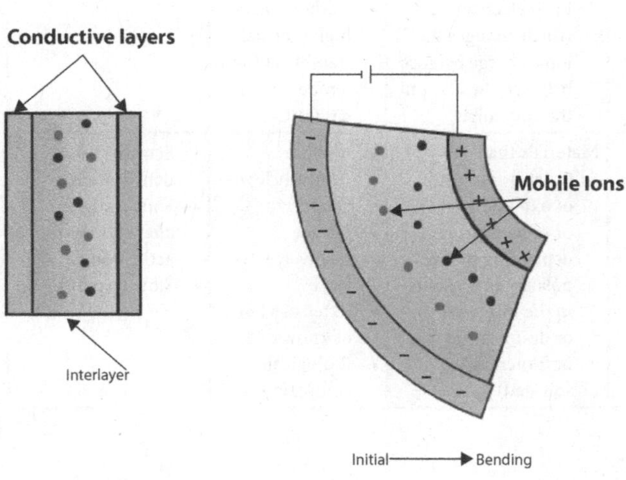

Figure 10.2 Ionic polymer actuator [7].

Table 10.1 Major benefits and drawbacks of ionic EAP.

Advantages	Disadvantages
Low voltage actuation	Low actuation force in bending
Bidirectional operation with voltage polarity	Electrolyte require humid condition
Extensive bending	Show response time range in second
	Unstainable strain under DC
	Electrolysis occur at >1.23V when a system involves water

Different types of Electronic EAPs:

Types of EAPs	Principle	Advantages	Disadvantages	References
Ionic polymer gel	Application of voltage causes movement of hydrogen ions in or out of the gel	• Application of voltage causes movement of hydrogen ions in or out of the gel	• Very slowly operates	[9]
Carbon nanotubes	Charge injection causes a change in the length of the carbon-carbon bond in nanotubes (NT) suspended in an electrolyte, which changes the ionic charge balance between the NT and the electrolyte.	• Potentially provide superior work/ cycle and mechanical stresses • Carbon offers high thermal stability at high temperatures <1000°C	• Expensive to produce • Difficult to mass produce	[10]
Conductive polymer	Materials that expand as a result of oxidation or reduction, depending on the polarity that results in the insertion or de-insertion of (potentially solvated) ions	• Require relatively low voltage • Induce relatively large force. • Extensive body of knowledge • Biologically compatible	• Sensitive to dehydration • Suffer fatigue after repeated activation • Slow response	[11]

(*Continued*)

(*Continued*)

Types of EAPs	Principle	Advantages	Disadvantages	References
Ionic polymer metal composite	When an electrical potential is applied to the metal electrodes, the hydrated mobile ions in the ion-exchange membrane migrate by means of electrostatic force, resulting in the actuation of the IPMCs	• Require low voltage • Produce significant bending	• Low frequency response (effectively below 1 Hz) • dc causes permanent deformation • Extremely sensitive to dehydration • Subject to hydrolysis above 1.23 V	[12, 13]

10.1.2 Electronic Polymer

Since Coulomb forces power electronic EAPs, they can be built to hold the induced displacement while running at a DC voltage. The use of these EAP materials in air is not significantly constrained and they have a high mechanical energy density. They call for an extremely strong activation field (>30 MV/m), which could cause painful electric shocks. Electronic EAPs have the advantages of quick response and the transmission of strong mechanical forces. They don't need to be coated to be protected, and they use only a small amount of electricity to stay in place. In the following Table 10.2, the main benefits and drawbacks of electronic EAPs are outlined:

Table 10.2 Main benefits and drawbacks of electronic EAPs.

Advantages	Disadvantages
Induce relatively large actuation forces.	Independent of the voltage polarity, it produces mostly nonpolar actuation due to associated electrostriction effect.
Exhibit rapid response.	Require high voltage.
Electronic EAPs can operate for a long time in a room condition.	
It can hold strain under DC activation.	
Exhibit high mechanical energy density.	

Different types of Electronic EAPs:

Types electronic EAPs	Principle	Advantages	Disadvantages	References
Dielectric Elastomer	By using DE, electric energy is converted to mechanical energy. Coulomb forces between the electrodes squeeze the material, causing it to expand in the plane of the electrodes.	• Very large strain reaching levels of 380%. • Inexpensive to produce. • Rapid response.	• Rquire high voltage. • Obtaining large displacement compromises the actuation forces.	[14]
Liquid crystal Elastomer	polymers can form either by dissolving a polymer in a solvent or by heating it over its glass or melting transition point.	• High temperature resistance. • Accuracy and remote control. • Long term cost effective.	• It consumes a lot of electricity which produces a lot of heat.	[15]
Ferroelectric elastomer	polymers that alter their form over time in a non-centrosymmetric manner in response to an electric field. These polymers include several that are ferroelectric due to spontaneous electric polarisation.	• Induce relatively large strain (5%) • Offer high mechanical energy density resulting from the relatively high elastic modulus • Permit ac switching with little generated heat • Rapid response (ms levels)	• Require high voltage. • Making thin multilayers is still a challenge	[16]
Electrostrictive Graft Elastomer	The pendant group comprised of graft crystalline elastomers connected to the backbone's electric field leads the molecules in the group to align.	• Relatively large force. • Strain levels of 5%. • Cheaper to produce. • Rapid response.	• Require high voltage.	[17]

10.2 Challenges in EAPs

Although many technologies still have issues, most of them can be fixed by changing the material's properties. The synthesis and application of EAP provide several difficulties.

1) **Improvement of mechanical strength:** Electroactive polymers, compared to metals or ceramics, have lower mechanical strengths. To improve the mechanical strength of electroactive polymers, novel formulations and processing techniques can be developed by scientists and engineers. These may involve structural modification of the polymer or incorporation of reinforcing elements to improve strength properties.
2) **Stability and strength:** Electroactive polymers may be made stronger and more stable by such strategies as providing protective coatings, improving material composition, and developing encapsulating procedures. Additionally, extensive testing and long-term evaluation will allow the degradation pathways to be identified and will guide the development of more sturdy materials.
3) **Low power consumption:** Researchers can concentrate on creating electroactive polymers with lower energy consumption by refining their molecular structures, investigating alternate actuation mechanisms, and improving electrode designs. The overall power consumption of these materials can also be decreased by integrating energy-harvesting technologies.
4) **Reducing manufacturing costs:** Ongoing research and development initiatives might result in less expensive methods for producing electroactive polymers. This might entail looking into other raw materials, refining production techniques, and employing mass production to benefit from economies of scale.
5) **Large-scale production:** Working together, manufacturers, industrial partners, and research scientists may be able to resolve scaling challenges. The mass manufacture of electroactive polymers without compromising quality or performance may be made possible by standardized procedures, the sharing of best practices, and investments in cutting-edge manufacturing technology.

6) **Better integration with electronics:** Collaboration between electronics engineers and material scientists can resolve integration problems, such as designing complementary interfaces, improving signal-processing technologies, and constructing flexible electronic components that would not interfere with the operation of electroactive polymers.
7) **Expanding material options:** More research and development is required to increase the selection of electroactive polymers that are now available. This might entail researching new chemical compositions, using creative synthesis techniques, and utilizing nanotechnology advancements to produce materials that are specifically suited for particular uses.
8) **Reduction of environmental effects:** Scientists can concentrate on creating environmentally friendly electroactive polymers by using sustainable raw materials and reducing the usage of hazardous chemicals. Establishing suitable disposal and recycling protocols can help lessen the impact of these items on the environment.
9) **Development of standards and specifications:** By working with researchers, industry associations, and regulatory bodies, standardized testing protocols, performance metrics, and design guidelines for electroactive polymers may be created. As a result, these materials will be easier to compare and assess, increasing their acceptance across sectors.
10) **Acceptance and awareness:** Outreach and educational initiatives can create a substantial impact for the increased usage of electroactive polymers. In this, conferences, workshops, and seminars could be organized to create awareness and consciousness about the potential applications and benefits of these materials. Other approaches for closing the gap between research and commercialization would include participating in knowledge transfer programs and working with industry partners.

10.3 Applications of EAPs

Because EAPs have the flexibility to transform their dimensions, form, or mechanical characteristics upon electrical stimulation, their applications are found in several fields like

1) **Robotics**
   Electroactive material-based artificial muscles can be used to construct robotic systems. These materials allow for more natural and flexible robotic movements and provide actuation capabilities that mimic human muscles.
2) **Soft robotics**
   Electroactive materials are particularly friendly to be used in soft robotics, where compliance and flexibility are of utmost importance. These materials can be used to make soft, flexible robots that can safely interact with humans and find their way into challenging settings [18].
3) **Prosthetics**
   The limbs and other medical devices may be manufactured using electroactive materials. The new prosthetics made from electroactive materials will have the ability to be more sensitive to the user's movements as well as more comfortable to wear [19].
4) **Biomedical devices**
   Electroactive materials could be used for the development of biomedical devices. These include drug delivery systems and tissue engineering scaffolds. They can be activated by electrical stimuli to release controlled amounts of drugs or to foster tissue growth and regeneration.
5) **Energy harvesting**
   Electroactive materials can also be used in energy-harvesting applications. These materials are able to convert mechanical or thermal energy into electrical energy and hence can power small electronic devices or sensors, hence eliminating the use of an external power source [20].

Apart from robotics and prosthetics, future applications would include smart textiles that change their breathability depending on temperature and moisture and window shutters that automatically open and close to save energy.

There are various challenges associated with these applications of EAPs such as:

10.3.1 Challenges in Artificial Muscle

One of the challenges in making electroactive polymers that can mimic human movement is to make artificial muscles. Artificial muscles, which control the generation of force and regulation of motion, are essential in allowing lifelike movements in robots [21]. Creating artificial muscles using electroactive polymers poses several major challenges. Getting the required amount of force and power production is one of them. It is very difficult to mimic the enormous force that human muscles can generate in artificial muscles. The researchers need to optimize the design and chemical composition of the polymer so that it can produce enough force while remaining light and flexible [22].

Another challenge is that of controlled, precise motions. Human muscles are capable of precision movements; one can write or play musical instruments. In the case of artificial muscles, achieving this kind of control will require advanced sensor integration and feedback mechanisms. Scientists have to design sensors that can recognize muscle contractions with high accuracy and convert these into movements for the robot.

Longevity and durability are also very important factors. Human muscles can withstand prolonged repeated movements without sustaining critical damage. Electroactive polymers need to be developed so that they can last as long as human muscles, considering the reliability of the robot. Last but not least, one of the greatest problems is safety. The artificial muscles have to be safe enough to be employed in proximity to humans because robots that mimic human movement can occasionally touch them [23].

The researchers involved in the field of materials science, robotics, biomechanics, and other similar fields must collaborate to come up with a solution for these problems. Through the integration of expertise from various fields, technological challenges can be overcome, and electroactive polymers that precisely mimic human movement and appearance can be developed [24].

10.3.2 Challenge in Soft Robotics

An emerging area of study called "soft robotics" focuses on developing soft, flexible robots that can mimic the abilities of living things. With their ability to change shape and size upon receiving electrical stimulation, electroactive polymers (EAPs), a class of materials, have been extremely promising for application in soft robotics. Researchers are working to overcome many challenges in soft robotics [25]. Some of these challenges are:

1) **Actuation:** Soft robots often require specialized actuators that can mimic the flexibility and elasticity of biological muscles. Traditional stiff actuators are not suitable for soft robots because they limit their ability to navigate complex

environments and interact with fragile objects. To give soft robots, the required flexibility and compliance, scientists have developed novel types of actuators, including pneumatic or hydraulic systems, shape memory alloys, and electroactive polymers.

2) **Sensing and control:** To better interact with their environment, soft robots must possess reliable and flexible sensing and control systems. Because soft robots are fundamentally soft, classical sensors and control algorithms used for rigid robots may be inappropriate. Innovative sensing technologies such as flexible sensors and soft strain gauges are under research by researchers to correctly detect and react to changes in the shape and environment of the robot. Advanced control algorithms capable of taking into account the non-linear and dynamic behavior of soft robots are also under development.
3) **Durability and reliability:** Soft materials tend to get fatigued, wear out, and tear down after a period, which degrades performance and lifetime. To make soft robots more robust and reliable, researchers are exploring new materials and manufacturing methods. This includes developing self-healing materials that can heal their damage as well as inventing innovative manufacturing techniques that ensure reliable and long-lasting soft robot structures.
4) **Power and energy supply:** Soft robots often require continuous power supply to operate. However, it could be problematic to integrate conventional power sources, for example, batteries in the context of soft and flexible structures. Alternative power sources including energy harvesting devices for example, solar or kinetic energy or even lightweight, flexible batteries are being considered to satisfy the power requirements of soft robotics.
5) **Scalability and complexity:** Because they deform and require several degrees of freedom, soft robots are typically more complex than rigid robots. Major fabrication, control, and actuation issues could be present when scaling up soft robots to larger sizes or more complicated configurations. Coupled with developing innovative fabrication processes, such as 3D printing that simplify the production process, researchers are coming up with scalable and modular

designs that will allow simple assembly and integration of soft robot components.

Researchers continued to overcome these technical challenges and were now able to realize the full potential of soft robotics with continuous research and technology advancement [26]. Such endeavors thus far have produced soft robots that are more dexterous and flexible as well as safer, which opens their applications to medicine, manufacturing, and space exploration.

10.3.3 Challenges in Prosthetic

In addition to the challenges mentioned above, there are specific challenges associated with the prosthetics are:

1) **Biocompatibility:** The EAPs must be biocompatible to be safely implemented inside the human body without causing any adverse response or tissue damage; this involves considering factors such as materials used, surface characteristics, and long-term stability in the body.
2) **Sensory feedback:** For various prosthetic limbs, a natural and intuitive experience for its user is the need for sensory feedback. Currently, the sensing capabilities of EAPs are so limited that it is not easy to provide consumers with specific and prompt sensory feedback.
3) **Controlling and Coordinating EAP-based Prosthetic Limbs:** They have to exactly replicate the coordination and range of motion of actual limbs. For most EAP actuators, coordinating and controlling a large number of them in precision is tough and requires some of the more advanced control algorithms and feedback mechanisms.
4) **Power supply and energy efficiency:** To offer extended operating periods, and reduce the frequent need to charge or replace batteries, prosthetic limbs must be powered efficiently. EAPs tend to have high power consumption, thus challenging the design of energy-efficient prosthetic devices [27].
5) **Fit and Customization:** For ensuring comfort, functionality, and aesthetics of prosthetic limbs, each user needs a set of unique ones properly fitted to them. The complicated

process of building might make fit and customization more challenging with EAP-based prosthetics than with traditional prosthetic materials [28].

10.3.4 Challenges in Biomedical Devices

Besides the above challenges, there are particular challenges associated with the Biomedical devices mentioned below as

1) **Reliability and durability:** To withstand the conditions they would be exposed to in the body, biomedical equipment must be reliable and rugged. This includes factors such as corrosion resistance, wear resistance, and fatigue resistance [29].
2) **Biocompatibility:** Any type of EAPs implemented in biomedical devices, including prostheses, has to be biocompatible to allow safe integration in the human body without causing adverse reactions or damage to tissues. This would be more important for equipment that is in direct contact with body tissue or fluid [30].
3) **Miniaturization:** Many biomedical devices have to be small and portable in order to be implanted or used inside the body. Miniaturization can be quite challenging when all the necessary parts have to fit into a tight space while retaining functionality.
4) **Power Source:** Biomedical equipment frequently needs a power source to function, which can be batteries or an external power supply. It would be hard to ascertain a reliable power source that will be safe to use inside the body.
5) **Connectivity and Data Management:** Increasingly, today most biomedical devices are designed to be networked with other devices or systems in order to collect, analyze, and communicate with data. This can be difficult to ensure continuity of smooth connectivity and efficient management of data with respect to the compatibility of hardware and devices.
6) **Regulatory compliance:** Biomedical equipment needs to adhere to strict regulatory requirements to ensure that they are effective and safe. It can be challenging to operate in the regulatory environment, particularly for novel or advanced technologies.

7) **Cost-efficiency:** The development, manufacturing, and maintenance of medical devices can be expensive. It is very challenging to ensure cost-efficiency while maintaining quality and security, particularly with small businesses or organizations with limited resources.

10.3.5 Challenges in Energy Harvesting

1) **Performance:** One of the major challenges is improving the energy conversion efficiency of electroactive polymers in many energy harvesting applications. Electroactive polymers convert thermal or mechanical energy into electrical energy; however, they still do so with lower efficiency compared to other energy harvesting methods that exist at the moment [31].
2) **Stability and durability:** In order to endure the extreme circumstances frequently found in energy harvesting applications, electroactive polymers must be stable and robust. They ought to be able to execute at a high level for an extended period without deteriorating or losing capability. Due to significant advancements in self-healing materials over the past ten years, equipping gadgets with this feature has become a particularly promising way to significantly improve their usefulness and durability [32].
3) **Scalability:** Increasing the manufacture of electroactive polymers for commercial use presents another difficulty. Since most research and development efforts are now concentrated on small-scale prototypes, scalable manufacturing procedures must be created in order to generate electroactive polymers in large quantities.
4) **Integration:** Electroactive polymers are not easily integrated into already-existing systems or devices. For a smooth integration into different applications, they should be compatible with many materials, electronics, and interfaces.
5) **Environmental impact:** The environmental impact of electroactive polymers should be considered, just like any other technology. These materials should be produced and disposed of in an environmentally responsible and sustainable manner.
6) **Cost-efficacy:** To make the electroactive polymer commercially more favorable for broad uses, its manufacturing cost must become lower. For this purpose, it includes reducing

the raw-material cost, facilitating the production procedure, and therefore the total production cost.

7) **Regulatory compliance:** Law and standards can be a benchmark for the efficient commercialization of energy harvesting solutions using electroactive polymers. It may pose a challenge when trying to ensure that the produced technology meets performance and safety regulatory requirements.

Achievement of successful commercialization of energy harvesting solutions using electroactive polymers would indeed rely heavily on conformation to the laws and standards. This may not be easy to meet the safety and performance standards required by regulatory agencies, mainly for emerging technology [33].

10.3.6 Material Used to Overcome these Challenges

Graphene is one modern material that has been used to solve some of the difficulties associated with electroactive polymers. Basic studies show the enormous potential of graphene and its derivatives in numerous significant applications, such as tissue engineering, drug transport, and sensing. A single layer of carbon atoms organized in a hexagonal lattice makes up the two-dimensional substance known as graphene. It is highly electrically conductible, has great mechanical qualities, and has a large surface area [30]. EAPs can be made with graphene to increase their electrical conductivity and mechanical stability. The degradation problems related to ESPs are lessened by adding graphene to the polymer matrix, which produces a composite material with improved mechanical strength and longevity. Due to its high electrical conductivity, the material can effectively move charges across it, which improves actuation.

The processing issues are easily eliminated by adding graphene to EAPs prepared by solution mixing or melt blending. The other advantage of it is its compatibility with various substrates and filler materials. Currently, graphene is very costly; however, studies on it are currently focused on developing scalable manufacturing techniques in order to reduce its cost. As part of the effort to address the environmental concerns associated with the production and waste management of graphene, green synthesis routes and recycling methods are being explored. In general, graphene is a material that can help EAPs overcome some of their challenges and open up a wider scope of potential applications. Despite its promising prospects as the material for long-lasting artificial muscles, graphene still has to address the problems that limit it and fully unlock its potential. The problems graphene has included:

1) **Cost:** Graphene production is, comparatively speaking still very expensive. Because graphene costs so much to produce, several applications may prove too expensive and not viable in many applications.
2) **Scalability:** Currently, graphene of high quality cannot be made on a large scale. There is a high demand for green materials in most industries, so the scalability of graphene production should be improved.
3) **Toxicity:** Many studies have questioned the toxicity of graphene and its derivatives. Further research is required to fully understand and address any potential health and environmental issues associated with the use of graphene-based products.
4) **Integration:** Graphene cannot easily be added to three-dimensional things, like artificial muscles, due to its two-dimensional nature. The challenging development of effective production processes and techniques for adding graphene to complex structures is still current [34].
5) **Stability:** Although graphene is highly robust, some environmental factors, such as moisture or high temperatures, may still damage it. For real applications, the stability and toughness of graphene-based products would need to be ensured through additional protective coatings or encapsulation techniques [35].
6) **Carbon nanotubes (CNTs):** Carbon nanotubes, similar to graphene, possess high electrical conductivity and mechanical strength. These could be incorporated in EAPs to improve electrical and mechanical properties, enhancing the actuation performance [36].
7) **Conductive polymers:** These types of polymers, such as polyaniline and polypyrrole, can exhibit high deformations but also electrical conductivity. They could be used in EAPs as the active element since they have properties of both electric conductivity and actuation.
8) **Metal nanoparticles:** Adding metal nanoparticles, such as gold or silver nanoparticles, to EAPs can enhance the electrical conductivity of EAPs. Such nanoparticles can lead to more effective actuation by improving charge transfer across the material.
9) **Ionic liquids**: These salts are organic and, therefore, liquid at room temperature. Ionized liquids can be used as electrolytes in EAPs, providing ions that allow charges to pass through the material. This improves the overall actuation performance of the artificial muscles [37].

10) **Hydrogels:** Water-swollen polymer networks, known as hydrogels, can respond to electrical stimuli by exhibiting large volume changes. When used as the active component in EAPs, they can be employed for applications in biomimetic systems or soft robotics for the offer of actuation and biocompatibility [38].

These applications constructed from electroactive polymers (EAPs) have been developed using a variety of materials in addition to graphene. These materials include, among others:

In addition, graphene may help EAP materials in improving the mechanical stability, electrical conductivity, and ease of processing. The performance and utility of artificial muscles that are developed from electroactive materials are continually being improved as new materials and methods are being researched.

10.3.7 Challenges in Conducting Polymers

Some of the challenges faced by conducting polymers are

1) **Low conductivity:** Though polymers can conduct electricity, they have a lesser conductivity than inorganic or traditional metals. Because of this, their usage will be limited in situations where high conductivity is required.
2) **Stability and degradation**: Polymers of conductivity can degrade after some time, especially if exposed to heat, light, or moisture. This may pose a challenge in their use in long-term applications because of how this may affect their dependability and performance [39].
3) **Fabrication and processing**: Generally, conducting polymers require specific processing methods, which can be troublesome and time-consuming. Some examples of such methods include chemical synthesis and electrochemical deposition. It may be hard to find regulated and uniform quality across the board [40].
4) As with EAPs, the environmental impacts of conducting polymers are unknown. The suspicion that the preparation and disposal of these materials will involve hazardous chemicals or produce hazardous waste creates environmental and waste disposal concerns.

5) **Cost:** The cost of producing conductive polymers can be higher than that of traditional materials. Their widespread acceptability in various industries is restricted by the high manufacturing costs brought on by the specialist materials and processing techniques required for their production.

Conducting Hydrogel Materials for Neural Interfacing Electrodes
Conducting hydrogels (CHs) are a new technology in the fields of medical electrodes and brain-machine interfaces. The main challenge to the fabrication of CH electrodes is the formation of the interpenetrating polymer networks (IPNs) necessary to develop soft and electroactive materials by hybridizing several polymers (conductive polymer and hydrogel) [41].

10.4 Sustainability of EAPs Material

This makes the sustainability of electroactive polymers (EAPs) a matter of great importance since they are widely used in most applications. The following are the aspects that one should consider concerning the sustainability of EAPs:

1) **Raw material availability:** To have long-term sustainability, it would be essential to consider the raw materials used for EAPs. A sustainable choice could not be available if the material is rare or not renewable.
2) **Durability and long-term:** You may raise the life span of your EAPs if you design them for durability. In that way, you shall reduce the number of times you need to replace them, which can reduce trash production and other resource consumption.
3) **Waste management:** Proper waste management is crucial for the sustainability of EAPs in the long term. Always make an effort to recycle or dispose of any kind of waste produced in a proper manner to minimize waste generation.
4) **Environmental impact:** The extraction and processing of materials on the environment should be assessed. This includes evaluating variables of the manufacture of the materials; such as energy, water use, emission of greenhouse gas and generation of waste.

5) **End-of-life:** To make EAPs a sustainable end-of-life alternatives have to be incorporated. EAPs after usage environmental impact can be reduced by making them recyclable or biodegradable.
6) **Energy usage:** Manufacturing procedures for EAP should try to minimize the usage of energy and reduce greenhouse gases. Energy-efficient manufacturing processes and renewable energy can be utilized in this process.
7) **Toxicity and health hazard:** Toxicity and potential health hazards of the material must be evaluated. Sustainable use of substances must not cause harm to the environment or human health.
8) **Recyclability or biodegradability:** EAPs are to be made recyclable or biodegradable. This reduces the environmental impact of the process of disposal and makes it possible to recover valuable resources.
9) **Lifecycle analysis:** An LCA of the chosen material can disclose information regarding its overall sustainability. This comprises assessing its environmental impacts through all phases-from extraction to production, use, and final disposal.
10) **Alternative materials:** Exploring more environmentally friendly substitute materials could help increase the overall sustainability of EAPs. This might require research and development of new materials or considering already existing materials with stronger sustainability characteristics.

A holistic approach that considers all dimensions of sustainability is what would make EAPs truly sustainable, and it must be remembered that the sustainability of EAPs depends not only on the material but many other aspects which, as mentioned in the previous answer, include energy consumption, waste disposal, and end-of-life management [42].

10.4.1 Platinum

Ancient electroactive material: platinum. Platinum, due to its excellent chemical stability and catalytic activity, is widely used in electrochemistry. This material is commonly used as an electrode within the fuel cell, where it accelerates electrochemical reactions for chemical-to-electrical energy conversion [43]. Platinum has high electrical conductivity with low corrosion rates, making it a suitable choice for the manufacture of such cells.

Platinum is not usually a sustainable electroactive material for the following reasons:

1) **Limited Availability:** Platinum is very rare and very expensive. Platinum has a limited number of deposits in the world. The increasing demand for electroactive materials poses a challenge to scale up the manufacturing process due to their scarcity.
2) **Environmental Impact:** Platinum mining and extraction can significantly harm the environment, causing habitat destruction, water pollution, and carbon emissions. Moreover, the refining process requires a lot of energy and may lead to greenhouse gas emissions.
3) **Recycling Hurdles:** Recycling platinum is feasible but involves a labour-intensive and expensive procedure. Recovery of platinum from electronic waste or used electrochemical devices is a challenging process owing to the expense of recycling which limits its practicality on a large scale.
4) **Alternative Materials:** Investigators are actively seeking other electroactive materials that would be as effective or even more effective than platinum. These alternatives include more easily available and reasonably priced substances like organic polymers, transition metal oxides, and carbon-based electrodes.

Although platinum shows good electrochemical characteristics, the finite supply and environmental effect together with difficulties of recycling and the existence of other materials make it unsuitable as a sustainable electroactive material.

10.4.2 Graphite

Graphite has been the material that carbon electrochemical applications depend on since time immemorial. Graphite is a most widely used material for electrode compositions of lithium-ion batteries. Being electrically conducting with a large capacity for electricity, it would allow it to hold and yield electricity effectively. Being lightweight yet of high specific capacity, the graphite is excellent material for the portable electrical device. Supercapacitors, which are energy storage devices that can deliver high power quickly, also make use of graphite as an electrode material. The uses of supercapacitors include renewable energy storage and regenerative braking systems in automobiles [44]. Graphite can also be used as a catalyst support material in various electrochemical processes. It can improve stability and catalytic activity of the other materials used, such as metal nanoparticles in procedures like water splitting for hydrogen production or reactions in fuel cells. Graphite is widely available, which happens to be the main justification behind its sustainability.

Carbon is the most abundant element on Earth and is the major constituent of graphite. Graphite is also highly recyclable and reusable, thus minimizing the demand for fresh extraction and manufacture. Additionally, in comparison to other materials, graphite has a minimum environmental impact. It has no toxic components and does not release any negative emissions when used. Therefore, it is an environmentally friendly choice for electrochemical applications [45]. Traditional electroactive materials like graphite have good performance and sustainability. Its availability, recyclability, and minimal environmental effect make it a viable candidate for many electrochemical applications [46].

10.4.3 Graphene

This is a major problem of typical ionic-type artificial muscles under prolonged excitations: the interior electrolyte and hydrated cations can seep out through fissures in the metallic electrodes. Using hydrophobic reduced graphene oxide sheets in this instance, we created a highly sustainable electroactive artificial muscle. The unique qualities of graphene, including its great strength, flexibility, and conductivity, make it a viable material for the creation of artificial muscles. It may prove to be a very crucial component in building artificial muscles and imparting to them the mechanical characteristics, which will make them operational. Using actuators comprising graphene is the most widely followed technique to fabricate artificial muscles with this material [47]. The actuators of graphene-based muscles are reported to respond by inducing mechanical movement through electric-field variations or even by changes in temperature. Because of that, they are appropriate for applications in biomedical equipment, robotics, and prosthetics.

Graphene-based artificial muscles enable a higher strength-to-weight ratio, thus allowing a more compact and lightweight design. Because they are so flexible, they can closely mimic the movement of real muscles. Additionally, the effective control and actuation of the artificial muscles are achieved due to the high electrical conductivity of graphene [48].

Synthetic muscles can utilize graphene and thereby may have some beneficial effects, but at the same time detrimental to sustainability. Positively, the resource of graphene, being renewable might help in a reduction of the consumption of fossil fuels. Besides, it might be recycled through several processes thereby minimizing waste output. Some toxic chemicals could also be applied during the process of synthesis for graphene which would require energy for synthesis. Optimizing the processes of manufacturing can reduce energy usage and waste production. In addition, because these artificial muscles containing graphene may require specific recycling methods to recover important elements, its disposal is relevant. There are several advantages when using graphene as a sustainable material.

1) **Environmental friendliness:** Carbon is an element found naturally in high abundance, which is used in the production of graphene. Its production process does not involve rare or dangerous materials, making it a sustainable alternative for artificial muscle development.
2) **Energy efficiency:** Graphene-based artificial muscles require less energy to operate compared to materials based on more traditional materials. Thus, there can be significant savings in energy and less environmental impact.
3) **Lifespan:** Graphene's robustness and resistance to degradation ensure that artificial muscles made from it are long-lasting. This extended durability reduces the frequency of replacements, promoting more sustainable resource utilization.
4) **Recycling:** Graphene can be recycled and reused; this also saves waste and fosters a circular economy. This makes it an appropriate material for applications requiring artificial muscle [34, 35].
5) **Reduced carbon footprint:** Due to their small weight, high energy efficiency, and extended lifespan, graphene-based artificial muscles have the potential to diminish carbon emissions. This may help reduce climate change and increase sustainability across several industries.

Since the final life duration of graphene as a sustainable material depends on various variables, such as the specific application, usage circumstances, and maintenance procedures, this last durability cannot be identified [49]. Conclusion While graphene has significant benefits in the creation of artificial muscles, its sustainability as a material should be assessed and maintained appropriately. Artificial muscles that use graphene can also help toward a greener and more ecologically sustainable business that not only bases its operations on ethical sourcing and good manufacturing techniques but also has effective recycling strategies along with long-lasting performance.

10.4.4 Conducting Polymer

In the last few years, conducting polymers have drawn considerable attention as they offer electrical conductivity and properties of polymers. One component of the sustainability of conducting polymers is their environmental effect. Traditionally, most techniques used for the synthesis of conducting polymers, including PANI, PPy, and PTh, sometimes employ the use of dangerous chemicals and harmful solvents that are not safe for human health and the environment. However, there have been efforts toward developing greener synthesis techniques that use safer chemicals and more environmentally friendly solvents. Some examples of such friendly conducting polymers include:

1. **Poly(3,4-ethylenedioxythiophene) (PEDOT):** Being one of the most studied conducting polymers, it can be prepared using non-toxic, ecologically acceptable solvents. It can be used in organic solar cells and organic light-emitting diodes (OLEDs), among other uses. It exhibits relatively high electrical conductivity.
2. **Polyaniline (PANI) derivatives:** Although PANI is a common conducting polymer, its derivatives could be modified towards sustainability. Thus, for instance, PANI can be modified with biodegradable or renewable materials such as cellulose and lignin such that the increased environmental friendliness of the polymers is enhanced without losing any conductivity. Both the nature of the dopant acid and their concentration have dramatically affected the electric properties, stability at elevated temperatures, and eco-friendliness of PANI Films [50].
3. **Poly(3-hexylthiophene) (P3HT):** P3HT is an organic photovoltaic material that is often used as a conducting polymer. It has shown promising effectiveness in renewable energy applications and can be prepared with non-toxic solvents.
4. **Polypyrrole (PPy) derivatives:** As with PANI, PPy can be functionalized with environmentally friendly components to reduce its environmental impact. For example, less toxic conducting composites may be prepared by combining PPy with biopolymers such as chitosan or starch [51].
5. **Polythiophene derivatives:** The synthesizing process may be either water-based or solvent-free with the aid of conducting polymers based on PTh. Considering these compounds over the conventional process of synthesizing PTh, they offer enhanced sustainability.

The main targets involved in developing a more sustainable conductive polymer will be developing better synthesis techniques by using environmentally friendly methods such as renewable or biodegradable reagents, followed by decreasing harmful chemicals and solvent usage. Considering their improved sustainability, synthesis simplification, and friendliness toward the environment, conducting and redox polymers have been evaluated as promising storage solutions for energy. In terms of quinones, AQ stands out as the most promising molecule from an environmental standpoint, because of its processability, sustainability, high electrochemical applicability, and versatility in that it can also be used as an anode as well as a cathode [52].

10.4.5 Heteroatom (N and S)-Doped Carbon (ATC) from Aesculus Turbinata

Using biomass and biowaste as carbon precursors in the design of environmentally friendly, effective energy storage and conversion materials According to the research, ATC produced from Aesculus turbinata seeds might be employed as a supercapacitor's high-performance, reusable electroactive material. Additionally, this study presents a convincing illustration of recycling waste biomass into a useful energy-storage medium, offering practical suggestions for handling biowaste/biomass for enhanced environmental preservation [53].

10.4.6 Ionic Gel Polymers

For example, a sustainable ionic polymer is, of course, gel. This particular hydrogel contains a polymer network along with ionic groups. The ion conduction is due to the capability of these gels to absorb and retain enormous volumes of electrolyte solution. They were researched for sensors, fuel cells, and batteries owing to the properties like strong ionic conductivity and mechanical flexibility of these compounds, which are compatible with a wide variety of materials for electrode use. What makes them a greener alternative to traditional conducting polymers is that potentially they can be prepared through a green process using non-toxic solvents and renewable material resources. The following examples demonstrate the diversity of environmentally friendly, non-toxic, and sustainable ionic polymer gels that can be produced using renewable materials and non-toxic solvents.

1. **Ionic polymer gel based on chitosan:** Chitosan is a biodegradable and regenerative polysaccharide obtained from crab shells. It may be coupled with ionic groups to create a

gel that can be utilized for a variety of purposes, including tissue engineering and drug delivery systems.

2. **Cellulose-based ionic polymer gel:** Cellulose is a biopolymer occurring naturally within plant cell walls, and hence is considered an eco-friendly, renewable source of material. Ionic groups may be attached to cellulose in order to make the gel applicable in biosensors, wound healing, and drug delivery [54].
3. **Alginate-based ionic polymer gel:** Alginate is a polysaccharide derived from brown seaweed, which is biocompatible and biodegradable. This material can be conjugated with ionic groups to form a gel that has been studied as an application for tissue engineering, drug delivery, and wound healing [54].
4. **Ionic polymer gel based on gelatin:** Gelatin is a collagen-based protein found in animal connective tissues. Ionic groups can be attached to this gelatin to form a gel material that can be potentially used in tissue engineering, medication delivery, and wound healing applications [55].

10.4.7 Nitinol and Magneto

Nitinol, a sustainable electroactive material, has drawn interest due to its unique characteristics and potential applications. The shape memory effect of nitinol makes it an excellent candidate for use as an electroactive material. Nitinol can "remember" and return to its original shape after distortion due to a reversible phase change between two distinct crystal structures. This property makes nitinol suitable for use in actuators, sensors, and smart materials [56]. Due to such properties, Nitinol, though tough, possesses many biomedical applications, including stents, orthopedic implants, and dental braces. It is highly corrosion-resistant; hence, it has enabled strength and durability in various environments. This makes the alloy suitable for challenging working conditions in aerospace and marine work environments with different temperatures and humidities. The sustainability of Nitinol is in its highly recyclability; it is often remelted and reused without losing its unique properties, therefore requiring fewer new raw materials and thus reducing waste. These properties make nitinol viable as a sustainable electroactive material with extensive applications. Versatile and sustainable, it is suitable for a variety of applications and techniques [57].

Recently, an innovative method has been proposed for the encoding of different shapes and forms with magneto-electroactive shape memory polymer composite (SMPC) structures that combine sustainability with 4D printing (4DP) technology. The 4DP composite actuators with the potential to minimize waste of material and energy/effort for the application offer plenty of promise toward shape morphing and sustainability [58].

Magneto is a material responsive to the magnet, which means its viscosity or stiffness can be altered when a magnetic field is presented. On the other hand, nitinol is popularly known for the shape memory effect and super elasticity.

10.4.8 Organic Electroactive Materials for Aqueous Redox Flow Batteries

In contrast to currently available commercial inorganic materials, organic electroactive materials benefit from structural tunability and possibly sustainable manufacture. Due to their inherent safety, organic electroactive materials in aqueous redox flow batteries (ARFBs) have attracted considerable attention recently for low-cost and sustainable energy storage devices. In ARFBs, the primary reaction types of organic electroactive materials are categorized in order to give a general overview of how to control their solubility, potential, stability, and viscosity. Next, the organic anolyte and catholyte in ARFBs are summarized with the various quinones, viologens, nitroxide radicals, hydroquinone, etc., in mind and stress ways to improve solubility through the creation of different functional groups [59].

10.5 Conclusion

To conclude, electroactive polymers (EAPs) are an interesting and rapidly developing family of materials with immense potential in a great variety of technical fields. These new possibilities have been made possible by the singular capacity of these polymers to undergo big changes in form, size, or conductivity under the action of electrical stimulation in various areas, such as robotics, sensors, actuators, and energy harvesting, among others. On the other hand, their environmental difficulties and sustainability cannot be disregarded. Many of these polymers are produced from fossil resources, which is a source of greenhouse gases and harmful to the environment. So, for production, sustainable renewable sources must be established. Other hand, most of these polymers are tough to biodegrade and have a tendency to stay in an environment for such a long period. This will cause damage and pollution to the ecosystem. As a consequence, develop efficient recycling and waste management techniques for these polymers. Another concern is the amount of energy used in the production of electroactive polymers. We are getting closer to a day when electroactive polymers will interpose positively towards the performance and capabilities of a wide spectrum of devices and systems. On this long journey from discovery to demystified reality, we expect exciting outcomes. Priorities for resolving these issues include developing sources for electroactive polymers that are as sustainable and renewable as possible, that have a greater capacity to biodegrade, and that optimize energy use in their production. It is also necessary to encourage interdisciplinary cooperation between academia, industry, and governments to enable innovation and the acceptance of sustainable practices in the arena of electroactive polymers. In essence, this said, whilst there are exceptional expectations for electroactive polymers in particular applications, the environmental and sustainability concerns surrounding it must be resolved in ushering a greener and more sustainable future.

AI disclosure statement: The author confirms that no artificial intelligence (AI)-generated content or AI-assisted writing tools were used in the preparation of this chapter (including summaries and the abstract). All text, analysis, and conclusions are the result of the author's original work.

References

1. Pipkin, A.C. and Rivlin, R.S., Electrical conduction in deformed isotropic materials. *J. Math. Phys.*, *1*, 2, 127–130, 1960.
2. Broadhurst, M.G., Davis, G.T., Gerhard-Multhaupt, R., Gross, B., Mascarenhas, S., Sessler, G.M., West, J.E., *Electrets*, vol. 33, Springer Science & Business Media, New York, 2005.
3. Kawai, H., The piezoelectricity of poly (vinylidene fluoride). *Jpn. J. Appl. Phys.*, *8*, 7, 975, 1969.
4. Fukada, E., Piezoelectricity of wood. *J. Phys. Soc. Jpn.*, *10*, 2, 149–154, 1955.
5. Hanson, D. and Bar-Cohen, Y., *The coming robot revolution: Expectations and fears about emerging intelligent, humanlike machines*, Springer New York, 2009.
6. Yang, Y., Tse, Y.A., Zhang, Y., Kan, Z., Wang, M.Y., A low-cost inchworm-inspired soft robot driven by supercoiled polymer artificial muscle, in: *2019 2nd IEEE International Conference on Soft Robotics (RoboSoft)*, 2019, April, IEEE, pp. 161–166.

7. Wang, Y. and Sugino, T., Ionic polymer actuators: Principle, fabrication and applications, in: *Actuators*, pp. 55–86, IntechOpen, United Kingdom, 2018, March.
8. Park, S.W., Kim, S.J., Park, S.H., Lee, J., Kim, H., Kim, M.K., Recent Progress in Development and Applications of Ionic Polymer–Metal Composite. *Micromachines, 13*, 8, 1290, 2022.
9. Chang, L., Liu, Y., Yang, Q., Yu, L., Liu, J., Zhu, Z., Hu, Y., Ionic electroactive polymers used in bionic robots: a review. *J. Bionic Eng.*, 15, 765–782, 2018.
10. Alekseyev, N.I., Khmelnitskiy, I.K., Aivazyan, V.M., Broyko, A.P., Korlyakov, A.V., Luchinin, V.V., Ionic eap actuators with electrodes based on carbon nanomaterials. *Polymers, 13*, 23, 4137, 2021.
11. Ruano Torres, G., Conducting polymers and hybrid materials for technological applications, 2021.
12. Wang, H.S., Cho, J., Song, D.S., Jang, J.H., Jho, J.Y., Park, J.H., High-performance electroactive polymer actuators based on ultrathick ionic polymer–metal composites with nanodispersed metal electrodes. *ACS Appl. Mater. Interfaces, 9*, 26, 21998–22005, 2017.
13. ul Haq, M. and Gang, Z., Ionic polymer–metal composite applications. *Emerging Mater. Res., 5*, 1, 153–164, 2016.
14. Hajiesmaili, E. and Clarke, D.R., Dielectric elastomer actuators. *J. Appl. Phys., 129*, 15, 151101–151110, 2021.
15. Kularatne, R.S., Kim, H., Boothby, J.M., Ware, T.H., Liquid crystal elastomer actuators: Synthesis, alignment, and applications. *J. Polym. Sci., Part B: Polym. Phys., 55*, 5, 395–411, 2017.
16. Sikulskyi, S., Mekonnen, D.T., El Atrache, A., Divo, E., Kim, D., Effects of ferroelectric fillers on composite dielectric elastomer actuator. *Actuators*, 10, 7, 137, 2021, June, MDPI.
17. Su, J., Electrostrictive Graft-Elastomers (G-Elastomers)-Materials, Devices and Applications: A Review, in: *Seminar at Chiba University* (No. NF1676L-33711), 2019, August.
18. Cianchetti, M., Laschi, C., Menciassi, A., Dario, P., Biomedical applications of soft robotics. *Nat. Rev. Mater., 3*, 6, 143–153, 2018.
19. Ning, C., Zhou, Z., Tan, G., Zhu, Y., Mao, C., Electroactive polymers for tissue regeneration: Developments and perspectives. *Prog. Polym. Sci., 81*, 144–162, 2018.
20. Elvin, N. and Erturk, A., *Advances in energy harvesting methods*, Springer Science & Business Media, New York, 2013.
21. Chen, Y., Chen, C., Rehman, H.U., Zheng, X., Li, H., Liu, H., Hedenqvist, M.S., Shape-memory polymeric artificial muscles: Mechanisms, applications and challenges. *Molecules, 25*, 18, 4246, 2020.
22. Mirvakili, S.M. and Hunter, I.W., Artificial muscles: Mechanisms, applications, and challenges. *Adv. Mater., 30*, 6, 1704407, 2018.
23. Haines, C.S., Li, N., Spinks, G.M., Aliev, A.E., Di, J., Baughman, R.H., A new twist on artificial muscles. *Proc. Natl. Acad. Sci., 113*, 42, 11709–11716, 2016.

24. Zhang, J., Sheng, J., O'Neill, C.T., Walsh, C.J., Wood, R.J., Ryu, J.H., Yip, M.C., Robotic artificial muscles: Current progress and future perspectives. *IEEE Trans. Rob.*, *35*, 3, 761–781, 2019.
25. Rus, D. and Tolley, M.T., Design, fabrication and control of soft robots. *Nature*, *521*, 7553, 467–475, 2015.
26. Trivedi, D., Rahn, C.D., Kier, W.M., Walker, I.D., Soft robotics: Biological inspiration, state of the art, and future research. *Appl. Bionics Biomech.*, *5*, 3, 99–117, 2008.
27. Rasmussen, L., Rodriguez, S., Bowers, M., Smith, D., Martino, G., Moy, L., d'Almeida, J., Synthetic Muscle Electroactive Polymer (EAP) based actuation and pressure sensing for prosthetic and robotic gripper applications, in: *Electroactive Polymer Actuators and Devices (EAPAD) XXI*, 2019, March, vol. 10966, SPIE, pp. 306–314.
28. Tryggvason, H., Starker, F., Lecomte, C., Jonsdottir, F., Modeling of stiffness characteristics in a prosthetic foot, in: *Smart Materials, Adaptive Structures and Intelligent Systems*, 2017, September, vol. 58264, American Society of Mechanical Engineers, p. V002T03A014.
29. Nezakati, T., Seifalian, A., Tan, A., Seifalian, A.M., Conductive polymers: opportunities and challenges in biomedical applications. *Chem. Rev.*, *118*, 14, 6766–6843, 2018.
30. Reina, G., González-Domínguez, J.M., Criado, A., Vázquez, E., Bianco, A., Prato, M., Promises, facts and challenges for graphene in biomedical applications. *Chem. Soc. Rev.*, *46*, 15, 4400–4416, 2017.
31. Kornbluh, R.D., Pelrine, R., Prahlad, H., Wong-Foy, A., McCoy, B., Kim, S., Low, T., From boots to buoys: promises and challenges of dielectric elastomer energy harvesting, in: *Electroactivity in Polymeric Materials*, pp. 67–93, 2012.
32. Chen, D., Wang, D., Yang, Y., Huang, Q., Zhu, S., Zheng, Z., Self-healing materials for next-generation energy harvesting and storage devices. *Adv. Energy Mater.*, *7*, 23, 1700890, 2017.
33. Mori, T. and Priya, S., Materials for energy harvesting: At the forefront of a new wave. *MRS Bull.*, *43*, 3, 176–180, 2018.
34. Lv, W., Li, Z., Deng, Y., Yang, Q.H., Kang, F., Graphene-based materials for electrochemical energy storage devices: opportunities and challenges. *Energy Storage Mater.*, *2*, 107–138, 2016.
35. Lee, S.M., Kim, J.H., Ahn, J.H., Graphene as a flexible electronic material: mechanical limitations by defect formation and efforts to overcome. *Mater. Today*, *18*, 6, 336–344, 2015.
36. Chen, I.W.P., Yang, M.C., Yang, C.H., Zhong, D.X., Hsu, M.C., Chen, Y., Newton output blocking force under low-voltage stimulation for carbon nanotube–electroactive polymer composite artificial muscles. *ACS Appl. Mater. Interfaces*, *9*, 6, 5550–5555, 2017.
37. Lu, C. and Chen, X., Flexible and electroactive ionogel graphene composite actuator. *Materials*, *13*, 3, 656, 2020.

38. Tungkavet, T., Seetapan, N., Pattavarakorn, D., Sirivat, A., Graphene/gelatin hydrogel composites with high storage modulus sensitivity for using as electroactive actuator: Effects of surface area and electric field strength. *Polymer, 70*, 242–251, 2015.
39. Green, R.A., Lovell, N.H., Wallace, G.G., Poole-Warren, L.A., Conducting polymers for neural interfaces: challenges in developing an effective long-term implant. *Biomaterials, 29*, 24-25, 3393–3399, 2008.
40. Yoon, H. and Jang, J., Conducting-polymer nanomaterials for high-performance sensor applications: issues and challenges. *Adv. Funct. Mater., 19*, 10, 1567–1576, 2009.
41. Goding, J., Gilmour, A., Martens, P., Poole-Warren, L., Green, R., Interpenetrating conducting hydrogel materials for neural interfacing electrodes. *Adv. Healthc. Mater., 6*, 9, 1601177, 2017.
42. Bar-Cohen, Y. and Leary, S.P., Electroactive polymers (EAP) charaterization methods, in: *Smart Structures and Materials 2000: Electroactive Polymer Actuators and Devices (EAPAD)*, vol. 3987, pp. 12–16, SPIE, 2000, June.
43. Mechtler, C. and Marschner, C., Polymerization of Silanes with Platinum Metal Catalysts, in: *Organosilicon Chemistry Set: From Molecules to Materials*, pp. 554–557, 2005.
44. Jabbour, L., Gerbaldi, C., Chaussy, D., Zeno, E., Bodoardo, S., Beneventi, D., Microfibrillated cellulose–graphite nanocomposites for highly flexible paper-like Li-ion battery electrodes. *J. Mater. Chem., 20*, 35, 7344–7347, 2010.
45. Min, J., Jung, Y., Ahn, J., Lee, J.G., Lee, J., Ko, S.H., Recent Advances in Biodegradable Green Electronic Materials and Sensor Applications. *Adv. Mater.*, New York, 35, 52, 2211273, 2023.
46. Pandey, R.K., Chen, L., Teraji, S., Nakanishi, H., Soh, S., Eco-friendly, direct deposition of metal nanoparticles on graphite for electrochemical energy conversion and storage. *ACS Appl. Mater. Interfaces, 11*, 40, 36525–36534, 2019.
47. Kim, J., Jeon, J.H., Kim, H.J., Lim, H., Oh, I.K., Highly sustainable electroactive artificial muscle with graphene, in: *2014 IEEE/ASME International Conference on Advanced Intelligent Mechatronics*, 2014, July, IEEE, pp. 1442–1445.
48. Azman, N.H.N., Mamat@ Mat Nazir, M.S., Ngee, L.H., Sulaiman, Y., Graphene-based ternary composites for supercapacitors. *Int. J. Energy Res., 42*, 6, 2104–2116, 2018.
49. Tiwari, A. and Syväjärvi, M., *Graphene Materials*, John Wiley & Sons, Inc., Hoboken, NJ, USA, 2015.
50. Ansari, R. and Keivani, M.B., Polyaniline conducting electroactive polymers thermal and environmental stability studies. *J. Chem., 3*, 202–217, 2006.
51. Ansari, R., Polypyrrole conducting electroactive polymers: synthesis and stability studies. *E-J. Chem., 3*, 4, 186–201, 2006.
52. Zarren, G., Nisar, B., Sher, F., Synthesis of anthraquinone-based electroactive polymers: a critical review. *Mater. Today Sustain., 5*, 100019, 2019.

53. Perumal, S., Chandra Kishore, S., Atchudan, R., Sundramoorthy, A.K., Alagan, M., Lee, Y.R., Sustainable synthesis of N/S-doped porous carbon from waste-biomass as electroactive material for energy harvesting. *Catalysts*, *12*, 4, 436, 2022.
54. Gebeyehu, E.K., Sui, X., Adamu, B.F., Beyene, K.A., Tadesse, M.G., Cellulosic-based conductive hydrogels for electro-active tissues: A review summary. *Gels*, *8*, 3, 140, 2022.
55. Lee, J. and Aida, T., "Bucky gels" for tailoring electroactive materials and devices: the composites of carbon materials with ionic liquids. *Chem. Commun.*, *47*, 24, 6757–6762, 2011.
56. Swensen, J.P. and Dollar, A.M., Optimization of parallel spring antagonists for Nitinol shape memory alloy actuators, in: *2014 IEEE International Conference on Robotics and Automation (ICRA)*, 2014, May, IEEE, pp. 6345–6349.
57. Taşkan, E., Bulak, S., Taşkan, B., Şaşmaz, M., El Abed, S., El Abed, A., Nitinol as a suitable anode material for electricity generation in microbial fuel cells. *Bioelectrochemistry*, *128*, 118–125, 2019.
58. Lalegani Dezaki, M. and Bodaghi, M., Sustainable 4D printing of magneto-electroactive shape memory polymer composites. *Int. J. Adv. Manuf. Technol.*, *126*, 1–2, 35–48, 2023.
59. Yang, G., Zhu, Y., Hao, Z., Lu, Y., Zhao, Q., Zhang, K., Chen, J., Organic Electroactive Materials for Aqueous Redox Flow Batteries. *Adv. Mater.*, 35, 33, 2301898, 2023.

Index

Printed in the USA/Agawam, MA
July 7, 2026

907425.005